Competition in Agriculture
The United States in the World Market

Pre-publication
REVIEWS,
COMMENTARIES,
EVALUATIONS . . .

"Colyer and his colleagues have put together a handy guidebook about the economic forces that determine the competitiveness of U.S. agriculture across a wide array of products. The contributors to this volume characterize trade patterns and disentangle the comparative costs and policy interventions affecting traditional and emerging trade flows. This compendium highlights the competitive pressures faced by agriculture, even for commodities where the United States has long dominated exports, and demonstrates the importance of further productivity-enhancing R&D investments to maintain a competitive position in dynamic world markets. The distributional impacts of free trade, as well as its net benefits, are well documented, and the editors have tied the analysis of specific commodities together into a valuable overview of the forces shaping the future of this important sector of the American economy."

David Orden, PhD
Professor,
Virginia Polytechnic Institute
and State University,
Blacksburg, VA

More pre-publication
REVIEWS, COMMENTARIES, EVALUATIONS . . .

"This book addresses the question of how U.S. agriculture is performing in a global context at the dawn of the twenty-first century. It begins with a review of the issue of competitiveness and the forces that impact a sector's or nation's competitiveness. The introductory chapters are followed by a series of commodity-specific chapters covering the major agricultural commodities produced in the United States. The final chapters are broader, as they consider NAFTA and the emerging biotechnology issues.

The book is of value to researchers and decision makers seeking a broad overview of the competitiveness issue. It brings together a diverse literature on U.S. agricultural trade with a rich commodity-by-commodity summary of U.S. export performance versus competing countries and how that performance is linked to trade and farm policies."

Philip L. Paarlberg, PhD
Associate Professor,
Purdue University,
West Lafayette, IN

Food Products Press®
An Imprint of The Haworth Press, Inc.

NOTES FOR PROFESSIONAL LIBRARIANS AND LIBRARY USERS

This is an original book title published by Food Products Press®, an imprint of The Haworth Press, Inc. Unless otherwise noted in specific chapters with attribution, materials in this book have not been previously published elsewhere in any format or language.

CONSERVATION AND PRESERVATION NOTES

All books published by The Haworth Press, Inc. and its imprints are printed on certified pH neutral, acid free book grade paper. This paper meets the minimum requirements of American National Standard for Information Sciences-Permanence of Paper for Printed Material, ANSI Z39.48-1984.

Competition in Agriculture
*The United States
in the World Market*

FOOD PRODUCTS PRESS
Agricultural Commodity Economics,
Distribution, and Marketing

Andrew Desmond O'Rourke, PhD
Senior Editor

New, Recent, and Forthcoming Titles:

Marketing Livestock and Meat by William H. Lesser

Understanding the Japanese Food and Agrimarket: A Multifaceted Opportunity edited by A. Desmond O'Rourke

The World Apple Market by A. Desmond O'Rourke

Marketing Beef in Japan by William A. Kerr, Kurt K. Klein, Jill E. Hobbs, and Masaru Kagatsume

Effects of Grain Marketing Systems on Grain Production: A Comparative Study of China and India by Zhang-Yue Zhou

Competition in Agriculture: The United States in the World Market edited by Dale Colyer, P. Lynn Kennedy, William A. Amponsah, Stanley M. Fletcher, and Curtis M. Jolly

Competition in Agriculture
The United States in the World Market

Dale Colyer
P. Lynn Kennedy
William A. Amponsah
Stanley M. Fletcher
Curtis M. Jolly

Editors

Food Products Press®
An Imprint of The Haworth Press, Inc.
New York • London • Oxford

Published by

Food Products Press®, an imprint of The Haworth Press, Inc., 10 Alice Street, Binghamton, NY 13904-1580

© 2000 by The Haworth Press, Inc. All rights reserved. No part of this work may be reproduced or utilized in any form or by any means, electronic or mechanical, including photocopying, microfilm, and recording, or by any information storage and retrieval system, without permission in writing from the publisher. Printed in the United States of America.

Cover design by Marylouise E. Doyle.

Library of Congress Cataloging-in-Publication Data

Competition in agriculture : the United States in the World Market / Dale Colyer ... [et al.] editors.
    p. cm.
  Includes bibliographical references and index.
  ISBN 1-56022-892-X (hard : alk. paper) — ISBN 1-56022-893-8 (pbk. : alk. paper)
  1. Produce trade—Government policy—United States. 2. Agriculture and state—United States. 3. United States—Commercial treaties. 4. International trade. 5. Competition, International. I. Colyer, Dale.

HD9006 .C66 2000
338.1'873—dc21

99-462142

CONTENTS

About the Editors	xi
Contributors	xiii
Foreword	xv
Thomas H. Klindt	
Chapter 1. Introduction	1
Dale Colyer	
P. Lynn Kennedy	

Agricultural Exports   1
Competitiveness   4
Plan of Book   8

Chapter 2. Agricultural Competitiveness Issues and Concepts   **11**
   P. Lynn Kennedy

Issues in Agricultural Competitiveness   11
Definitions of Competitiveness   15
Factors Influencing Competitiveness   17
Indicators of Competitiveness   22

Chapter 3. The Competitiveness of U.S. Agriculture   **25**
   Munisamy Gopinath
   Terry L. Roe

Introduction   25
A Conceptual Framework   26
The Case of U.S. Agriculture   29
Comparison of Agricultural Growth in the United States
   and European Union   33
Conclusions   35

Chapter 4. Corn   **41**
   Paul W. Gallagher

Introduction   41
The Record: Changing Competitiveness in the U.S. Corn
   Industry   41
Market Developments Influencing Competitiveness   43

U.S. Policy and Competitiveness	46
U.S. Trade Agreements and Corn Exports	49
Indirect Effects of International Policy Changes	51
Biotechnology and Competitiveness in the Corn Industry	52
Export Performance and Technology Adoption	53
Market Structure	54
Conclusions	56

Chapter 5. Wheat   59
Won W. Koo

Introduction	59
World Wheat Industry	60
Trade Agreements and Policies	66
Regional and National Competitiveness	70
Conclusions	75

Chapter 6. Rice   79
Gail L. Cramer
James M. Hansen
Kenneth B. Young

Introduction	79
United States Relative Cost of Production and Milling	80
United States Agriculture Policy	81
Rice Trade Agreements, Rules, and Regulations	82
Impact of El Niño on Global Rice Production	85
The Arkansas Global Rice Model	86
Conclusions	96

Chapter 7. Soybeans   99
Donald W. Larson

Introduction	99
Soybean Trade Research	103
Competitive Position of the U.S. Soybean Industry	107
Trade Agreements	112
Conclusions	115

Chapter 8. Cotton — 119
Darren Hudson
Don Ethridge

Introduction	119
Imports and Exports	119
Trade Research in Cotton	125
Competitiveness of U.S. Cotton	128
Conclusions	132

Chapter 9. Peanuts — 139
Stanley M. Fletcher

Production	140
Trade	141
Competitive Position of the U.S. Peanut Industry	143
Agriculture and Trade Policy	146
Conclusions	151

Chapter 10. Fruits — 155
Dale Colyer

Introduction	155
Fruit Production	155
Citrus Fruits	157
Noncitrus Fruits	161
Agricultural and Trade Policy Impacts	164
Conclusions	166

Chapter 11. Vegetables — 169
John J. VanSickle

Introduction	169
Vegetable Trade Research	170
Competitive Position of the U.S. Vegetable Industry	172
Conclusions	175

Chapter 12. Miscellaneous Crops — 177
Glenn C. W. Ames

Wine	177
Sugar	182
Tobacco	188
Implications for U.S. Trade in Wine, Sugar, and Tobacco	194

Chapter 13. Beef   **199**
Rudy M. Nayga Jr.
Flynn Adcock
Parr Rosson

Introduction	199
The Changing Dynamics of World Beef Demand	199
Overview of World Meat and Fish Consumption, 1985-1997	202
Trends in World Beef Production, 1985-1997	203
Trends in World Beef Consumption, 1985-1997	205
Trends in World Beef Trade, 1985-1997	207
U.S. Competitiveness in World Beef Markets	209
Conclusion	214

Chapter 14. Poultry   **217**
Dale Colyer

Poultry Trade Research	219
Competitive Position of the U.S. Poultry Industry	221
Conclusions	230

Chapter 15. Pork   **235**
William A. Amponsah
Xiang Dong Qin

Introduction	235
Structural Change in the U.S. Pork Industry	237
The Changing Dynamics of Pork Consumption	241
U.S. Competitiveness in Global Pork Markets	243
Conclusions	248

Chapter 16. The Food Processing Industry   **251**
Michael Reed

Introduction	251
U.S. Exports of Processed Foods	252
U.S. Imports of Processed Foods	255
The Competitive Environment for Processed Foods	258
Competitive Position of the United States	260
Trade Agreements, Rules, and Regulations	263
Conclusions	266

Chapter 17. North American Free Trade and U.S. Agriculture — 269
Parr Rosson

Introduction — 269
CUSTA and NAFTA Implementation — 270
Overview of Agricultural Trade in NAFTA Countries — 271
NAFTA Impacts on Selected Commodities — 272
Animals and Animal Products — 273
Grains and Oilseeds — 279
Horticultural Products — 283
Other Commodities — 286
Conclusions — 289

Chapter 18. Biotechnology and Competitiveness — 293
Nicholas Kalaitzandonakes

Introduction — 293
Agrobiotechnology and Value Creation — 293
Value Appropriation and Firm and Industry
 Competitiveness — 297
Network Competitiveness — 301
Impacts of Agrobiotechnology on National
 Competitivencss — 303
Conclusions — 304

Chapter 19. Findings and Implications — 307
Dale Colyer
Curtis M. Jolly

Introduction — 307
Importance of Agricultural Exports — 307
Market Shares — 310
Factors Affecting Competitiveness — 311
Technology — 317
Effects of Trade Agreements — 318
Conclusions — 322

Index — 325

ABOUT THE EDITORS

Dale Colyer, PhD, is Professor of Agricultural and Resource Economics and Interim Associate Director of the West Virginia Agricultural and Forestry Experiment Station at West Virginia University in Morgantown. Dr. Colyer has served as a consultant to several countries, including Ecuador and Guatemala. His current research projects include investigating the global dynamics of the poultry trade.

P. Lynn Kennedy, PhD, is Associate Professor of Agricultural Economics and Agribusiness at Louisiana State University in Baton Rouge. Dr. Kennedy has been a presenter at over twenty international, national, and regional agricultural economics meetings. His current research interests involve analyzing how trade and economic welfare impacts domestic and multilateral policies on trade-in commodities that are of importance to the United States.

William Amponsah, PhD, is Associate Professor of International Trade and Economic Development and Coordinator of the Center of Excellence in International Trade at North Carolina A&T State University in Greensboro. Dr. Amponsah currently conducts research in the areas of macroeconomic policy, trade, and foreign investment policies.

Stanley M. Fletcher, PhD, is Professor of Agricultural and Applied Economics at the University of Georgia in Griffin. Dr. Fletcher is the only agricultural economist to receive the National Peanut Council's Research and Education Award. His current research interests include the domestic and international trade aspects of peanuts.

Curtis M. Jolly, PhD, is Professor of Agricultural Economics/International Trade and Economic Development in the Department of Agricultural Economics and Rural Sociology at Auburn University in Alabama. Dr. Jolly serves on the Technical Committee of Collaborative Research Program and works in Haiti on peanut marketing.

CONTRIBUTORS

Flynn Adcock, MS, is Research Associate and Assistant Director, Center for North American Studies, Department of Agricultural Economics, Texas A&M University, College Station, Texas.

Glenn C. W. Ames, PhD, is Professor of Agricultural and Applied Economics, The University of Georgia, Athens, Georgia.

Gail L. Cramer, PhD, is L. C. Carter Chair Professor of Agricultural Economics, University of Arkansas, Fayetteville, Arkansas.

Don Ethridge, PhD, is Professor and Chairman, Department of Agricultural and Applied Economics, Texas Tech University, Lubbock, Texas.

Paul W. Gallagher, PhD, is Associate Professor, Agricultural Economics, Iowa State University, Ames, Iowa.

Munisamy Gopinath, PhD, is Assistant Professor of Agricultural and Resource Economics, Oregon State University, Corvallis, Oregon.

James M. Hansen, BA, is Research Associate, Agricultural Economics, University of Arkansas, Fayetteville, Arkansas.

Darren Hudson, PhD, is Assistant Professor of Agricultural Economics, Mississippi State University, Mississippi State, Mississippi.

Nicholas Kalaitzandonakes, PhD, is Associate Professor of Agricultural Economics, University of Missouri, Columbia, Missouri.

Won W. Koo, PhD, is Professor of Agricultural Economics, North Dakota State University, Fargo, North Dakota.

Donald W. Larson, PhD, is Professor, Department of Agricultural, Environmental, and Development Economics, The Ohio State University, Columbus, Ohio.

Rudy M. Nayga Jr., PhD, is Associate Professor of Agricultural Economics, Texas A&M University, College Station, Texas.

Xiang Dong Qin, PhD, is Assistant Professor of International Trade and Agricultural Economics, North Carolina A&T State University, Greensboro, North Carolina.

Michael Reed, PhD, is Professor of Agricultural Economics, University of Kentucky, Lexington, Kentucky.

Terry L. Roe, PhD, is Professor of Applied Economics and Director of the Center for Political Economy, University of Minnesota, St. Paul, Minnesota.

Parr Rosson, PhD, is Professor and Director, Center for North American Studies, Department of Agricultural Economics, Texas A&M University, College Station, Texas.

John J. VanSickle, PhD, is Professor of Food & Resource Economics and Director of the International Agricultural Trade and Policy Center, Institute of Food & Agricultural Sciences, at the University of Florida, Gainesville, Florida.

Kenneth B. Young, PhD, is Senior Research Associate, Agricultural Economics, University of Arkansas, Fayetteville, Arkansas.

Foreword

The importance of international trade for the country's agricultural sector has been underscored during the past two years, with declines in international demand due to the Asian and Russian economic crises. These have contributed to the current very low prices of major farm products and a consequent economic crisis for U.S. farmers. This book is a synthesis of research carried out under successive Southern Regional Research Committees on international trade in agricultural commodities and by other research economists at land grant universities in the United States. It presents both theoretical and applied research results. But it is primarily oriented toward applied, practical problems and issues faced by producers and processors of food and fiber products as they face increasingly strong competition from producers in other areas of the world; it is intended to provide information to assist U.S. producers to remain competitive.

Agricultural research has led to the development and implementation of new technologies that have contributed to the ability of the nation's farmers to be highly efficient, productive, and competitive. This ongoing process will become more important as it becomes easier to transfer technology, as more of the world's farmers possess increased skills and education, and as freer trade in agricultural products is extended. The development of global markets and multinational corporations in food processing and marketing is changing the nature of the food and fiber industries. The resulting struggle for competitiveness becomes even more important since such businesses must seek the least-cost sources of high-quality products to compete in retail food and fiber markets. In addition, international trade agreements, including the World Trade Organization (WTO), created during the last round of the General Agreement of Tariffs and Trade, and the North American Free Trade Agreement strongly influence agricultural trade. In the past, agricultural trade tended to be omitted from trade agreements, and each country was free to follow an agricultural policy agenda without considering its impacts on other nations. This is no longer the case, as U.S. producers are now affected by what happens in competitor countries such as Brazil, France (the European Union), and China, as well as by policies of importing countries such as Japan. In the approaching round of negotiations under the WTO, agricultural trade is likely to be one of the more contentious issues, as it was in the last round. But as agricultural trade has increased in importance and as the issue of domestic food security becomes less important with the demise of the Cold War, agricultural trade has become a part of the general, but slow, trend toward freer trade that has existed since the end of World War II. Despite the movement toward freer trade regimes, many tariff and nontariff barriers to trade still exist and new ones are being created as policymakers respond to producer complaints about perceived unfair trade practices.

Agricultural trade is a major source of trade disputes. Examples include the banana dispute between the European Union and the United States together with banana producers from Central and South America, the issue of exports of beef produced with feed fortified with hormones, and a number of other disputes that are now wending their way through the WTO's dispute settlement process. Negative consumer reactions to incorporating genetically modified organisms into the food production process promises to yield contentious issues that will affect competitiveness, import restrictions or prohibitions, labeling requirements, and other barriers to trade.

The editors of this book are members of the Southern Regional Research Committee on Impacts of Trade Agreements and Economic Policies on Southern Agriculture and predecessor regional research committees. Thus, they have several decades of combined experience in trade research and have worked jointly on research activities related to agricultural trade and competitiveness. The authors of the individual chapters include members of the regional research committee and other university researchers who were selected for the depth and quality of their research activities on agricultural trade and competitiveness.

Important issues and problems related to the competitiveness of U.S. agriculture are addressed in this book. Research on these topics is synthesized and presented in nontechnical terms. The degree of competitiveness and factors affecting it vary widely from commodity to commodity and was a major factor in opting for a commodity approach to competitiveness. The material covered in the work is divided into: (1) general analyses of competitiveness issues; (2) commodity-specific research results as the main focus of the book; and (3) special topics including trade agreements, processed food products, and biotechnology. The authors of the individual chapters are trade experts who have written many research articles and papers in their areas of expertise. Nonetheless, they have utilized the research of other analysts to provide a more complete coverage of each topic. Consequently, the book will provide ample coverage for those who are interested in understanding agricultural competitiveness in general terms, as well as for those who want an in-depth analysis of specific commodities and issues.

Thomas H. Klindt
Associate Dean, College of Agriculture
University of Tennessee
and Administrative Advisor
to the Southern Regional Research Committee
on Impacts of Trade Agreements
and Economic Policies
on Southern Agriculture

Chapter 1

Introduction

Dale Colyer
P. Lynn Kennedy

The agricultural sector of the United States has been a significant contributor to the nation's export earnings and has consistently produced a favorable sectoral balance of trade. This favorable situation has been the result of a highly productive and internationally competitive agricultural industry. However, the international situation is undergoing significant changes due to globalization and liberalization of world markets and economies. As a consequence, shifts are occurring in trade patterns, with some products benefitting from the new situations while others lose out in this new world order. The ability of the United States to maintain a growing and favorable agricultural trade regime depends on the capacity to improve productivity, on the willingness to adapt to changing forces in the demand and supply of agricultural products, and on the continued evolution of more trade-oriented policies and programs, i.e., on the elimination of trade barriers. Many universities, federal government agencies, and private sector entities are engaged in research on various aspects of international trade and competitiveness in agriculture, and there has been a large volume of publications on these issues. However, this research has been published in widely scattered sources and there is a need to consolidate and systematize the results of this monumental research endeavor. This book is designed to help fill that gap.

AGRICULTURAL EXPORTS

Agriculture in the United States has become increasingly dependent on international trade, with around $60 billion worth of exports per year in 1995-1997 compared to $7 billion in 1970, $42 billion in 1980, and $45 billion in 1990—see Figure 1.1 for trends in agricultural trade (FAO, 1998). More than 20 percent of the total value for all farm products is exported each year (USDA, ERS, 1997, p. 28). For primary crops the export share is about 30 percent, while it averages around 10 percent for livestock production. The percentage of particular products exported varies considerably, being nearly zero for some products whereas virtually all of others are exported; e.g., the export share is around 97 percent for almonds. Generally, over half of the wheat crop is exported, while around 40 percent of rice and cotton production is exported. Among livestock products, about 17 percent of broiler production is exported along with over 8 percent of beef and about 6 percent of pork.

Agricultural exports benefit the nonfarm as well as the farm sector. It is estimated that each $1 million of agricultural exports creates around 17 jobs; in 1996 U.S. agricultural exports were more than $60 billion, indicating that over one million jobs were supported by this activity (USDA, FAS, 1998). Some three-fifths of

FIGURE 1.1. Values of U.S. Agricultural Exports and Imports

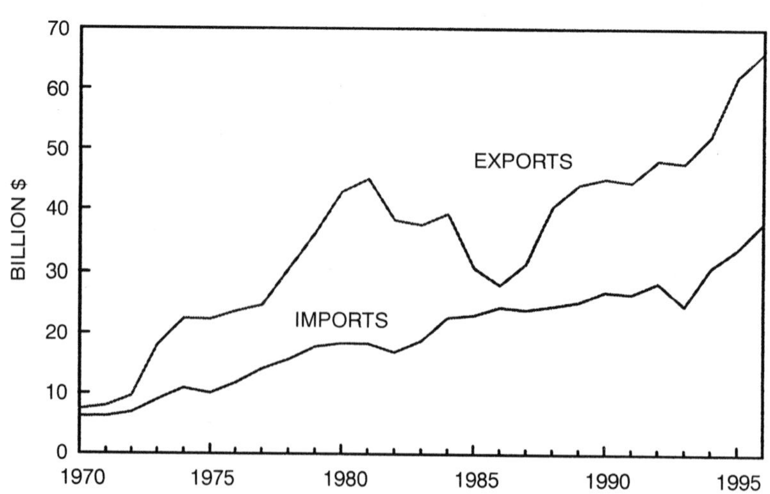

Source: USDA ERS, 1999.

these are off-farm jobs, most of which are located in urban-suburban areas. Processed high-value exports, which have been growing more rapidly than bulk exports in recent years, contribute even more to job creation.

The United States also imports many agricultural products, with a substantial share (about one-third) being tropical products not produced in significant amounts in the United States; others provide variety, help fill seasonal gaps, or reduce costs to consumers. Despite large volumes of imports, the agricultural sector produces a positive balance of payments, which helps to reduce the size of the unfavorable balance that results from large importations of industrial and other products (see Figure 1.1).

Although the total value of U.S. agricultural exports has been rising in recent years, because of worldwide growth in agricultural exports the U.S. share has been constant at around 14 percent of the world total (see Figure 1.2). This share increased to around 18 percent of the total during the international food crisis of the

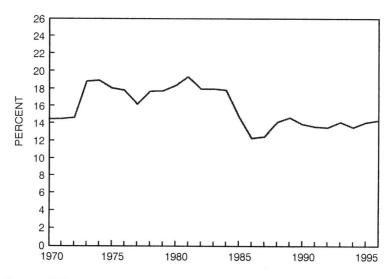

FIGURE 1.2. U.S. Share of World Agricultural Exports

Source: FAO, 1999.

early 1970s, when U.S. farm programs were changed to encourage production and exports. During the international debt crisis and worldwide recession of the early 1980s, the U.S. share of world agricultural exports dropped back to previous levels of about 14 percent. However, world trade in agricultural products has increased, rising from about $52 billion in 1970 to more than $463 billion in 1996.

COMPETITIVENESS

This book addresses issues affecting international trade in U.S. agricultural products, a set of factors that are frequently referred to as competitiveness. There are many definitions of competitiveness and considerable controversy about the use of the term, but it essentially refers to the ability to sell products in international markets. Tweeten (1992, p. 27) defines competitiveness as "a nation's ability to maintain or gain market share by exploiting competitive advantage in world markets through increasing productivity from technological advances or other sources." Other analysts prefer to tie competitiveness to the individual firm and make it more amenable to measurement. Krugman (1994), however, claims that it is not meaningful to talk about the competitiveness of a country, since nations are not the same as corporations, which compete with one another. Thus, Harrison and Kennedy (1997) define competitiveness from the viewpoint of a firm as the ability to profitably create and deliver value at prices equal to or lower than those offered by other sellers in a specific market. Hudson (1993), in a publication for the USDA, indicates that competitiveness has three aspects: economics and policy; geography; and history. However, while economics and policy are related, it might be more appropriate to consider them as separate forces that influence the competitive position of agricultural products.

The economic issues in competitiveness involve, in general, the costs of production and other costs for delivering a product to a market as well as the prices received for the product. The costs of production are a function of the amounts of resources used and the prices of those resources. A very important factor in determining costs, then, is efficiency in the use of resources, which is, in turn, a function of technology. Primarily due to highly effective research in the United States, agricultural productivity has improved continuously, enabling the country's agricultural sector to remain competitive despite higher input costs, especially higher labor costs. U.S. agricultural productivity has increased rapidly since World War II; aggregate output has more than doubled since 1950, while aggregate inputs have actually declined, especially in recent years (see Figure 1.3). However, in addition to production costs, the ability to export to particular markets is affected by transportation, storage, and processing costs as well as by losses and changes in quality of the product. Thus, efficiency and its improvements are essential to attaining and retaining competitiveness.

A wide variety of economic, agricultural, and related policies, such as health, sanitary, and environmental policies, of both the exporting and importing country, also affect the ability to export agricultural products. These include agricultural policies that support prices, subsidize production or exports, or directly or indirectly tax agriculture or agricultural exports; of course the support of agricultural research and education (or a lack of such support) is a

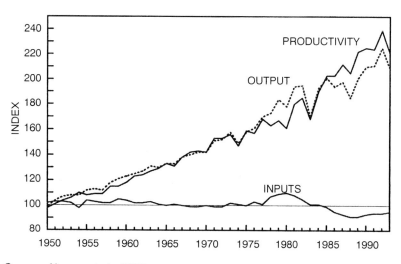

FIGURE 1.3. Growth in U.S. Agricultural Productivity

Source: Ahearn et al., 1998.

major factor impacting on competitiveness, especially in the long term. However, macroeconomic policies that affect the general circumstances of an economy, such as inflation, taxes, exchange rates, etc., are often more important influences on capacity to export; favorable sectoral policies can be easily offset by unfavorable macroeconomic policies (see, e.g., Krueger, 1992). Of course, very important impacts derive from specific trade policies, including tariff and nontariff barriers in potential importing nations, export or import subsidies and taxes, and phytosanitary requirements. In addition, costs of production and efficiency can be affected by environmental, health, and related policies that impose varying restrictions on producers in different countries.

Trade agreements, trading blocks, free trade areas, and similar arrangements are important factors in determining who trades with whom. The Uruguay Round of the General Agreement on Tariffs and Trade (GATT), which created the World Trade Organization (WTO), included agriculture for the first time since this process was started about 50 years ago. Although the initial impacts on agricultural trade are modest, it does begin to incorporate such trade within the GATT process, which has moved gradually toward free trade and which has been accompanied by a large expansion in international trade. In addition to GATT (WTO), many regional trade agreements facilitate trade within a restricted group of member countries, often while erecting barriers to trade with nonmembers. The European Union is one of the more important of these, especially as it has affected agriculture and trade in agricultural products through its Common Agricultural Policy (CAP). However, the North American Free Trade Agreement (NAFTA), the Common Market of the Southern Cone Countries (MERCOSUR), and agreements among several Central American countries, Andean countries, Baltic countries, etc. have very important impacts on trade within and between those blocks of countries.

What Hudson (1993) called geography is actually a combination of geography and climate, which essentially determines the types of products that can be produced within a particular country (or within specific areas of a country or region). The broadest of these are the tropical and temperate zones, although some products are semitropical. Since, except for Hawaii, the United States is in the temperate and semitropical zones, there is little competition with respect to tropical products, which must be imported and are classified as noncompetitive (USDA, ERS, 1997, p. 74). Although natural conditions tend to set the parameters that determine what can be produced efficiently within an area, it is possible to change these through interventions and/or changes in technology. For example, the development of irrigation has enabled production of many crops in arid or semiarid areas that could not be produced under rain-fed agriculture. Similarly, plant (and animal) breeding has enabled the expansion of production into areas where, say, for reason of day length a particular crop had not been economically feasible. An example is the development of hybrid corn varieties in the United States that are suited to longer day lengths in the South and shorter day lengths in the North.

Marketing and market development activities by governments, trade associations, and individual firms also are important in determining trading patterns, in developing new markets, and in retaining existing markets. For markets to operate effectively, information about products, their availability, and prices is essential. Such information can be provided through market development activities such as promotions, fairs, advertising, and other forms of information dissemination. Governments and trade associations often promote a type of product or provide generic information about product availability, characteristics, and prices while individual firms more often promote their particular brand or variety. Thus beef, pork, milk, or poultry (or even Washington apples) might be promoted by governments or trade associations, but corporations will advertise their own brands as, for example, United Brands promotes Chiquita bananas or Perdue Farms promotes their golden-skinned chicken, and so on.

Historical conditions and patterns of trade can influence the ability to sell in a particular market, especially in the short term, since there tends to be some inertia in the process. Existing facilities, contacts between importers and exporters, familiarity with products and shippers, brand names and loyalties, and other factors tend to cause established trading patterns to continue. However, such historical determinants generally will not prevail if producers in other countries become more efficient and can profitably sell their product at a lower price (or if government subsidies are used to sell at lower prices). Thus, trading patterns are constantly shifting as the underlying forces of costs, yields, policies, and other circumstances change. Understanding these forces and changes are essential to remain or become competitive. A purpose of this book is to provide information that will help meet the challenges facing the U.S. agricultural sector in existing and emerging global markets.

PLAN OF BOOK

This book is directed at evaluating the competitiveness of specific U.S. agricultural products since the competitiveness of the country's agricultural sector is a function of the competitiveness of each individual product; some products are competitive and others are not. For many, international trade is an important factor in their economic returns. The United States produces a large number of products, and it is impossible to adequately evaluate all of them completely in a single volume. Thus, the major but not exclusive emphasis is on those which contribute more to the nation's exports. Chapters 2 and 3 present general information on competitiveness and approaches to measuring and understanding the concept. The next nine chapters evaluate the competitiveness of major crops or groups of crops. Chapters 13, 14, and 15 examine the situation for international trade in the major livestock products of the United States, while chapters 16, 17, and 18 develop additional information on the competitiveness of further processed crop and livestock products (which are becoming increasingly important components of U.S. agricultural exports), on the impact of biotechnology on competitiveness, and the effects of NAFTA on U.S. agricultural trade. The last chapter summarizes the findings and presents a set of conclusions and implications from the individual product analyses.

REFERENCES

Ahearn, M., J. Yee, E. Ball, and R. Nehring (1998). *Agricultural Productivity in the United States.* Information Bulletin No. 740, Washington, DC: Economic Research Service, U.S. Department of Agriculture.

FAO (1998). *FAOSTAT Database:* <http: //apps.fao.org/lim500/nph-wrap.pl?Trade.CropsLivestockProducts&Domain=SUA>. Food and Agriculture Organization, United Nations.

FAO (1999). *FAOTSTAT Database.* Food and Agriculture Organization, United Nations. <http://apps.fao.org/lim500/nph-wrap.pl?Trade.CropsLivestockProducts&Domain=SUA>.

Harrison, R.W. and P.L. Kennedy (1997). "A Neoclassical Economic and Strategic Management Approach to Evaluating Global Agribusiness Competitiveness." *Competitiveness Review: An International Business Journal* 7(1):14-25.

Hudson, W.J. 1993. *The Basic Elements of Agricultural Competitiveness,* Miscellaneous Publication 1510. Washington, DC: Economic Research Service, U.S. Department of Agriculture.

Krueger, A.O. 1992. *The Political Economy of Agricultural Pricing: A Synthesis of the Political Economy of Developing Countries*, Volume 5. Baltimore, MD: The Johns Hopkins University Press.

Krugman, P.R. 1994. "Competitiveness: A Dangerous Obsession." *Foreign Affairs* 73(2):28-44.

Tweeten, L. 1992. *Agricultural Trade: Principles and Policies.* Boulder, CO: Westview Press.

USDA ERS (1997). *Foreign Trade of the United States: January, February, March 1997,* Final Issue. Washington, DC: U.S. Department of Agriculture, Economic Research Service.

USDA ERS (1999). "Total Value of U.S. Agricultural Trade." U.S. Department of Agriculture, Economic Research Service. <http://www.econ.ag.gov:80/briefing/agtrade/htm/Data.htm>.

USDA FAS (1998). "Fast Track and Agriculture: What's at Stake for U.S. Workers." FAS Online: <www.fas.usda.gov/itp/fast_track/5jobs.htm> (accessed June 16). U.S. Department of Agriculture, Foreign Agricultural Service.

Chapter 2

Agricultural Competitiveness Issues and Concepts

P. Lynn Kennedy

ISSUES IN AGRICULTURAL COMPETITIVENESS

The economic, political, and technological environment of the 1980s and early 1990s has contributed to the recent focus on competitiveness. Various issues, such as the U.S. budget and trade deficits, have led to an emphasis on the overall competitiveness of the United States economy. A fear of losing competitive advantage to European and Pacific Rim countries has contributed to the investment of time and resources in an attempt to retain and enhance our competitive edge. The agricultural sector has been no exception.

The competitiveness of the U.S. agricultural sector is evidenced by recent agricultural trade surpluses (USDA, ERS, various issues). These surpluses have been particularly significant given the chronic trade deficits experienced by the rest of the U.S. economy. The argument could be made that, given the contribution of agriculture to the trade position of the nation, enhancing the competitiveness of the U.S. agricultural sector benefits the overall economy. Advocates of this position might propose agriculture-specific research and development or export promotion as means to maintain the competitive position of U.S. agriculture. This raises the question as to whether policies of this nature increase the welfare of the nation as a whole.

Although increasing competitiveness appears to be a useful pursuit at first glance, it has been suggested that an obsession with competitiveness at the national level can be detrimental to a country's welfare. Krugman (1994) and Porter (1990) note that it is firms that compete, not nations. Attempts to enhance competitiveness at the national level with disregard to the specific advantages of firms or industries may not yield positive welfare consequences for the nation as a whole. To maximize the welfare of the nation, resources should be directed toward those firms or industries in a way that will maximize the overall welfare of the nation, a concept that hints at the law of comparative advantage.

The development of strategies that benefit the nation as a whole requires an awareness of the interrelationships between factors that influence competitiveness and the welfare of various interest groups. At the same time, several contemporary issues have influenced, and will continue to influence, the competitiveness of U.S. agriculture. To facilitate the commodity analyses presented in this book, this chapter will discuss four key issues and their relationship with agricultural competitiveness. These include domestic agricultural policies, agricultural trade agreements, processed and differentiated products, and biotechnology. Several definitions of competitiveness will be considered. In addition, factors that influence competitiveness will be identified and discussed. Finally, potential indicators of competitiveness will be reviewed.

Domestic Agricultural Policy

The U.S. agricultural sector has faced a turbulent policy environment in recent years. Changes in domestic and international policy mechanisms have forced producers to adapt to a new playing field. Central to these changes is the Federal Agriculture Improvement and Reform Act of 1996 (FAIR). The reforms that stem from FAIR are consistent with global trends in agricultural policy that include increased market orientation, decreased government regulation, and the desire to lower the costs of agricultural programs.

This trend in domestic agricultural policy toward increased market orientation has the potential to impact the competitiveness of U.S. agriculture in a number of ways. On the surface, decreased production incentives lower the effective commodity prices received by producers, resulting in reduced profits and, thus, competitiveness. However, these decreased production incentives could be a catalyst that causes domestic producers to tighten their belts, adopt state-of-the-art technologies, and reduce their cost of production. This, in turn, will enhance their competitive position relative to other domestic sectors and the rest of the world. It is, thus, important to account for the dynamic effects of various factors throughout analyses of agricultural competitiveness.

Agricultural Trade Agreements

In addition to changes in domestic policies, the rules governing the international trade of agricultural products are rapidly changing as institutions, such as the World Trade Organization (WTO) and the North American Free Trade Agreement (NAFTA), seek to lower trade barriers and increase market access. The course international agricultural policy will take as governments prepare for the next round of WTO agricultural negotiations is a critical issue, as indicated by the failure to reach an agreement during the negotiations held in Seattle in December, 1999.

Of importance to the competitiveness of U.S. agriculture is the type and degree of trade liberalization that occurs. Multilateral trade liberalization, such as that proposed within the WTO, has the potential to create a "level playing field." The removal of protection will have differing effects, depending on the initial levels of support. However, this trend toward freer trade will increase the clarity of world price signals. As a result, agricultural production will increasingly be based on comparative advantage rather than domestic or international agricultural policies.

Processed and Differentiated Products

The world market for agricultural products has historically been composed of commodity trade. The United States has a strong tradition in this market. In recent years the share of processed and differentiated agricultural products has increased. However, the United States has not shared in this market expansion to the extent of several European countries. This raises questions as to why the United States has failed to keep pace with this expanding market. Just as important, if not more, is the question of whether it is in the best interest of the United States to pursue this expanding market. If the country does not possess an advantage in the processed foods sector, is it in our best interest, through whatever means, to encourage the development of an advantage in the processed food products area?

Hughes (1992) examines the argument that, given the increasing competition from newly industrializing countries in low-technology products, maintenance of international competitiveness requires advanced countries to specialize and become competitive in higher technology sectors. Although this proposition may be true for the manufacturing and services sectors, it requires careful evaluation with respect to the agricultural sector. This issue is examined to some extent by Gopinath, Roe, and Shane (1996), who discuss the two-way transfer of efficiency gains between primary agriculture and the processed food sector. Given this symbiotic relationship, strategic policy should aim at coordination between sectors rather than specialization in only one.

Also of importance to the evaluation of competitiveness in processed and differentiated agricultural products is the framework used. Traditional concepts, such as comparative advantage, were useful in examining competitiveness when agricultural economists were, for the most part, dealing with commodities. The increased quantity and importance of processed and differentiated agricultural products necessitates the use of, at the very least, a modified concept. Firms are increasingly able to differentiate their products and themselves, thus affecting their ability to provide quality to the consumer. As a result, analysts must consider quality issues as they evaluate agricultural competitiveness.

Development and Adoption of Biotechnology

An additional issue facing the U.S. agricultural sector involves recent trends in the development and adoption of biotechnology. For example, how will the development of herbicide-resistant plant varieties by U.S.-based multinational firms affect the competitiveness of U.S. agriculture? A host of side issues, including consumer acceptance of the resulting products, will frustrate the quesiton.

A major consideration with respect to the development and adoption of biotechnology is related to the concept of the "agricultural treadmill." As more and more producers adopt technologies designed to improve their operational efficiency, the supply curve shifts to the right. Given typical demand elasticities for agricultural products, this results in a downward trend in producer prices. Unless producers possess proprietary technology to which access can be limited, care should be taken to ensure that the call to competitiveness does not adversely affect all producers. The paradox of this is that firms not aggressively adopting new technologies may ultimately find themselves squeezed out of the industry.

DEFINITIONS OF COMPETITIVENESS

Competitiveness has been addressed from a number of different perspectives in the literature. Researchers focusing on the national level have defined competitiveness as the ability to sustain an acceptable growth rate and real standard of living for their citizenry while efficiently providing employment and maintaining the growth potential and standard of living for future generations (Landau, 1992). This definition is linked to a nation's employment and, consequently, the standard of living of its citizens. The level of national employment, growth of employment, and the standard of living in an economy, however, depend on the competitiveness of firms within the country. Hence, a nation's competitiveness depends on the underlying factors that influence the competitiveness of individual firms and industries.

Other definitions contrast competitiveness with comparative advantage. The law of comparative advantage suggests that trade flows occur as the result of relative cost differentials between countries. Barkema, Drabenstott, and Tweeten (1991) contend that this theory does not apply to a world with market-distorting government policies. They assert that competitiveness takes a more realistic view of the world. Their definition, similar to that above, views competitiveness from a national perspective. It also implies that government policy affects competitiveness. However, it fails to provide insight into the underlying sources of competitiveness or account for demand-side factors, such as product differentiation. Thus, a description of the linkages between the sources and indicators of competitiveness must account for the effect of government policies and consumer demand.

Porter (1990) advances the notion that firms, rather than nations, compete with one another in international markets. When considering competitiveness, the emphasis must not be placed on the economy as a whole, but on specific industries and industry segments. Competitive advantage results from the difference between the value a firm is able to create for its buyers and the cost of creating that value. Superior value results from offering lower prices than competitors for equivalent benefits or by providing unique benefits that more than offset a higher price.

Firm-level definitions of competitiveness have been put forward by various economists. For example, competitiveness is defined as the ability to deliver goods and services at the time, place, and form sought by buyers at prices as good as or better than other suppliers while earning at least opportunity costs on resources employed (Sharples and Milham, 1990; Cook and Bredahl, 1991). This definition, although viewing competitiveness from the perspective of the firm, fails to address the sources that give firms the ability to deliver goods or services at "competitive" prices. Still other definitions view competitiveness as the sustained ability to profitably gain and maintain market share in domestic or foreign markets (Van Duren, Martin, and Westgren, 1991). This firm perspective explains competitiveness in terms of performance indicators (e.g., net worth, profitability, and market share).

These definitions contrast the differing approaches used to analyze competitiveness. The strategic management school defines competitiveness as the ability to profitably create and deliver value through cost leadership or product differentiation (Kennedy et al., 1997). This approach implies that competitiveness is directly related to factors that influence a firm's cost and demand structure. Other schools of thought place greater emphasis on the indicators of competitiveness, described as the sustained ability to profitably gain and maintain market share (Van Duren, Martin, and Westgren, 1991). Both approaches can be useful for evaluating competitiveness, depending on the objectives of the researcher. However, neither demonstrates a clear linkage between the factors that influence the cost and demand structure of the firm and possible measures of competitiveness.

FACTORS INFLUENCING COMPETITIVENESS

Analyzing a nation's competitiveness requires that the underlying factors influencing individual firms and industries be examined (Porter, 1990). Firms become more competitive by creating value through cost leadership or product differentiation (Porter, 1980). More specifically, technology, attributes of purchased inputs, product differentiation, production economies, and external factors are primary sources of competitiveness (Harrison and Kennedy, 1997). Each of these factors affect a firm's costs and the degree to which it can differentiate its products.

Technology

Development and adoption of technology affects the firm in several ways. The operational competitiveness of a production unit is a critical component of its overall competitiveness (Parkan, 1994). Cost advantage can be achieved through proprietary technologies that affect the productivity of labor and capital. The impact of employing new methods depends, to a large extent, on firm behavior and industry structure. For example, some technologies enable firms to lower production costs for a given quantity of outputs. Other technologies allow the firm to increase its quality of output given an initial set of inputs.

A technology is *productivity enhancing* if its adoption allows firms to decrease costs per unit of output. Conversely, a technology is *quality enhancing* if it increases quality per unit of input. Despite the tendency to categorize technology as either productivity enhancing or quality enhancing, many technologies fall into both classifications. The existence of technologies that are simultaneously productivity and quality enhancing, combined with the effects of firm behavior, imply that both cost and quality factors must be utilized in an analysis of competitiveness.

A firm's adoption of productivity enhancing or quality enhancing technologies will cause movement in the supply or demand curves. Although various technologies affect production in different ways, supply and demand link technology with profits. These linkages are useful in analyzing the relationship between sources of competitiveness and profits.

Input Costs

Costs are also influenced by the price, quality, and dependability of purchased inputs. This is one of the most direct and obvious sources of competitiveness. Even so, it is difficult for a firm to attain an advantage in this area. To illustrate this point, consider firms producing an identical product. Assume that inputs are proportionate for the companies and that the cost of inputs declines. This decrease in cost shifts the supply curves of the firms to the right. However, it does not change any firm's cost of production relative to others. Although the firms incur lower production costs, none gain competitive advantage relative to others.

To gain a competitive edge in this area, a firm must lower input costs relative to those of rival firms. Suppose one of the firms in the previous scenario has the capacity to obtain more accurate information regarding future input availability. Knowledge of this type permits the firm to better coordinate its input procurement. The resulting cost advantage influences the relative competitive advantage of the firms.

Production Economies

As mentioned previously, cost advantage is a major determinant of competitiveness, allowing a firm to gain a competitive edge over rivals and deter entry of new firms. One way that cost advantage can be attained is through economies of size and product scope. Economies of size occur when plant size is adjusted in a way that decreases average costs of production. When a firm is able to capture a larger share of the market, the resulting increase in production in the short run allows fixed costs to be spread over increased output, thus reducing average costs. Perhaps more important, the sustained ability to maintain market share allows the firm to adjust plant size and attain economies of size.

One of the arguments for the efficiency of the U.S. meat packing industry is its evolution from a large number of medium-sized packers to a few large firms that control most of the market. The increased size of these firms allows them to reduce average cost through a greater division of labor. However, Roy (1997) points out that excess profits, resulting from imperfect competition, are generated at the expense of payments to labor and capital. New entrants into the industry increase competition which, in turn, increases the payments to both labor and capital. Capital accumulation increases in response to the increased returns to capital, which speeds up growth and, in turn, competitiveness.

Economies can also be achieved by broadening the scope of products that a firm produces. The firm's scope can be adjusted to produce a wide variety of products that are close substitutes in the production process. Expansion of a firm's product line can allow it to utilize excess capacity. Thus, economies of scope permit the firm to spread the cost of its fixed assets over additional product lines.

Also relevant to this topic is the influence of externalities in industry development. The impact of spillovers from one sector to another can result in symbiotic benefits that would not accrue if the industries functioned in isolation. Going beyond the simple identification of optimal firm or industry size, attention to this concept raises the issue of how policymakers can best facilitate the development of industry networks. To attain strategic improvements in competitiveness, Grupp (1995) advocates increased cooperation in science, technology, and innovation studies based on regional differences in comparative advantage. The formation of strategic industrial alliances based on a region's relative strengths can result in a formidable advantage that is difficult for competitors to overcome.

Product Quality and Enterprise Differentiation

Another factor influencing the firm's competitiveness is its ability to differentiate itself. Firms differentiate their products from those of their competitors to increase market share and develop consumer loyalty. Product differentiation is the degree to which the products of competing sellers substitute for one another in consumption (Marion, 1986). A primary way to achieve product differentiation is to provide superior product quality. Research and development, quality control, and the use of higher quality inputs are among the sources that affect product quality. Another factor that affects a firm's competitiveness is enterprise differentiation: the ability to distinguish oneself from rivals. By providing superior services, firms can enhance the reputation of their company and product lines.

Of particular importance to the competitiveness of U.S. agriculture is the issue of reliability. As various segments of the U.S. agricultural sector attempt to differentiate themselves and their products from those of other countries, the dependability of export supply becomes a factor. Reliability and the timely delivery of goods and services are as much a part of quality as the physical attributes of the product. Given the history of U.S. agricultural exports, including episodes involving grain embargoes, producers must strive to nurture and maintain a reputation of dependability in the world market.

Promotion

Although physical differences between products contribute to the degree of product differentiation, in many cases promotional strategies are sufficient to differentiate products in the mind of the consumer. Promotional strategies influence the consumer's perception of a product, thus affecting demand. These include brand advertising and export promotion.

Brand advertising is one means by which a firm can distinguish its products from those of other firms. A successful advertising strategy establishes a barrier to market entry by creating brand loyalty. This loyalty is based on the customer's perception that the preferred product conveys greater value relative to close substitutes. Brand loyalty allows a firm to pursue one of two strategies. The firm can sell the same amount of its product at prices higher than competitors, or it can sell more of its product at prices equal to competitors. In either case, demand for the firm's product and its relative competitiveness in the market will increase.

Firms pursuing the first strategy will maintain market share, while those following the second strategy will increase market share. However, the effect of advertising on profitability is ambiguous. For example, a firm pursuing the first strategy will capture a larger portion of the market, but this expansion in market share may result in a loss of profits as increased advertising will also increase short-run costs. As a result, an increase in short-run profits will occur only if the marginal return from advertising outweighs its marginal cost. Yet, the firm may choose to incur these losses in order to gain economies of size in the long run. In this case, the firm's long-run profits will increase.

Discussions regarding the impact of promotion on competitiveness must consider the differentiated product and commodity distinctions that exist within the agricultural sector. The previous discussion of advertising and promotional strategies has clear implications for producers of differentiated products. Just as important, although perhaps not as immediately visible, are the domestic and export promotional activities that occur on behalf of various commodity groups. As with the formulation of any plan to enhance competitiveness, developing strategic programs of this nature must simultaneously account for long-run costs and benefits. In particular, the development of foreign market demand, although successfully created, is futile if that market is later captured by competitors. Successful strategic planning must consider these contingencies.

External Factors

A number of external factors influence the competitiveness of firms and industries. Among these, government policies affect competitiveness in both domestic and international markets. This linkage is such that changes in the real agricultural price consist of a world price component, a real exchange rate component, and a sector-specific price intervention component (Quiroz and Valdés, 1993). Policies that subsidize the production of raw agricultural commodities directly affect the prices that food processors pay for inputs. Lowering the price of agricultural commodities leads to lower costs for downstream firms and an increase in their competitiveness relative to foreign rivals.

Government policies also affect a firm's ability to obtain world market share. Export subsidies lower the world price at which domestic industries are willing to sell various quantities of their product. As a result, exporters can sell their products at a discounted price on the world market while maintaining, or increasing, their effective price per unit. This acts to expand the world market share of the subsidized firm or industry.

Macroeconomic variables, such as exchange rates, consumer incomes, and population growth, also influence the competitiveness of the firm. For example, a devaluation of the U.S. dollar has the effect of lowering the price of U.S. goods in foreign markets. Although individual firms have little influence on the exchange rate, they benefit from increased profits and market share. Thus, government policies and other factors beyond firms' control influence competitiveness.

Competitiveness is also influenced by publicly funded programs. For example, the land grant system is responsible for a large portion of the productivity of U.S. agriculture. Publicly funded research and development have resulted in a variety of technological innovations that have affected U.S. agricultural competitiveness over the past several decades. In addition, the Cooperative Extension Service has served to facilitate the transfer of information and the adoption of new technology. According to Mowery and Oxley (1995), inward technology transfer typically depends on the presence of successful national innovation systems that enhance "national absorptive capacity." Investments in scientific and technical training and the existence of an economic policy environment that fosters competition among domestic firms are critical to the development of this capacity. In the presence of publicly funded research by foreign competition, a laissez-faire response may, at the very least, be extremely costly (Milberg, 1991).

INDICATORS OF COMPETITIVENESS

The previous discussion identified a number of sources that influence competitiveness. These sources can be grouped into two categories: those which affect the firm's relative cost of production and those which affect the quality, or perceived quality, of its product or business enterprise. As firms gain advantage through the various sources of competitiveness, relative market share and profits increase. In situations where firms are able to decrease production costs or improve their products relative to other firms in the industry, market share will increase.

The ability of existing firms to profitably gain and maintain market share indicates a competitive advantage. Yet knowledge of a firm's profitability or market share does not provide information regarding any specific source of competitiveness. An increase in the profitability of a firm or industry may indicate an increase in competitiveness but not whether this result stems from decreased cost, increased quality, or some external factor. Similarly, relative advantage in any individual source of competitiveness does not guarantee profitability or a sustained share of the market. For example, cost-reducing technologies that adversely affect product quality do not necessarily increase competitiveness. As a result, the measures and indicators used to evaluate competitiveness must be selected based on the circumstances of the unit of analysis.

Broad measures, such as market share and profitability, provide useful insights into overall competitiveness. On the other hand, the individual sources of competitiveness provide information with respect to specific strengths and weaknesses. Used separately, these tools provide a valuable indication of the firm's competitive position. Utilized together, they provide information regarding the strengths to be maintained and exploited or the weaknesses that are prime targets for improvement.

From an international perspective, agricultural competitiveness is reflected by the ability to profitably gain and maintain world market share. An increase in market share typically indicates an increase in competitiveness, while a decrease in market share would indicate the opposite. However, it must be remembered that the factors influencing competitiveness are not identical to those affecting comparative advantage. If the enhancement of societal welfare is an objective of the policymakers, each determinant of competitiveness must be considered in the formulation of strategic agricultural policy.

REFERENCES

Barkema, A., M. Drabenstott, and L. Tweeten. (1991). "The Competitiveness of U.S. Agriculture in the 1990s in Agricultural Policies." In *Agricultural Policies in the New Decade*, edited by K. Allen, Washington, DC: Resources for the Future, National Planning Association, pp. 253-284.

Cook, M. and M.E. Bredahl. (1991). "Agri-business Competitiveness in the 1990s: Discussion." *American Journal of Agricultural Economics* 73(5): 1472-1473.

Gopinath, M., T.L. Roe, and M.D. Shane. (1996). "Competitiveness of U.S. Food Processing: Benefits from Primary Agriculture." *American Journal of Agricultural Economics* 78(4): 1044-1055.

Grupp, H. (1995). "Science, High Technology and the Competitiveness of EU Countries." *Cambridge Journal of Economics* 19(1): 209-223.

Harrison, R.W. and P.L. Kennedy. (1997). "A Neoclassical Economic and Strategic Management Approach to Evaluating Global Agribusiness Competitiveness." *Competitiveness Review: An International Business Journal* 7(1): 14-25.

Hughes, K.S. (1992). "Technology and International Competitiveness." *International Review of Applied Economics* 6(2): 166-183.

Kennedy, P.L., R.W. Harrison, N.G. Kalaitzandonakes, H.C. Peterson, and R.P. Rindfuss. (1997). "Perspectives on Evaluating Competitiveness in Agribusiness Industries." *Agribusiness: An International Journal* 13(4): 385-392.

Krugman, P. (1994). "Competitiveness: A Dangerous Obsession." *Foreign Affairs* 73(2): 28-44.

Landau, R. (1992). "Technology, Capital Formation and U.S. Competitiveness." In *International Productivity and Competitiveness,* edited by B.G. Hickman, New York: Oxford University Press, pp. 299-325.

Marion, B.W. (1986). *The Organization and Performance of the U.S. Food System*. Lexington, MA: D.C. Heath and Company.

Milberg, W.S. (1991). "Structural Change and International Competitiveness in Canada: An Alternative Approach." *International Review of Applied Economics* 5(1): 77-99.

Mowery, D.C. and J.E. Oxley. (1995). "Inward Technology Transfer and Competitiveness: The Role of National Innovation Systems." *Cambridge Journal of Economics* 19(1): 67-93.

Parkan, C. (1994). "Operational Competitiveness Ratings of Production Units." *Managerial and Decision Economics* 15(3): 201-221.

Porter, M. (1980). *Competitive Strategy: Techniques for Analyzing Industries and Competitors*. New York: The Free Press.

Porter, M. (1990). *The Competitive Advantage of Nations*. New York: The Free Press.

Quiroz, J. and A. Valdés. (1992). "Agricultural Incentives and International Competitiveness." *Food Policy* 18(4):342-354.

Roy, U. (1997). "Intra-Industry Competitiveness and Economic Growth." *Journal of Economics and Business* 49(2): 117-125.

Sharples, J. and N. Milham. (1990). *Long-Run Competitiveness of Australian Agriculture*. Foreign Agricultural Economics Report No. 243. Washington, DC: United States Department of Agriculture, Economic Research Service.

USDA ERS (various issues). *Foreign Agricultural Trade of the United States*. Washington, DC: United States Department of Agriculture, Economic Research Service.

Van Duren, E., L. Martin, and R. Westgren. (1991). "Assessing the Competitiveness of Canada's Agrifood Industry." *Canadian Journal of Agricultural Economics* 39(4): 727-738.

Chapter 3

The Competitiveness of U.S. Agriculture

Munisamy Gopinath
Terry L. Roe

INTRODUCTION

The widening trade deficit and stagnant gross domestic product (GDP) growth rates of the early 1990s raised concerns about the "competitiveness" of the U.S. economy. The popular view of competitiveness has focused on the ability to sell abroad (export), while academic circles have debated the existence, merits, and underpinnings of the concept in economic theory (Krugman, 1996). The debate and the ensuing controversy led Robert Reich, an early proponent of the concept of competitiveness, to remark, "rarely has a term in public discourse gone so directly from obscurity to meaninglessness without an intervening period of coherence" (Reich, 1992, p. 1). Despite the confusion on the concept of competitiveness, some lessons have been learned. The purpose of this chapter is to show how competitiveness can be measured within the bounds of economic theory, and usefully analyzed from a public policy perspective.

U.S. agriculture provides a growing array of food of increasing variety and quality. Agriculture's share of the U.S. GDP is only slightly over 1 percent, but agricultural exports accounted for almost

---

This chapter is based on a two-year research project seeking empirical insights into how U.S. agriculture's competitiveness can be sustained or increased, conducted in collaboration with the Economic Research Service, USDA.

9 percent of total U.S. exports of goods and services in 1997 (Gopinath and Roe, 1996). This export performance and the increases in variety and quality of food have come about despite the negative terms of trade with the rest of the economy faced by the agricultural sector, i.e., real prices received by farmers show a long-term declining trend (Gopinath and Roe, 1996). Although U.S. agriculture was a trade surplus sector, the concept of competitiveness found itself associated with foreign market shares and opportunities. The U.S. share of global agricultural trade, worth over $381 billion in 1994, has fallen from 22 percent in 1962 to less than 15 percent in 1994 (Gehlhar and Vollrath, 1996). Coupled with the fact that processed (high-value) products accounted for less than 40 percent of U.S. agricultural exports, compared with a global share of over 60 percent, U.S. agriculture was suspected of losing competitiveness. As we discuss in detail below, the performance of U.S. agriculture in the past was largely due to a high rate of growth in its total factor productivity. Although policies that give access to foreign markets are a prerequisite, concerns relating to U.S. agriculture's competitiveness should focus on investments that sustain and enhance productivity.

A CONCEPTUAL FRAMEWORK

The concept of competitiveness has been used in a broad set of contexts. For instance, competitiveness has been related to growth in the real standard of living (Landau, 1992) or rate and determinants of productivity growth (Porter, 1990). In other cases, competitiveness has been viewed as a combination or continuum of economic theories and subdisciplines including macroeconomics, trade theory, and business strategies (Abbott and Bredahl, 1994). Although these definitions had a broad focus, a major shortcoming in them is the lack of a unified theory or framework to analyze competitiveness. In what follows, we define competitiveness, as it relates to a growth decomposition exercise. Our focus is at the sectoral level rather than that of the aggregate economy, as it is a relative concept with two dimensions: domestic and international. If within an economy, say the United States, the rate of growth in agriculture's real GDP exceeds that of the economy, i.e.,

$$d(\ln GDP_A)/dt > d(\ln GDP)/dt \tag{1}$$

then, we say that agriculture's (A) domestic competitiveness is growing relative to the rest of the economy. The derivatives in equation (1) are total rather than partial, and suggest that the sources of growth can be decomposed into effects of prices and inputs, and technology or total factor productivity (TFP) effects. That is, trade models (Hecksher-Ohlin type) have defined GDP, $G(p,v,\tau)$, as a function of output prices (p), factor endowments or inputs (v), and the level of technology (τ). Growth arising from each of these three sources can be significantly affected by public policy. For example, a price support policy can bring about faster growth in agriculture relative to other sectors; acreage reduction and other conservation programs can affect the availability of land, and thus, agricultural growth; investments in research and development (R&D) and infrastructure can bring about increases in the level of technology.

Now, consider a comparison of agricultural sectors of two countries, say that of the United States and country "X":

$$\frac{d(\ln GDP_{A, US})/dt}{d(\ln GDP_{US})/dt} > \frac{d(\ln GDP_{A, X})/dt}{d(\ln GDP_X)/dt} \tag{2}$$

If the real GDP of agriculture relative to nonagriculture of one country is growing compared with that of another, then we say the first country is gaining bilateral agricultural competitiveness over the second. In the case of equation (2), U.S. agriculture is growing relative to country X agriculture and is said to be gaining bilateral competitiveness. Note this is again a function of the underlying sources of growth in agricultural GDP.[1]

For the United States to be globally competitive, it has to be the case that:

$$\frac{d(\ln GDP_{A, US})/dt}{d(\ln GDP_{US})/dt} > \frac{d(\ln GDP_{A, W})/dt}{d(\ln GDP_W)/dt} \tag{3}$$

where $GDP_{A, W} = \Sigma_x GDP_{A, X}$ is world agricultural GDP. In a competitive economy with no trade distortions this result implies that in the aggregate and on average over a period, U.S. farmers are competing more successfully for world consumers, including U.S. consumers, than are the rest of the world's farmers.[2]

The distinction between price/input effects and TFP effects has implications for sustaining the competitiveness of a sector.[3] For instance, assume country X has high price support policies, while U.S. agricultural growth is dominated by TFP effects or growth in TFP. The price and input effects result in one-time benefits and have to recur periodically to sustain growth, but the TFP effects are long-run in the sense that they do not perish in one time period. Further, annual increases in price supports can increase growth in real GDP, but this source can be artificial and not sustainable when prices are supported above world market levels. In this example, U.S. agriculture will maintain its competitiveness in the long run, while country X may have high growth rates during the period of increasing price supports. However, increasing price supports only come about through growing budget support, which over time becomes increasingly difficult to sustain.

Krugman (1996) points out that the productivity of a sector per se has little, if anything, to do with international competitiveness. Instead, it is relative sectoral efficiency gains (gains in U.S. agriculture relative to nonagriculture compared with that of its major competitors) that determine trade performance. Although productivity growth of a sector or an economy is vital to a country's standard of living, absolute productivity comparisons across countries alone provide no insights into competitive advantage. The productivity of agriculture relative to nonagriculture in the United States compared with that of its major competitors determines international competitiveness or, as Krugman suggests, "the success of a country depends not on absolute but on comparative productivity advantage" (p. 272).

In what follows we draw upon our previous work in evaluating equations (1) and (2) for comparing U.S. agriculture with major European agricultural sectors.[4] The results from previous work are used here to address concerns about the competitiveness of U.S. agriculture.

THE CASE OF U.S. AGRICULTURE

Our analysis of U.S. agriculture's competitiveness begins by decomposing growth in the U.S. GDP into its various sources, and then analyzing these sources in more detail. Row 1 of Table 3.1 shows the sources of growth in U.S. GDP for the period 1959-1991 (Gopinath and Roe, 1996). Growth in total resources, labor, and capital (row 1, column 3) explains about 75 percent of the 2.92 percent annual rate of growth in U.S. GDP. Favorable changes in the terms of trade for the services sector relative to agriculture and manufacturing sectors account for a mere 5.4 percent, with about 19.6 percent caused by technological change, which we refer to here as growth in TFP. More simply, but a bit inaccurately,[5] had there been no change in the level of resources available to the economy, and had prices remained unchanged, the U.S. economy would have grown by about 0.6 percent per year due to technological change alone.

Now, consider a similar decomposition of growth for U.S. agriculture (Gopinath and Roe, 1997). The negative number associated with price effects (column 4, row 2; Table 3.1) reveals the declining terms of trade faced by agriculture within the U.S. economy. This decline in real prices (terms of trade) would have caused agriculture's GDP to decrease by about 1.19 percent per year on average over the period 1959-1991, holding all else constant. In light of the declining trend in real agricultural prices, some resources fled agriculture, as the smaller negative number indicates in column 3, row 2 of Table 3.1. Despite the negative changes in price and input levels, the growth in agricultural GDP is positive due to the growth in TFP, which averaged over 2.3 percent per year over the period 1959-1991 (column 5, row 2; Table 3.1). Had the levels of prices and inputs remain unchanged, agricultural output and GDP would have grown by 2.3 percent per year.

These aggregate effects of prices, inputs, and productivity on agricultural growth conceal the reallocation of resources among its various subsectors (Gopinath and Roe, 1995). While the outputs of grain, other crops, and livestock have all grown, the bias of productivity growth has been toward grains and crops relative to livestock sectors (see Table 3.2). Since grain production is far less labor-

TABLE 3.1. Growth Decomposition of U.S. Economy, Agriculture, and Food Processing, 1959-1991

Year	GDP Growth	Input Effect	Price Effect	TFP Growth
Economy	2.92	2.36	0.09	0.47
Agriculture	0.97	-0.15	-1.19	2.31
Food Processing	1.04	1.46	-0.83	0.41

Sources: Gopinath and Roe, 1996, 1997; Gopinath, Roe, and Shane, 1996.

TABLE 3.2. Productivity Growth Rates at the Subsectoral Level, U.S. Agriculture, 1949-1991

	Grains	Other Crops	Meat	Other Livestock	Overall Agriculture
TFP Growth	2.8	2.9	1.4	1.9	2.3

Source: Gopinath and Roe, 1995.

intensive than livestock, rising real wages caused by the growth of the services sector places larger upward pressures on cost of livestock production than on grain production. Moreover, the higher rates of productivity growth in the grain and crop subsectors have, at the margin, helped them to pull resources from the livestock sectors. As the grain and crop subsectors are more intensive in the use of intermediate inputs (e.g., fertilizers and pesticides), they also have benefitted from technological change embodied in these inputs.

The sources of growth in agricultural TFP are shown in Figure 3.1 (Shane, Roe, and Gopinath, 1998). The height of the bars are the measures of agriculture's rate of TFP growth after random and noise factors have been removed. The partition of the bars provides the explanation for their height. Notice the important role played by investments in public R&D in agriculture. This source accounts for nearly one-half of the average rate of growth in TFP (2.3 percent per year), and reached its peak contribution in the early 1980s. Investments in infrastructure, such as roads, rural electrification,

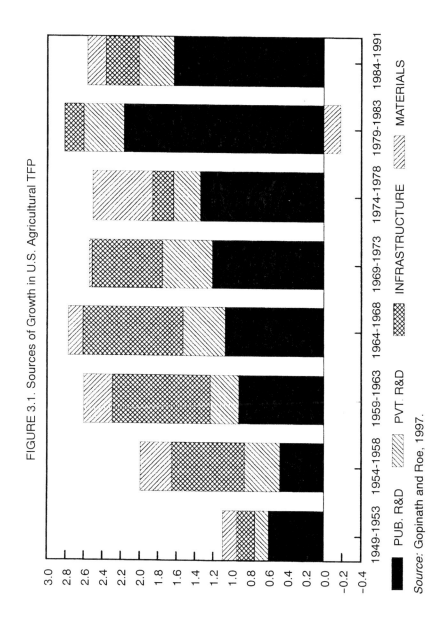

FIGURE 3.1. Sources of Growth in U.S. Agricultural TFP

Source: Gopinath and Roe, 1997.

and sanitation, have played an important role in agricultural TFP growth. This source reached its peak in the early 1960s, and now plays a smaller role as most current investments are for maintenance. Private investment in agriculturally related R&D is almost equal in importance to the embodied sources of technological efficiencies found in modern inputs of machinery, chemical and biological technologies not accounted for by public and private R&D.

So, the growth of agriculture relies mostly on those investments that increase the sector's TFP. Growth in TFP, which helps farmers to compete with others for economy-wide resources, also helps to keep the sector competitive in world markets. Of course, policies that give access to these markets are a prerequisite. Otherwise, farmers would face even more rapid declines in prices. Still, growth in agriculture's GDP is less than growth in its TFP. This implies that much of the efficiency gains that agriculture has achieved in the past have been passed on to the rest of the economy, both domestic and rest of the world. To see how these gains are passed on, consider the decomposition of growth in the U.S. food processing sector.

The growth in real GDP of the food processing sector averaged slightly more than primary agriculture, at 1.04 percent per year (row 3, Table 3.1), over 1959-1991 (Gopinath, Roe, and Shane, 1996). Intermediate inputs were the major contributors to the growth of GDP in the food processing sector. All else being constant, the contribution from all inputs to growth in its real GDP averaged 1.46 percent per year, with intermediate inputs alone (which include primary agricultural output) contributing 1.01 percent per year. Growth in TFP averaged 0.41 percent per year. These two sources, together, yield an average annual rate of growth in the sector's output of about 1.87 percent. However, like primary agriculture, this sector was subject to declining real prices (0.83 percent per year), resulting in an average (net) growth in GDP of 1.04 percent per year.

Primary agriculture accounted for, on average, about 26 percent of the intermediate inputs used by the food processing sector over the 1959-1991 period. So, approximately a fourth of the contribution from inputs to growth in the food processing sector's GDP is attributable to the primary agricultural sector. The drop in the primary agricultural price index of 1.19 percent per year translates into a decline of 0.32 percent per year in the total procurement costs of the food processing sector. Thus, productivity growth in primary agriculture is passed on to the food processing sector through the reduced cost of primary inputs. However, prices for processed foods declined, on average, 0.83 percent per year. The net decline (0.51 percent) comes from within the sector and exceeds the rate of productivity growth (0.41 percent) in food processing.

In summary, a major factor causing GDP growth rates in U.S. agriculture and food processing sectors lower than those of the overall economy is the effect of declining real prices. This should not be viewed as a decline in competitiveness but as transfer of efficiency gains from agriculture and food processing to the rest of the economy. Investments that enhance productivity growth in these sectors not only maintain their competitiveness, but also benefit the broader economy.

COMPARISON OF AGRICULTURAL GROWTH IN THE UNITED STATES AND EUROPEAN UNION

Growth in European agricultural output has been relatively high over the past few decades, coincident with the European Union's (EU) support for agriculture (Arnade, 1997). Growth in productivity, particularly in the 1970s, has been higher in European agriculture than in U.S. agriculture. In the 1980s, the gap between the rates of productivity growth declined (Ball et al., 1996). There are two hypotheses about growth in EU agriculture: (1) growth has been stimulated by high and stable prices that producers received under the European Union's Common Agricultural Policy (CAP); and (2) growth has been the result of technical change. Our results suggest both factors have contributed to agricultural growth in the EU.

Growth in real agricultural GDP for four countries, Denmark, France, Germany, and the U.K., was decomposed for the period 1974-1993, similar to that for U.S. agriculture. The results (Gopinath, Roe, and Shane, 1996) indicate that productivity is the primary source of growth in agricultural output in these countries (see Figure 3.2), as it was for the United States. Overall, the decline in real agriculture prices in Denmark, France, and the U.K. was smaller than that experienced by U.S. producers, and thus, the negative price effects on their agricultural GDPs were lower than that of the

FIGURE 3.2. Components of Agricultural GDP Growth in the United States and Major European Countries, 1974-1993

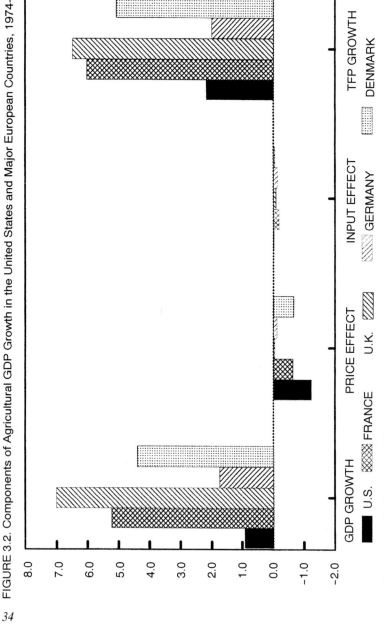

Source: Gopinath et al., 1997.

United States.[6] Since 1988, the effect of declining prices on the EU's agricultural GDP growth has been relatively large, but still smaller than that for the United States.[7] Accompanying the negative price effects is a fall in the EU's rate of agricultural TFP growth. Together, these two effects have substantially lowered the four countries' growth in agricultural GDP, particularly in recent years.

All else being constant, the level of a country's exports depends not on absolute but on comparative productivity advantage.[8] The assessment of changes in comparative advantage between two countries entails a comparison of the ratio of growth in agriculture to growth in the rest of the economy. Productivity growth is the dominant factor explaining output growth in both U.S. and European agriculture. The ratio of agricultural to nonagricultural productivity growth in the United States is about 10 (2.17 percent productivity growth in agriculture and 0.21 percent productivity growth in the entire economy) during the 1974-1991 period (see Figure 3.3). For the European countries, agricultural productivity growth ranged from about 7 percent for the U.K. to 2 percent for Germany. Economy-wide productivity growth rates vary between 1.7 percent for the U.K. to 2.9 percent for France (Boskin and Lau, 1992). For other countries, such as Australia, Brazil, Canada, Japan, and Mexico, the relative agricultural TFP growth, computed using a variety of sources (Arnade, 1997; Nehru, Swanson, and Dubey, 1995), is lower than that of U.S. agriculture. This suggests that the ratio of agricultural to nonagricultural productivity in major EU and other countries is between 1 and 4. Based on these comparisons, U.S. agriculture appears to have a fundamental comparative advantage in world markets relative to the EU and other countries, largely because of its relative technological progress.[9]

CONCLUSIONS

Growth in U.S. agriculture is primarily driven by growth in its TFP, which benefits the broader economy in three ways. First, productivity growth is the dominant factor explaining growth in agricultural output in the United States. TFP growth has offset the negative effects of agriculture's declining terms of trade and, thus, has sustained returns to agricultural factors of production. Lack of

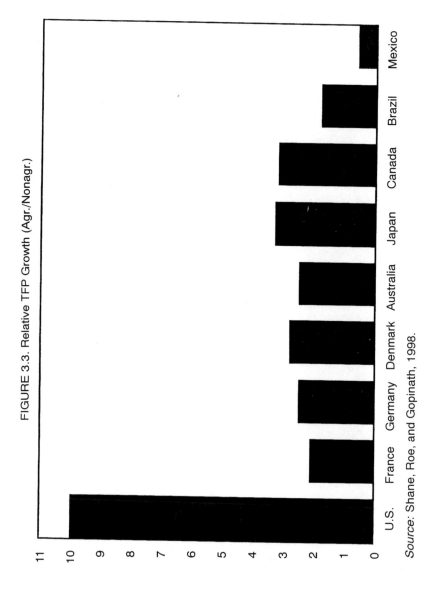

FIGURE 3.3. Relative TFP Growth (Agr./Nonagr.)

Source: Shane, Roe, and Gopinath, 1998.

growth in returns would almost surely have encouraged more resources to exit agriculture, which, in turn, would tend to lower growth in returns to family labor, farm buildings, land, and other sector-specific resources.

Second, real agricultural prices have declined at a rate of 1.19 percent per year since 1959. This has been passed on to consumers through the food processing sector. Growth in the food processing sector arises largely from its use of intermediate inputs. The share of primary agricultural output in the intermediate inputs of the food processing sector averaged 26 percent over the last three decades. The 1.2 percent fall in the real prices of primary agricultural outputs led to a decline of 0.32 percent per year in the procurement cost of intermediate inputs. This cost advantage should allow the processed foods sector to compete more effectively for export shares in the growing world markets for processed food. However, real prices of processed foods have also declined (-0.83 percent) more than the sector's rate of productivity growth (0.41 percent). Thus, food processing passes on almost all of its productivity growth and the gains from primary agriculture to downstream sectors, such as wholesale and retail food industries, and eventually to consumers.

Third, although the rate of European agricultural TFP growth is larger than that in the United States, U.S. agriculture has a higher relative productivity—the ratio of agricultural TFP to nonagricultural TFP—suggesting that it has the comparative advantage in agriculture. Hence, sustaining productivity growth in U.S. agriculture is crucial for maintaining its share of world agricultural markets.

Investments in public agricultural R&D and public infrastructure accounted for 75 percent of the growth in agricultural TFP. However, the rate of TFP growth in agriculture has stagnated in recent years. Due to the time lag between new investments in public agricultural R&D and productivity growth, the current slowdown in TFP growth may be explained, in part, by the decline in public investments in the 1980s. More problematic is the implication of continuing stagnation of public investment in agriculture on the future growth of the sector.

With declining real prices, growth in output and returns to factors of production in agriculture depends crucially on productivity growth.

Moreover, TFP growth in primary agriculture helps maintain the competitiveness of the food processing sector by lowering its costs for primary agricultural commodities. The decline in real prices of agricultural commodities and processed food products are passed on to consumers. Thus, the social rates of return to public agricultural R&D, which leads to productivity growth, are found to be relatively high. This suggests that positive growth of public agricultural R&D investments should increase the living standards of farm households, increase the welfare of consumers, and sustain the U.S. competitive edge in foreign markets.

NOTES

1. Note that equation (2) can relate to the competitiveness indices developed by FAO, AgCanada, and other agencies at the individual commodity level. Their indices look at the ratio of export share to production share of a country for a single commodity. The larger the index, the higher the competitiveness in that commodity. Our index focuses, at the aggregate level, on the rate of change of relative share of production of a commodity in the overall economy (GDP) and its sources.

2. Of course, at the individual commodity level, some countries may be more competitive than the United States. Moreover, as the denominator of equation (3) is an estimate of mean growth rates, some countries included in the aggregate may have larger growth rates than the United States.

3. Note, these definitions of competitiveness are not normative.

4. We refer readers interested in these techniques to Gopinath and Roe (1997) and Gopinath et al. (1997) for a detailed exposition.

5. The inaccuracy here is that changes in prices, resource levels, and technological change affect one another, so that holding any one constant will likely affect the levels of the others.

6. Although the EU has a CAP, it is not necessary for the CAP to have the same effect on all EU economies. As the prices are relative to the other sectors of the economy, the price effects can be different for countries within the EU.

7. See Gopinath, Roe, and Shane, 1996 for detailed tables.

8. Input-driven growth is inevitably limited because mere increases in inputs must run into diminishing returns (Krugman, 1996). Growth through increases in efficiency/productivity is sustainable.

9. This is consistent with the results of Gehlhar and Vollrath (1996) on U.S. export performance in agricultural markets. The U.S. share of world agricultural markets has grown by 1.3 percent per year beginning in 1987, while the EU share has stayed constant.

REFERENCES

Abbott, P.C. and M.E. Bredahl. (1994). "Competitiveness: Definitions, Useful Concepts and Issues." In *Competitiveness in International Food Markets*, edited by M.E. Bredhal, P.C. Abbott, and M.R. Reed, Boulder, CO: Westview Press, pp. 11-35.

Arnade, C.A. (1997). *Productivity Growth in 76 Countries*. ERS Staff Paper, U.S. Department of Agriculture, Washington, DC.

Ball, V.E., A. Barkaoui, J.C. Bureau, and J.P. Butault. (1996). "Agricultural Productivity in Developed Countries: A Comparison Between the United States and the European Community." Paper presented at *The Global Agricultural Science Policy for the 21st Century*, Melbourne, Australia.

Boskin, M.J. and L.J. Lau (1992). "Post-war Economic Growth in the Group of Five Countries: A New Perspective." Technical Paper, Department of Economics, Stanford University, Stanford, CA.

Gehlhar, M. and T. Vollrath. (1996). *U.S. Agricultural Trade Competitiveness in Foreign Markets*. ERS Technical Paper 1854, U.S. Department of Agriculture, Washington, DC.

Gopinath M., C. Arnade, M.D. Shane, and T.L. Roe. (1997). "Agricultural Competitiveness: The Case of the U.S. and Major EU Countries." *Agricultural Economics* 16(2): 99-109.

Gopinath, M. and T.L. Roe. (1995). *General Equilibrium Analysis of Supply and Factor Returns in U.S. Agriculture 1949-91*. Bulletin 95-8, Economic Development Center, University of Minnesota, St. Paul.

Gopinath, M. and T.L. Roe. (1996). "Sources of Growth in U.S. GDP and Economy-Wide Linkages to the Agricultural Sector." *Journal of Agricultural and Resource Economics* 21(2):152-167.

Gopinath, M. and T.L. Roe. (1997). "Sources of Sectoral Growth in an Economy-Wide Context: The Case of U.S. Agriculture." *Journal of Productivity Analysis* 8(3): 293-310.

Gopinath, M., T.L. Roe, and M.D. Shane. (1996). "Competitiveness of U.S. Food Processing: Benefits from Primary Agriculture." *American Journal of Agricultural Economics* 78(4):1044-1055.

Krugman, P. (1996). *Peddling Prosperity: Economic Sense and Nonsense in the Age of Diminished Expectations*. New York: Norton Publishers.

Landau, R. (1992). "Technology, Capital Formation and U.S. Competitiveness." In *International Productivity and Competitiveness*, edited by Bert G. Hickman, New York: Oxford University Press, pp. 299-325.

Nehru, V., E. Swanson, and A. Dubey. (1995). "A New Database on Human Capital Stock in Developing and Industrial Countries: Sources, Methodology, and Results." *Journal of Development Economics* 46(2): 379-401.

Porter, M. (1990). *The Competitive Advantage of Nations*. New York: The Free Press.

Reich, R. (1992). "Competitiveness." *The Wall Street Journal.*

Shane, M.D., T.L. Roe, and M. Gopinath. (1998). *U.S. Agricultural Growth and Productivity: An Economywide Perspective.* Agricultural Economics Report No. 758, Economic Research Service, U.S. Department of Agriculture, Washington, DC.

Chapter 4

Corn

Paul W. Gallagher

INTRODUCTION

During the last two decades, competitiveness has been an important policy and market development issue. In the U.S. corn industry, the issues have evolved with changing performance in international trade. This chapter reviews the sources of changing competitiveness in the U.S. corn industry and summarizes public and private strategies for improving the international market position. Domestic policy reactions have offset deterioration or improved the international position of the U.S. corn industry. Multilateral and regional trade agreements have also affected corn trade. Industry's strategy has emphasized improved product qualities for specific uses in the livestock feed industry. This review will help identify potential trends in changing performance and emerging issues regarding corn trade in international markets.

THE RECORD: CHANGING COMPETITIVENESS IN THE U.S. CORN INDUSTRY

Competitiveness refers to an industry's ability to price products attractively on the international market. Market share is a useful indicator of competitiveness. In the case of the U.S. corn industry, market

---

The data for this chapter come from various sources. Calculations of exchange rates for the dollar come from the International Monetary Fund (1998). Data on exports of corn gluten feed and corn gluten meal come from the U.S. Department of Commerce (1964-1998). Other information comes from the following USDA publications: price information from the *Feed Outlook* reports (USDA, 1994-1998) and U.S. and foreign livestock and grain data from the *PS&D View* database (USDA, 1999).

share computations should account for the growth of by-product feed exports and the dominant use of corn in foreign countries. Corn by-product feeds have grown from nothing to about 10 percent of corn exports by volume since the mid-1970s. By-products are like corn, except that there is no starch. Thus, there is some potential for substitution with corn exports. The by-products, corn gluten feed and meal, are included with raw corn exports in market share calculations.

The main foreign market for U.S. corn is livestock feed. The foreign feed market's size is approximated using USDA data. Total use of all grain for feed in foreign countries, total foreign use of protein feeds, and U.S. exports of corn by-product feeds are all included in calculations of total foreign feed consumption.

Figure 4.1 shows the evolution of U.S. corn's share of the foreign feed market. U.S. corn and by-products captured an increasing

FIGURE 4.1. U.S. Corn, Gluten Feed, and Meal Share of the Foreign Food Market

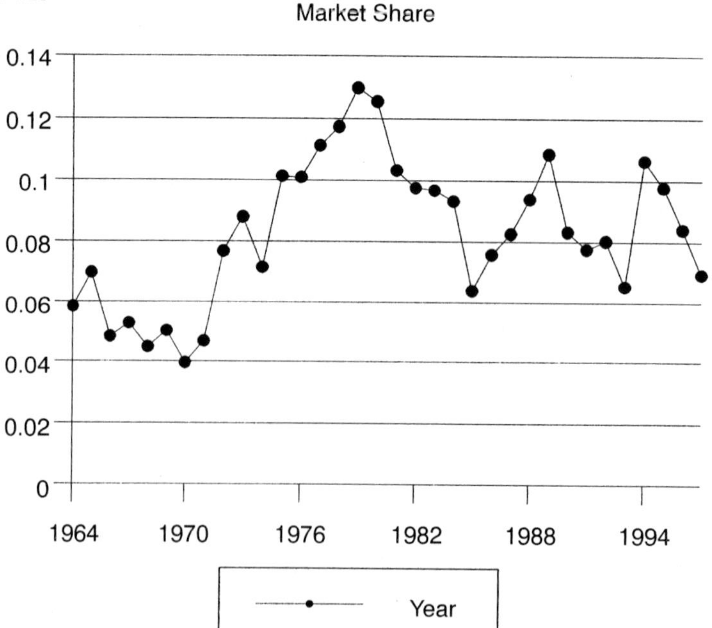

Source: U.S. Department of Commerce, 1964-1998.

share of the foreign feed market during the 1970s. The U.S. corn share peaked at 13 percent in 1980. Since then, the U.S. share has fluctuated widely, but trended downward. A market share of about 8 percent is typical for the late 1990s.

MARKET DEVELOPMENTS INFLUENCING COMPETITIVENESS

Shifting market conditions cause changing competitiveness. In turn, the events that produce shifts in supply and demand lie beyond the agricultural market and policy system. Changing cost conditions in competing countries are one fundamental cause of changing competitiveness. Rising foreign income and changing quality demand for feeds are also important determinants of corn market share.

Costs

Production costs are used in some macroeconomic analyses of competitiveness (Dornbush, 1980). But two important assumptions of the macroeconomic analysis, fixed proportions production technology and exogenous factor prices, do not hold for grain markets (Paarlberg et al., 1985). Instead, factor proportions and the intensity of land use varies with the application of fertilizer and other chemicals. Furthermore, land use is concentrated in the corn production region. Thus, land prices for grain production are flexible, responding to price conditions in the product market. For example, Midwest land prices have fallen by half from the peak to trough of past corn export cycles. In short, the cost dimension of competitiveness in the corn market is elusive. Consequently, indirect analyses of cost competitiveness for grain markets are cast in terms of the underlying factors that affect marginal costs in the United States and other competing countries.

Technology advance drives cost reductions for corn. Technology improvement is imperfectly measured by productivity increases. During the 1960s and 1970s, U.S. corn yield increases outpaced yields in other countries, leading to improving relative costs for the

United States. At the peak of the late 1970s, U.S. corn yields were four times the average of foreign grain yields (see Figure 4.2). The improving yield advantage for the United States reflected the U.S. role in technology development and the emphasis on yield-increasing technologies. The U.S. yield advantage has been declining recently; a typical yield ratio for U.S. corn and foreign grain is about 3 in the late 1990s. Possibly, U.S. advances of previous decades are being adapted to the conditions of other countries. Also, by 1980 the emphasis in technology development began moving away from yield increases for major grains in the United States (Office of Technology Assessment, 1991).

Feed Quality

During the 1970s, corn demand grew with expansions in the foreign feed market. Corn's market share grew as well. Since then,

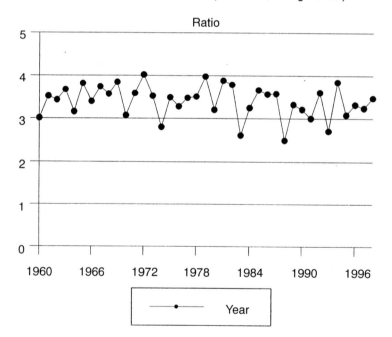

FIGURE 4.2. Corn Yield Ratio (U.S. Corn/Foreign Grain)

Sources: USDA, 1996-1998; 1999.

foreign feed use has grown more slowly. Further, feed demands have shifted toward the needs for poultry and pork rations. The feed mix has moved toward higher protein content in feeds. In addition, specific proteins and amino acids are limiting factors in hog and chicken diets.

Figure 4.3 shows that the share of pork and poultry in foreign meat production has increased steadily. The total market is defined as the sum of beef, pork, and poultry in foreign countries using USDA data. The share of pork and poultry has increased from about 45 percent in 1964 to nearly 75 percent in the late 1990s. The shift toward protein-using livestock has likely reduced U.S. corn's share of the foreign feed market.

FIGURE 4.3. Pork and Poultry Share of Foreign Market

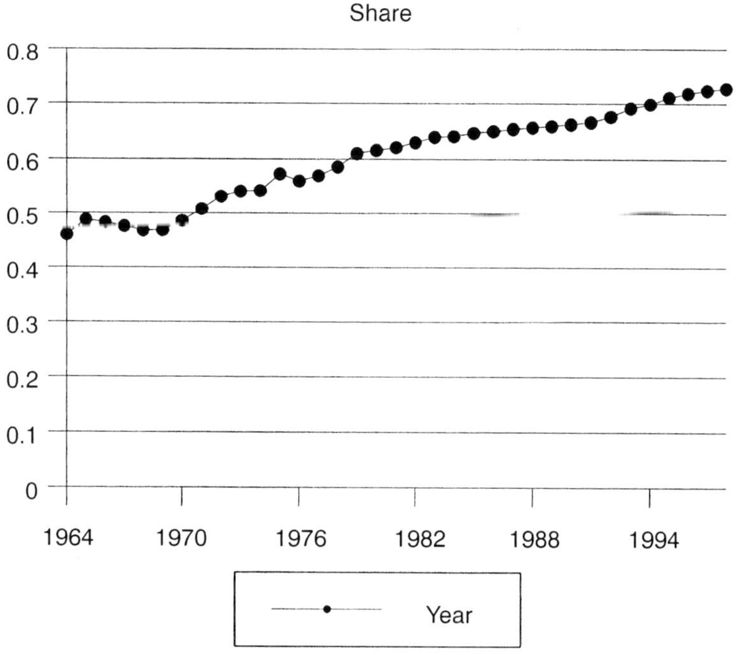

Source: USDA, 1999.

U.S. POLICY AND COMPETITIVENESS

In the United States, macroeconomic policy and corn market policy have had important effects on competitiveness in the corn industry. At times, passive U.S. policies have outlived their usefulness and erected impediments to competitiveness. At other times, policies have been designed with enhanced competitiveness for the corn market in mind.

Commodity Policy

When the market expansion phase of the foreign corn market ended in 1980, an escalating U.S. dollar and declining foreign corn demand produced international prices for U.S. corn that were not competitive. The taxlike consequences of a strong dollar for U.S. agricultural exports are well known (Schuh, 1974). In the corn market, export prices for U.S. corn increased by about 20 percent on foreign markets between 1979 and 1983 (Gallagher, 1986). At the same time, corn prices declined by about 10 percent on the domestic market (see Figure 4.4). Also, the U.S. corn loan rate supported market prices, because the U.S. government effectively purchased corn when the market price fell to the loan rate during the 1980s. The combination of strengthening dollar, declining foreign demand, and corn pricing on the loan rate floor caused corn pricing on international markets that was not competitive.

Subsequent revisions of U.S. agricultural policy gradually addressed the export-pricing issue. For instance, the 1985 Farm Bill included substantial cuts in the corn loan rate, from $2.55/bu in 1985 to $1.57/bu in 1990. But high target prices were retained until 1995 to support farm income during the export market collapse of the mid-1980s and to maintain a market position until the Uruguay Round of GATT negotiations was completed.

Beginning with the 1996 farm bill, the FAIR Act, the loan rate no longer supports corn market prices directly. Currently, the loan rate provides income support. After taking out a loan, farmers sell their corn at the prevailing market price. They also receive a loan deficiency payment from the government when the market price is less than the loan rate. The loan deficiency payment is equal to the

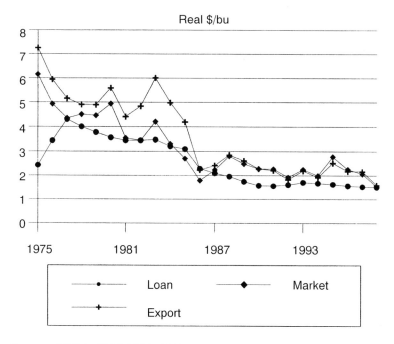

FIGURE 4.4. Real U.S. Corn Prices, 1975 to 1998

Sources: USDA, 1994-1998; 1999.

difference between the market price and the loan rate (Hoffman, 1996).

Administrators still have some discretion in setting the corn loan rate under the FAIR act. However, the law specifies minimum and maximum rates. The maximum loan rate for corn is $1.84/bu. The minimum loan rate is moderately below typical market conditions for recent years. It can be set at 85 percent of the average of market prices for the preceding five-year period, excluding the years with the highest and lowest price (Hoffman, 1996). So far, loan rate choices have held the maximum value. Nonetheless, the loan provisions were not effective until the end of the 1998 crop year (see Figure 4.4). However, the farm price averaged $1.70/bu for the first third of the 1999 crop year. Thus, loan deficiency payments totaled $1.4 billion through December 1999, with an average payment rate of $.29/bu.

The marketing loan facilitates competitiveness because the loan rate is not a price floor in weak markets. Under the old system, farmers defaulted on loans when the market price fell to the loan rate, P_l (see Figure 4.5); the government assumed ownership of grain, amount $Q_s - Q_d$; and demand fell, to Q_d. With the marketing loan, producers sell at the low market price, P_m, and receive a marketing loan payment of $P_l - P_m$; consumers use the amount produced, Q_s; and the government does not assume grain ownership.

The corn loan rate choice determines the government's exposure to marketing loan payments during weak market conditions. In particular, marketing loan payments could become substantial if the maximum rate held while the market price dropped to $1.50/bu. However, a guideline is given for adjusting the loan rate down. The loan rate could escalate beneath the market during a future export boom. But the problem with rigid loan rates experienced after the export booms of the 1950s and 1970s could not occur now; the law mandates loan rate reductions and defines the appropriate magni-

FIGURE 4.5. Corn Pricing: Price-Support Loan and Marketing Loan

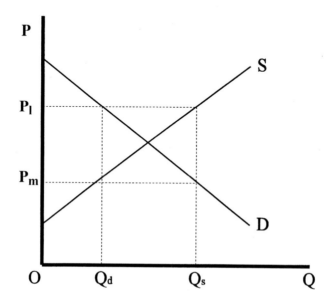

tudes. Overall, the farm policy revisions facilitate competitiveness and help to avoid prolonged periods of high government expenditures when exports decline.

Monetary Policy

The strong dollar episode of the early 1980s has not recurred, which is an important piece of the competitiveness formula for U.S. corn exports. U.S. monetary policy has shifted. A tight money position was implemented due to inflation concerns in the early 1980s. In contrast, monetary policy has accommodated productivity growth in the late 1990s. Also, agricultural interests are now more sensitive to the adverse consequences of a strong dollar. Combined export industries likely exert political-economic pressure against the recurrence of another strong-dollar episode.

Overview

Together, commodity programs and macroeconomic policies have removed most impediments to U.S. corn exports. The overseas value of U.S. corn has declined substantially since the end of the export expansion period of the late 1980s. However, corn market export performance has been less than spectacular during this period, despite steady economic growth in Europe and Japan prior to the mid-1990s. A remedy for sluggish export performance in the corn export market may lie elsewhere.

U.S. TRADE AGREEMENTS AND CORN EXPORTS

The U.S. government has participated in two trade treaties during the last decade. A multilateral trade agreement, the outcome of the Uruguay round of the GATT, was completed in 1994. The United States also participated in a regional trade agreement with Mexico and Canada, the North American Free Trade Agreement, completed in 1995. Both trade agreements include agriculture. Both agreements contain provisions for liberalizing corn imports by important markets, which are summarized below.

Under the recent GATT agreement, the EU has agreed to reduce its import tariff on maize by 36 percent (Josling et al., 1994). However, given the high initial level for prices, the tariff reduction should produce a moderate 15 percent reduction of corn prices on the European market during the six-year period ending in the year 2000.

Corn imports for livestock enter Japan duty-free. But corn used for starch processing has been subject to a quota. Similarly, processed starch imports are also restricted. Under the recent GATT agreement, Japan has converted its starch quota into a tariff quota. Up to 157,000 tons of starch can enter duty-free. Beyond the small minimum access quantity, the starch tariff is 140 ¥/kg. Japan will reduce its starch tariff by 20 percent by the year 2000 (Josling et al., 1994). Even so, the starch tariff will probably prohibit corn processing activity in Japan; typical U.S. corn and starch prices range from 11 ¥/kg to 30 ¥/kg.

Under NAFTA, concessions by Mexico on corn imports are probably the most important. Mexico eliminated import licenses for corn and set a tariff rate quota (TRQ) for duty-free corn imports up to 2.5 million tons. Above the TRQ, the import duty is 215 percent. An important feature of the agreement is that Mexico has agreed to eliminate the tariff by the year 2008 (USDA, 1997).

A system of price supports for small corn producers and a targeted subsidy program for low-income urban consumers of staple corn products (tortillas) does remain in Mexico. Together, the two subsidies are probably trade neutral since one subsidy encourages the consumption of the extra production created by the other subsidy. In recent years, the programs have been reasonably managed in the sense of committing only a moderate share of Mexico's domestic budget.

Overall, these trade concessions are a mixed bag that point toward a moderate growth in corn exports and competitiveness. The tariff reduction in the EU could account for moderate improvements in imports now. But expanded demand related to starch processing in Japan probably resides in the next decade and only if tariff concessions are obtained in future negotiations. Mexico's agreement to eliminate corn import barriers during the next ten years is promising. But realization may depend on moderate performance in their domestic economy. Another risk is that a changing domestic political situation could upset Mexico's balanced and moderate commitment to domestic subsidies.

INDIRECT EFFECTS OF INTERNATIONAL POLICY CHANGES

Some indirect effects of trade policy changes have significant effects on U.S. corn exports. Two cogent examples are macroeconomic effects of the NAFTA agreement and the combination of domestic reforms and most favored nation (MFN) trade status for China.

The emerging economic story about a trade agreement between a wealthy neighbor, such as the United States, and a developing neighbor, such as Mexico, starts with a processing investment by the wealthy country in the developing country for low-wage segments of a manufacturing process. Next, new processing wages induce multiplier growth in the developing country's income (Robinson et al., 1993). Indeed, computable general equilibrium models of the NAFTA that include investment and capital flows indicate a 10 percent GNP increase for Mexico in the trade agreement (Brown, 1993). Similarly, China's domestic reforms in the 1980s propelled incomes to new levels.

Income growth in Mexico and China ultimately increased U.S. corn exports because meat demand expanded and domestic corn production failed to fulfill the needs for producing the increased meat requirements. Furthermore, corn market liberalization was included in trade agreements; the corn agreement for NAFTA was discussed above, and MFN status was extended to China. This combination of policy change that increases developing country income growth and grain market reforms has been a significant source of growth in U.S. corn exports because meat is a relative luxury in these countries and has a high income elasticity. Indeed, some early projections showed that China's import/export status in the corn market could fluctuate, depending on the outcomes for income growth in the domestic reform process (Carter and Zhong, 1988). Generally speaking, trade agreements and domestic reforms in developing countries have strengthened the link between income growth in developing countries and U.S. corn export demand. Increasingly, this corn export market possesses good growth potential but has all of the instability that is associated with the economies of developing countries.

BIOTECHNOLOGY AND COMPETITIVENESS IN THE CORN INDUSTRY

Corn emerges at the forefront of technology change in agriculture because its relatively simple genetic structure is easier to modify than those of some other commodities; corn emerged with the first hybrids in the mid-1930s; and corn varieties provided steady yield increases through the next three decades (Bray and Watkins, 1964). The most recent advances have focused on product qualities instead of yield. Extensive control over quality is now possible using genetic engineering.

The new corn varieties can be placed in two different groups for purposes of economic analysis. One group provides more favorable production conditions—these varieties resist insects, pests, and chemicals. A second group has enhanced characteristics that are valuable to end users—these varieties have high oil, lysine, and starch content. Varieties in both groups are commercially available or are in advanced stages of development. Both groups of corn seeds have the potential to improve U.S. competitiveness in the corn market.

The first group reduces the variability of production because pests are more readily controlled. Hence, production risks should be reduced. Further, production costs are reduced for several reasons. First, smaller applications of less toxic herbicides provide more effective weed control, so yields increase (Schnittker, 1998). Second, the average costs over several years may be lower with less need to factor in occasional crop failures. Hoffman (1997) provides some evidence of reduced production costs with herbicide-resistant varieties.

The second group provides corn that is more desirable in specific processing or feed uses (Kalaitzandonakes and Maltsbarger, 1998). Thus, processing or feed ration costs should be reduced. Alternatively, feed efficiency could increase. High-starch corn will likely be useful to corn processors. High-lysine feed will be useful in swine rations, especially as world fishmeal supplies decline and this feed becomes more expensive. High-oil corn is useful in feed rations due to improved uniformity of the ration. High-oil corn may also be desirable in feed rations in countries such as Japan that import corn feed without tariffs but maintain tariffs on animal and vegetable oils. Specialized corn varieties may reverse the recent gains of protein feeds over U.S. corn in the international feed market.

EXPORT PERFORMANCE AND TECHNOLOGY ADOPTION

The new generation of corn seeds will likely improve the U.S. share of the export market because they are developed for U.S. growing conditions. In general, improvements developed for one set of growing conditions do not readily transfer to other growing conditions and locations (Barker and Pluchnett, 1991, p. 115). The current thinking is that conditions favor a prolonged advantage for the United States in biotechnology. First, technology-generating research is possible only in the United States, France, and Japan, and the United States is the only one of these countries that has refrained from product development or use restrictions. Further, the normal product-development cycle for new seeds, ten years, is effectively doubled due to regulations. Finally, large firms that can manage the long product development cycle own the technology (Office of Technology Assessment, 1991).

Nevertheless, many other countries have the capacity to adapt corn varieties that have been successful in the United States or Europe to local growing conditions (Office of Technology Assessment, 1991). Consequently, modifications that have been profitable in the United States will be scrutinized for use in other countries. Ultimately, the pattern of eroding U.S. advantage in yield-increasing technology (see Figure 4.2) may occur with quality-improving technologies. Hence, increases in corn market share that are driven by biotechnology during the next decade may be less dramatic or turn out to be transitory in the longer run.

MARKET STRUCTURE

In contrast to previous generations of technology for major field crops, ownership for the newest technologies is private. Increased private ownership has occurred due to the strengthening of U.S. patent laws. Critics have expressed two concerns that might affect U.S. export performance in the corn market.

First, major seed companies and grain companies have formed joint ventures and used producer contracts to market new varieties. Hence, a few firms control the modified corn varieties. Contracts have been used as an instrument of oligopoly control in other agriculture industries (Helmberger, Campbell, and Dobson, 1981, p. 616). Some are concerned about monopsony pricing for corn producers.

However, corn output (and exports) would likely increase if enhanced varieties improve feed efficiency and if commodity corn remains an alternative to grower contracts. For illustration, consider a local corn production area where the traditional and the enhanced variety have the same marginal costs, defined by supply curve S in Figure 4.6. Corn producers and feed users in a large market will arrive at competitive price P_c without the technology owner in the middle and produce amount Q_c. Feed users will pay a higher price for the enhanced variety, P_e, because feed efficiency improves. The technology owners will expand the corn up to the quantity Q_m, where the marginal expense on inputs to corn producers equals the enhanced feed price, MEI = P_e, where they seek maximum profits. Corn producers are then paid the monopsony price, P_m. In the case shown, producers choose the contract instead of the market because profits are higher by the area $A + B$. Further, the amount produced under contract, Q_m, exceeds the amount of market production with the traditional variety. Corn exports would likely increase as well. However, reductions in P_e would shrink producers' profits and contract volumes at first, and eventually induce farmers to decide on market production with traditional varieties.

Ultimately, the issue is whether the enhanced varieties dominate the corn market or just obtain a significant market share. Currently, enhanced corn varieties make up a moderate proportion of the corn production area. Specifically, 1998 corn production included 3.8 million acres of enhanced varieties, or about 5 percent of the total U.S.

FIGURE 4.6. Producer's Choice of Enhanced Corn Varieties in a Monopsony Market

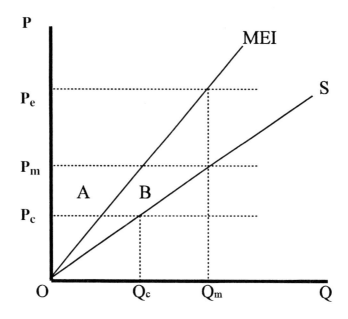

corn area (U.S. Feed Grains Council, 1999). Concern about the demise of the commodity corn market is not yet warranted. But manageable handling and transport costs for enhanced varieties will require geographical concentration. Consequently, public information should monitor local market areas on the extent of contracting and evidence of pricing problems.

Second, there is a group of poor countries that cannot adapt the successful new varieties to their local growing conditions. Past generations of publicly owned corn technology were transferred to these countries through international institutes such as CIMMYT. Some argue that these countries will be left out of the technology development if the direction is defined by profits (Barker and Pluchnett, 1991). Further, increasing agricultural production is crucial to economic development and income growth in these countries. Finally, income growth in poor countries is in the interest of corn-exporting countries such as the United States, because today's tortilla consumer is tomorrow's meat consumer.

CONCLUSIONS

This chapter has reviewed market developments, government policies, and industry strategies that affect the performance of U.S. corn on international markets. The market situation has shown mild erosion of our traditional cost advantage and shifting foreign demands that favor protein and limiting amino acids in feed rations. U.S. policy has been increasingly competitive in general. Specifically, revisions in loan rate functions and level-setting procedures have facilitated advantageous export pricing of U.S. corn. Further, U.S. monetary policy has been favorable to all export industries throughout the 1990s. U.S. trade negotiations have secured modest or slight concessions in corn market access for our traditional trade with Japan and Europe. In contrast, Mexico provides a substantial promise for corn market access. The indirect effect of trade agreements on corn demand will also be significant; GNP increases have stimulated livestock production in Mexico and China, and boosted U.S. corn exports. Private strategies for increasing corn's share of the foreign feed market include new varieties for reduced production costs and yield risk for U.S. corn producers. Other varieties meet the needs for low-cost and high-quality feed rations for the growing pork and poultry markets of developing countries. As long as these varieties do not completely replace commodity corn, they should contribute to increased U.S. corn production and exports.

As a result of trade policy concessions and technology development strategies during the last decade, the economic link between developing country incomes and foreign corn demand has strengthened. These markets now have growth potential due to a large income elasticity for meat. But now the risk that is closely associated with unstable growth rates in developing countries is magnified: Mexico, Taiwan, and Russia are meat-consuming, corn-using countries with fluctuating growth rates for GNP. Perhaps more attention to stability and diversification is now in order. Further geographical diversification of export markets might help if the risk of worldwide recession is only moderate. Also, human consumption of corn is relatively stable in developing countries (Hoffman, 1997). Some new varieties with enhanced taste and nutritional qualities might secure a higher share of the human consumption market for U.S. corn.

REFERENCES

Barker, R. and D. Pluchnett (1991). "Agricultural Biotechnology: A Global Perspective." In *Agricultural Biotechnology: Issues and Choices*, B.R. Baumgardt and M.A. Martin (Eds.). West Lafayette, IN: Purdue University Agricultural Experiment Station, pp. 107-109.

Bray, J.D. and P. Watkins (1964). "Technical Change in Corn Production in the United States, 1870-1960." *Journal of Farm Economics* 64(4): 751-765.

Brown, D.K. (1993). "The Impact of a North American Free Trade Area: Applied General Equilibrium Models." In *Assessing the Impact, North American Free Trade*, N. Lusting, B.P. Bosworth, and R.Z. Lawrence (Eds.). Washington, DC: The Brookings Institution, pp. 49-64.

Carter, C.A. and F. Zhong (1988). *China's Grain Production and Trade: An Economic Analysis*. Boulder, CO: Westview Press.

Dornbush, R. (1980). *Open Economy Macroeconomics*. New York: Basic Books.

Gallagher, P. (1986). "Competitiveness of U.S. Grain in International Markets: An Introduction." In *Proceedings of Symposium on Competitiveness of U.S. Grains in International Markets*, W.W. Wilson (Ed.). Agricultural Economics Report No. 97, Department of Agricultural Economics, North Dakota State University, Fargo.

Helmberger, P.G., G.R. Campbell, and W.D. Dobson (1981). "Organization and Performance of Agricultural Markets." In *A Survey of Agricultural Economics Literature, Volume 3*, L.E. Martin (Ed.). Minneapolis, MN: University of Minnesota Press, pp. 503-653.

Hoffman, L. (1996). "Title I: The Agricultural Market Transition Act." In *Provisions of the Federal Agriculture Improvement and Reform Act of 1996*, L. Hoffman (Ed.). U.S. Dept. of Agriculture, Agriculture Information Bulletin No. 729, Washington, DC.

Hoffman, L.A. (1997). "Enhanced Grains and Oilseeds: Issues and Prospects Relating to Producers, Marketing Systems, and Vertical Integration." Paper presented at the Annual Meeting of the American Agricultural Economics Association, Toronto.

International Monetary Fund (1998). *International Financial Statistics*. Washington, DC: International Monetary Fund.

Josling, T., M. Honma, J. Lee, D. MacLaren, B. Miner, D. Sumner, S. Tangermann, and A. Valdés (1994). *The Uruguay Round Agreement on Agriculture: An Evaluation*. Commissioned Paper Number 9. St. Paul, MN: International Agricultural Trade Research Consortium.

Kalaitzandonakes, N. and R. Maltsbarger (1998). "Biotechnology and Identity-Preserved Supply Chains." *Choices* Fourth Quarter: 15-18.

Office of Technology Assessment (1991). *Biotechnology in a Global Economy.* Washington, DC: Congress of the United States.

Paarlberg, P., A.J. Webb, J.C. Dunmore, and J.L. Deaton (1985). "The U.S. Competitive Position in World Commodity Trade." In *Agricultural and Food Policy Review: Commodity Program Perspectives*, Agricultural Economics Report 530. Washington, DC: ERS, USDA, July, pp. 93-121.

Robinson, S., M.E. Burfisher, R. Hinojosa-Ojeda, and K. Thierfelder (1993). "Agricultural Policies and Migration in a U.S.-Mexico Free Trade Area: A Computable General Equilibrium Analysis." *Journal of Policy Modeling* 15(5,6): 673-701.

Schnittker, J.A. (1998). "An Agricultural Revolution with Implications for Sustainability." *Choices* Fourth Quarter: 1-3.

Schuh, G.E. (1974). "The Exchange Rate and U.S. Agriculture." *American Journal of Agricultural Economics* 56(1): 1-13.

U.S. Department of Commerce (1964-1998). *U.S. Exports. Schedule B Commodity by Country.* Washington, DC: Bureau of the Census.

U.S. Feed Grains Council (1999). "1997-98 Value-Enhanced Corn Report." URL <www.grains.org>, January.

USDA (1994-1998). *Feed Outlook.* Washington, DC: Economic Research Service, U.S. Department of Agriculture.

USDA (1997). *NAFTA: Situation and Outlook Series.* Economic Research Service, International Agriculture and Trade Report, WRS-97-2. Washington, DC: United States Department of Agriculture

USDA (1999). *PS&D (Production, Supply & Distribution) View.* Economic Research Service, U.S. Department of Agriculture. Internet document: <http://jan.mannlib.cornell.edu/datasets/international/93002/>.

Chapter 5

Wheat

Won W. Koo

INTRODUCTION

Wheat is widely grown around the world. The total world wheat production has increased slightly, from 531 million tons in 1986-1987 to 582 million tons in 1996-1997. China was the largest wheat producer in 1996 (110 million tons), followed by the European Union (94 million tons) and the United States (62 million tons). Other major wheat-producing countries are the former Soviet Union (FSU), Canada, Australia, Turkey, India, and Argentina. These nine countries produce about 74 percent of the world wheat crop. Because of the concentration of production in a few countries, a large volume of wheat is traded in the world market. The total quantity of wheat traded in the world market was 95.4 million tons in 1996, which is about 17 percent of wheat produced in that year. Major exporting countries are the United States, Canada, the European Union, Australia, and Argentina.

The world wheat market has changed dramatically in the past decade. Farm support policies in exporting and importing countries have encouraged production, resulting in large stock buildups. Countries have used quotas, variable levies, tariffs, and other forms of import restrictions to protect domestic producers. As world trade decreased during the early 1980s due to a depressed world economy, major exporting countries expanded the use of export subsidies or promotion programs to maintain their grain market shares. The Uruguay Round (UR) of GATT negotiations, which was effective in 1995, has affected trade flows of wheat from exporting countries to importing countries. In addition, recent financial crises in several Asian countries, including South Korea, Thailand, Indonesia, and Taiwan, also have affected the world wheat market. Import demand for wheat from those countries has been reduced substantially, resulting in depressed wheat prices in the world market. The average export price of wheat at U.S. Gulf of Mexico ports decreased from $5.02 per bushel in 1996-1997 to $3.89 per bushel in 1997-1998.

The objective of this chapter is to assess U.S. competitiveness in exporting wheat to major importing countries. Special attention is given to evaluate the impacts of the UR agreement on trade flows of wheat from exporting countries to major importing countries and their implications for the U.S. wheat industry.

WORLD WHEAT INDUSTRY

World wheat trade is dominated by a few exporting countries: the United States, Canada, Argentina, Australia, and the European Union. These countries have handled over 80 percent of wheat traded in the world market. Although competition among exporting countries is strong, the world wheat market is not perfectly competitive. Australia and Canada use wheat boards to market their grain, while the United States and the European Union rely on export subsidies to increase their market share. In addition, all major wheat exporters use credit guarantees and long-term preferential trade agreements to promote their exports.

Wheat Classes

Most wheat varieties grown today belong to the broad category of common or bread wheat, which accounts for approximately 95 percent of world wheat production. The remaining 5 percent is durum wheat used to produce pasta and couscous. Common wheat is further divided into hard and soft wheat. Wheat varieties are highly differentiated in terms of their agronomic and end-use attributes. Based on criteria such as kernel hardness, color, growth habit, and protein content, wheat is divided into several classes. Color and hardness refer to physical properties of the wheat kernel.

Based on the color of the outer layer of the kernel, common wheat varieties are described as white, amber, red, or dark, while the hardness of the kernel is used to characterize them as hard or soft.

Growth habit is an important agronomic feature of wheat varieties. Winter wheat is planted in late summer or fall and requires a period of cold winter temperatures for heading to occur. After using fall moisture for germination, the plants remain in a vegetative phase or dormancy during the winter and resume growth in early spring. In contrast to winter wheat, spring wheat changes from vegetative growth to reproductive growth without exposure to cold temperatures. In temperate climates, spring wheat is sown in spring. Since yields for winter wheat tend to be higher than for spring wheat, spring wheat is produced primarily in regions where winter wheat production is infeasible, i.e., where hard frozen soil kills the wheat plants or where winters are too warm. Countries with mild winters, such as Argentina, Australia, and Brazil, produce spring wheat, but plant in the fall rather than in the spring.

Wheat Production

Because of differences in soil types and climates, wheat produced in one country generally differs from that produced in other countries. The United States produces hard, soft, and durum wheats. Hard wheat produced in the United States is further divided into hard red winter (HRW), and hard red spring (HRS) wheat and soft wheat into soft red winter (SRW) and white wheat. SRW wheat is produced in the Corn Belt and Southern states. HRS and durum wheat are grown in the Northern Plains, mainly North Dakota, which produces about 80 percent of durum wheat and 60 percent of HRS wheat grown in the United States. HRW wheat is grown primarily in the central plains, particularly Kansas and Oklahoma. White wheat, a type of soft wheat, is grown in the Pacific Northwest, Michigan, and New York. Average U.S. wheat production for the 1995-1997 period was 63.5 million tons, with 24.6 million tons in HRW, 14.6 million tons in HRS, 12.5 million tons in SRW, 9.2 million tons in white wheat, and 2.8 million tons in durum wheat (see Table 5.1).

The majority of Canadian wheat is produced in Saskatchewan, southwestern Manitoba, and southeastern Alberta. Canada produces primarily HRS wheat (Canadian Western Red Spring) and durum

TABLE 5.1. Wheat Production by Class, 1995 to 1997 Average Production

Country/Class	1995	1996	1997	Average	Share (%)
	(1,000 metric tons)				
Argentina					
Common	9,445	15,914	14,733	13,364	2.3
Australia					
Common	16,504	23,702	19,417	19,874	3.4
Canada					
All	25,009	29,802	24,298	26,370	4.5
Common	20,361	25,175	19,948	21,828	3.8
Durum	4,648	4,627	4,350	4,542	0.8
European Union					
All	87,709	99,724	94,798	94,077	16.2
Common	81,009	91,624	87,098	86,577	14.9
Durum	6,700	8,100	7,700	7,500	1.3
United States					
All	59,418	62,194	68,781	63,464	10.9
Hard Red Winter	22,455	20,713	30,512	24,560	4.2
Hard Red Spring	12,929	17,175	13,636	14,580	2.5
Soft Red Winter	12,412	11,486	13,174	12,357	2.1
White	8,846	9,662	9,118	9,209	1.6
Durum	2,776	3,157	2,341	2,758	0.5
Other Producers					
All	353,242	361,238	395,614	370,032	63.7
Total World					
All	542,822	586,474	612,760	580,685	

Sources: FAO, 1999; International Grains Council, 1997; Canadian Wheat Board, 1996-1998.

wheat. Average Canadian wheat production for the 1995-1997 period included 21.6 million tons of HRS and 4.5 million tons of durum wheat (see Table 5.1).

The EU produced an annual average of 86.6 million tons of soft wheat and 7.5 million tons of durum wheat during the 1995-1997 period. France accounted for 40 percent of soft wheat production in the EU in 1995. Germany and the United Kingdom are also major producers. The majority of durum is produced in Italy, Greece, and France. Italy accounted for nearly 60 percent of EU durum production in 1995, followed by Greece (22 percent) and France (13 percent).

Australia primarily produces a winter wheat which is similar to HRW in terms of quality and characteristics (Ortmann, Rask, and Stulp, 1989). Average Australian wheat production amounted to 19.9 million tons for the 1995-1997 period. Production is concentrated in the eastern Australian states of New South Wales and Victoria.

Argentina produces a wheat with characteristics of both soft and hard wheat (Harwood and Bailey, 1990). Argentina's average wheat production amounted to 13.4 million tons for the 1995-1997 period.

Wheat Consumption and Imports

Different wheat classes have preferred uses. Hard wheat flour has excellent bread baking properties; soft wheat flour is well-suited for cakes, cookies, and Asian noodles; and durum wheat is used for pasta products and couscous. However, since different types of wheat can be blended to produce flours with certain characteristics, some substitution among wheat classes is possible in flour milling.

Although wheat is used primarily for human consumption, it is also an excellent feed grain for poultry and livestock. Feed use of wheat tends to be highly variable and depends on the quality of the crop and on the price relationship between wheat and other feed grains. Generally, only lower quality wheat is used for feed, and differences among wheat classes are not important for feeding purposes. Wheat is a differentiated product only for human consumption.

Major importing countries include Algeria, Brazil, China, Egypt, Japan, Mexico, Morocco, South Korea, Taiwan, Tunisia, and Venezuela (see Table 5.2). Most of these countries use various types of barriers to restrict inflows of wheat. China has been the largest importer of wheat, followed by Brazil and Japan. However, China's wheat imports have been highly volatile, depending upon its domestic production and import policies. China recently reduced wheat imports substantially, from 12.6 million tons in 1995 to 2.8 million tons in 1997.

Trade flows of wheat from exporting to importing countries are shown in Figure 5.1. The European Union and the United States are

TABLE 5.2. Wheat Imports by Country, 1995 to 1997 Average

Country	1995	1996	1997	Average	Import Share (%)
	(1,000 metric tons)				
Algeria	3,507	1,972	2,974	2,818	2.8
Brazil	6,135	7,663	4,850	6,216	6.2
China	12,602	9,194	2,826	8,207	8.2
Egypt	5,070	5,121	4,500	4,897	4.9
European	2,832	2,187	3,775	2,931	2.9
Japan	5,965	5,928	6,315	6,069	6.0
Korea	2,342	2,223	3,325	2,630	2.6
Mexico	1,223	1,980	1,801	1,668	1.7
Morocco	2,549	2,240	2,054	2,281	2.3
Former Soviet Union	3,792	3,510	4,605	3,969	4.0
United States	1,519	1,313	2,216	1,683	1.7
Venezuela	1,037	939	939	972	1.0
Others	53,018	56,613	58,594	56,075	55.8
Total World	101,591	100,883	98,774	100,416	

Sources: FAO, 1999; International Grains Council, 1997; Canadian Wheat Board, 1996-1998.

FIGURE 5.1. Average Wheat Export Flows: 1993-1994 to 1995-1996 (1,000 metric tons)

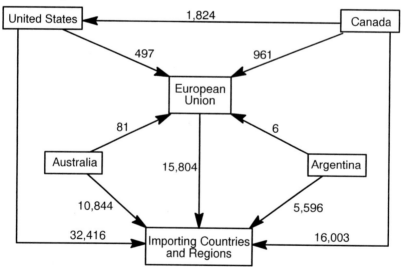

Source: FAO, 1999.

major exporters of wheat, but they also import considerable amounts. The United States imports wheat from Canada, while the European Union imports it from the United States, Canada, Argentina, and Australia.

Wheat Exports

The major wheat exporting countries, the United States, Canada, the EU, Australia, and Argentina, supply approximately 80 percent of the wheat traded in the world market. The United States is the largest exporter, followed by Canada and the EU (see Table 5.3).

TABLE 5.3. Wheat Exports by Class, 1995 to 1997 Average Production

Country/Class	1995	1996	1997	Average	Share (%)
		(1,000 metric tons)			
Argentina					
All	6,913	3,532	8,767	6,404	6.2
Australia					
All	7,818	14,568	19,378	13,921	13.5
Canada					
All	16,960	16,520	18,858	17,446	17.0
Common	13,739	12,426	14,205	13,457	13.1
Durum	3,221	4,094	4,653	3,989	3.9
European Union					
All	12,190	17,005	17,500	15,565	15.1
Common	11,890	16,505	17,000	15,132	14.7
Durum	300	500	500	433	0.4
United States					
All	33,778	27,246	29,260	30,094	29.2
Hard Red Winter	10,452	7,784	10,751	9,662	9.4
Hard Red Spring	8,982	8,165	6,260	7,803	7.6
Soft Red Winter	6,805	3,811	5,444	5,353	5.2
White	6,478	6,451	5,580	6,169	6.0
Durum	1,062	1,034	1,225	1,107	1.1
Others					
All	24,869	19,582	13,962	19,471	18.9
Total					
All	102,528	98,453	107,725	102,902	

Sources: FAO, 1999; International Grains Council, 1997; Canadian Wheat Board, 1996-1998.

The United States leads in exports of HRW and SRW wheats; an annual average of 30 million metric tons was exported in 1995-1997, of which nearly 16 million tons were HRW and SRW; another 7.8 million tons were HRS. The United States competes with the EU for market share of SRW wheat exports. Major U.S. markets for SRW wheat include China, West Asia, and North Africa. EU markets for SRW wheat include the FSU, China, West Asia, and North Africa.

Canada is the leader in exports of HRS and durum wheat. The United States also exports these wheats, competing with Canada. The EU competes with the United States and Canada for market share of durum wheat exports. Major U.S. markets for HRS wheat include Southeast Asia and East Asia, including Japan and South Korea. Major Canadian markets include China, the FSU, and the East Asian markets. The United States, Canada, and the EU compete intensely for the North African durum markets.

Australia and Argentina compete with the United States in exporting HRW wheat. Major U.S. markets for HRW wheat include the FSU, China, and East Asia. Argentina exports HRW wheat mainly to South America and West Asia. Australia's major markets are the North African countries, China, the FSU, and West Asia.

TRADE AGREEMENTS AND POLICIES

The major exporting countries use several export promotion policies, including export subsidies, credit arrangements, and long-term agreements, to protect or enhance their positions in the world market.

Export Subsidies

The EU and the United States are the primary users of direct export subsidies. The EU subsidy is equal to the difference between the EU market price and the world price. The EU's threshold and intervention prices keep domestic market prices well above the world price.

The EU uses two methods to establish export restitutions. First, refund tenders cover the majority of EU exports in which traders apply for refunds on specific quantities exported to specific markets. The exporter receives an export certificate, indicating the refund and a time period within which the certificate is valid. The second method is the "ordinary restitution" and is published regularly. These refunds are designed for particular destinations and often are used for stable import markets. Restitutions are the same for every origin of wheat in the EU, but may differ depending on destination. Subsidies to certain regions, such as Africa, Switzerland, or Scandinavia, are fixed by the commission and published in the EC journal. However, fixed export subsidies apply only to a small proportion of the total EU exports. Most grain is exported by weekly tenders. The commission publishes the quantities to be exported and grain trading firms are invited to submit bids.

The United States instituted export subsidies under the Export Enhancement Program (EEP) in May 1985 to regain lost market shares and compete with EU subsidies (Haley, 1989). The EEP uses a competitive bid process under which the U.S. Department of Agriculture targets a country for a specific quantity of a commodity. U.S. exporters then compete for sales to the targeted market and bonuses are awarded to the exporters whose sales price and bonus bid fall within an acceptable range. The bonus is calculated by taking the difference between the U.S. market price and world price. The exporters complete the sale, present proof of delivery, and receive a cash subsidy in the form of certificates. The exporter may sell these certificates or exchange them for Commodity Credit Corporation stocks. Sales targeted to the FSU and China accounted for half of EEP wheat sales and those to North Africa and Middle Eastern countries accounted for one-third of sales during the 1986-1995 period.

The UR agreement restricted export subsidies in terms of the quantity of grain subsidized and total expenditures under subsidy programs. The subsidized quantity under the program should be reduced by 21 percent of the average quantity for the 1986-1990 period by the end of the year 2000, and total expenditures should be reduced by 36 percent of the average expenditures for the 1986-1990 period. The EEP has not been used in the United States since 1996.

The Canadian rail subsidy under the Western Grain Transportation Act (WGTA) provided direct government payments to Canadian railroads for shipments of specified commodities, including wheat (U.S. International Trade Commission, 1990). Rail shipments subject to this subsidy include those from any point west of Thunder Bay, Ontario, or Armstrong, Ontario, to Thunder Bay or Armstrong for export or domestic use, and any port in British Columbia for exports (except to the United States).

The WGTA rail subsidy was estimated at $21.31 per metric ton, which was equivalent to 70 percent of the estimated freight rate of $30.31 per metric ton in 1989-1990 (U.S. International Trade Commission, 1990). However, the Canadian government eliminated the subsidy as of August 1, 1995. New regulations for Western grain transportation are contained in the 1996 Canada Transportation Act (CTA), which came into effect on July 1, 1996. A cap on freight rates still exists for Western grain in the CAP.

Export Credit

The United States, the Canadian Wheat Board (CWB), the Australian Wheat Board (AWB), and Argentina offer export credit to their buyers. The EU does not offer credit assistance as a group; however, some member countries do. France, for example, guarantees repayment through COFACE and certain commercial banks.

The credit terms and conditions vary widely across countries. Argentina's credit has primarily been granted to other Latin American countries, including Peru and Cuba, and has not exceeded twelve months. Australia extends credit through the AWB for up to three years. Egypt and Iraq are regular recipients. The Canadian government guarantees loan repayment on credit extended which does not exceed three years. Brazil is the largest credit buyer of wheat, followed by Iraq, Egypt, and Algeria. France's COFACE provides short-term credit and guarantees 85 percent of the credit if the purchaser is a private buyer and 90 percent if the purchaser is a foreign government. Medium- and long-term credit financing is provided through the Banque Francaise du Commerce Exterieur.

The U.S. Department of Agriculture operates two credit programs, GSM-102 and GSM-103. The GSM-102 program guarantees repayment of private credit extended to importers in specified countries for up to three years. The GSM-103 program covers private credit extended for between three and ten years.

Long-Term Agreements

Long-term agreements (LTAs) are advantageous to both exporting and importing countries. Exporting countries use LTAs to maintain export shares, attain new markets, and stabilize exports from year to year. Importers use LTAs to assure reliable supplies. LTAs include provisions for an upper and lower bound on purchases and, in some cases, financing arrangements; they may involve shipments over two or more seasons. Historically, 75 percent of the FSU's wheat imports were through LTAs. Canada and Australia have an advantage in negotiating LTAs because their grain boards can guarantee these trade commitments. Actual LTA shipments account for a small share of U.S. wheat exports (Harwood and Bailey, 1990).

North American Free Trade Agreement

The Canada-United States Free Trade Agreement (CUSTA) was signed in 1988 and implemented in 1989. The objective was to create a U.S.-Canada free trade area so that trade between the two countries would be uninhibited by border measures. The agreement called for conversion of nontariff border measures to tariffs, with all tariffs to be phased out over a ten-year period. The agreement was expanded to the North American Free Trade Agreement (NAFTA) by including Mexico in 1994.

There are no barriers in trading wheat between the United States and Canada under CUSTA. NAFTA reduces trade barriers among the United States, Canada, and Mexico. The United States is phasing out a 0.77 cents per kilogram tariff on durum wheat from Mexico over ten years; other wheat tariffs are being reduced to zero over five years. Mexico agreed to convert its license for wheat imported from the United States and Canada to tariffs. U.S. wheat exports to Mexico were subject to an initial 15 percent tariff, which is being reduced in equal installments over a ten-year period. In addition, Canada agreed to eliminate its import license requirement for wheat coming from Mexico.

Trade Barriers and Internal Supports

Importing countries have used various types of trade barriers to protect their domestic producers. Under the World Trade Organization (WTO), created by Uruguay Round (UR) agreements, nontariff trade barriers were to be converted into tariffs and reduced by 36 percent of the average level for the 1986-1988 period by 2000 in developed countries and 24 percent in developing countries by 2004. In 1995, the minimum access was to be 3 percent of base level consumption for the 1986-1988 period, and was to increase to 5 percent by 2000 in developed countries and by 2004 in developing countries.

In addition, both exporting and importing countries use various types of internal support to enhance their competitiveness in producing grain in the world market. Under the UR agreement, internal supports in terms of aggregate measures of supports (AMS) are to be reduced by 20 percent of the average support for the 1986-1988 period by 2000 in developed countries and by 13 percent by 2004 in developing countries.

REGIONAL AND NATIONAL COMPETITIVENESS

A country's competitiveness in a market can be measured in many different ways. One method is to compare the sum of production costs in exporting counties and marketing costs from exporting countries to an importing country. When considering several importing countries simultaneously, comparisons can be obtained by developing a mathematical programming model that has demand and supply constraints. Shadow prices associated with supply constraints of the model can be used to measure competitiveness in producing wheat among exporting countries (Golz and Koo, 1993; Koo, Golz, and Yang, 1993).

The model developed by Golz and Koo includes four classes of wheat: HRW, HRS, SRW, and durum. SRW wheat includes both soft and white wheat. The objective of the model is to minimize production costs in growing regions, inland transportation costs from production regions to utilization centers and ports for export, and ocean shipping costs from export to import ports. The model includes seven exporting countries and twelve importing regions representing fifty importing countries. Exporting countries are the United States, Canada, Australia, Argentina, the EU, Eastern Europe, and Mexico. Importing regions are Northeast Africa, Northwest Africa, China, South Asia, Southeast Asia, West Asia, East Asia, the FSU, Western Europe, Western South America, Northern South America, and Eastern South America. The objective function was constrained by (1) land constraints for wheat production in exporting countries or regions, (2) supply constraints of wheat in exporting countries or regions, and (3) import demand constraints for wheat in importing countries or regions.

The model by Koo, Golz, and Yang (1993) is similar to the wheat model. However, this model focuses on durum wheat production in major producing countries or regions, durum wheat trade from producing countries to importing countries, and trade of semolina from durum wheat mills to food processors in the United States and Canada.

Since the wheat model includes all classes and allows competition among classes for production, the results from the model are used to evaluate U.S. competitiveness in the world market in this analysis. In the Golz and Koo (1993) study, competitiveness in producing wheat is evaluated by using both base and free trade models. The base model optimizes wheat production and trade flows with existing trade policies of the exporting and importing countries. However, export promotion programs in exporting countries, trade barriers, and internal supports in both exporting and importing countries are entirely eliminated in the free trade model.

Optimal Wheat Production

The base model evaluates competition among exporting countries based on production costs, marketing costs, internal supports, export subsidies, tariffs, and import licenses. The optimal total wheat production by country or region is presented in Table 5.4. Comparing actual wheat production with that of the base model shows the United States is the only country with more wheat production in the model than is actually produced, indicating that the United States should produce more wheat under the given resource endowments and subsidy programs.

TABLE 5.4. Optimal Wheat Production by Class in Major Exporting Countries Under the Base and Free Trade Models

Country/Class	Base Model	Free Trade Model
	(1,000 metric tons)	
United States		
Soft Red Winter	14,812	18,230
Hard Red Winter	30,712	28,770
Hard Red Spring	9,444	9,605
Durum	3,230	4,493
Canada		
Hard Red Spring	20,190	19,920
Durum	4,005	4,317
European Union		
Soft Red Winter	83,536	74,217
Durum	7,671	4,643
Argentina	10,920	15,735
Australia	14,107	19,760

In the model, the EU is the leading SRW wheat producer, followed by the United States. Both Mexico and Eastern Europe produce enough wheat to meet domestic utilization. The United States produces more than actual levels and the EU less. SRW wheat production in the United States is about 28 percent more in the base model than actual, primarily because Australia and Argentina, the two competitors in the SRW wheat market, have relatively small domestic supports and export subsidies.

The U.S. and EU durum production in the model solution are more than their actual levels, while Canada's production is smaller than actual because subsidy levels are high in the United States and the EU relative to those in Canada. Although production practices in the United States and Canada are similar, marketing systems differ, with a government enterprise, the Canadian Wheat Board, marketing wheat in Canada while the United States has a private marketing system. The difference in marketing systems was not captured in this model and, thus, may contribute in part to the difference in durum wheat production between the actual and the base model amounts.

HRS wheat production in the model is less than actual production for both the United States and Canada. HRS and HRW wheat are substitutable in domestic utilization in the United States and Canada. Production cost per bushel for U.S. HRW wheat is lower than for HRS wheat in the United States and Canada (McElroy, 1987; Ortmann, Rask, and Stulp, 1989). This may account for lower HRS wheat acreage in the model than actual areas used for HRS wheat production.

The elimination of all trade barriers, domestic subsidies, and export promotion policies in the free trade model would decrease U.S. and EU production significantly, slightly increase Canada's production, and significantly increase production in Australia and Argentina. This indicates that Australia and Argentina would be major beneficiaries under a free trade policy regime for wheat.

HRW wheat production in the free trade model would decrease in the United States and increase in Australia, compared to the base model. Thus, under free trade Australia has a competitive advantage over the United States in producing HRW wheat. Australia has lower production and transportation costs to major HRW wheat import markets, such as China and Japan. Argentina's HRW and SRW wheat production is doubled in the free trade scenario, compared to the base model, due primarily to low production costs (Ortmann, Rask, and Stulp, 1989; Stanton, 1986). SRW wheat production in the EU would decrease 12 percent, while SRW production in the United States increases slightly. This indicates that the United States and Argentina have a competitive advantage over the EU in producing SRW.

U.S. production of HRS wheat under the free trade model increases, compared to the base model, while Canadian HRS wheat production is nearly the same as the base model. U.S. and Canadian production of durum wheat would increase in the free trade model compared with the base model, while durum production in the EU would decrease by 65 percent. This indicates that the United States and Canada have a competitive advantage over the EU in producing durum wheat.

Optimal Wheat Exports

In the base model, the United States is the leading exporter of HRW wheat, indicating that it has a competitive advantage in exporting HRW wheat (see Table 5.5). Canada is the largest exporter of HRS wheat and the EU is the largest exporter of SRW and durum

TABLE 5.5. Optimal Wheat Exports by Class in Major Exporting Countries Under the Base and Free Trade Models

Country/Class	Base Model	Free Trade Model
	(1,000 metric tons)	
United States		
Soft Red Winter	11,584	8,939
Hard Red Winter	10,050	8,112
Hard Red Spring	13,957	11,432
Durum	1,290	2,073
Canada		
Hard Red Spring	15,256	15,316
Durum	2,271	3,171
European Union		
Soft Red Winter	16,978	10,587
Durum	757	0
Argentina	4,252	8,073
Australia	9,269	14,733

wheats in the base model. Australia and Argentina do not have competitive advantage in exporting SRW and HRW wheat, respectively, in the world market, mainly because their subsidy levels are lower than those of the other exporting countries.

Under the free trade model, the United States loses its competitiveness in the hard wheat market, but maintains its competitiveness in exporting HRS, durum, and SRW wheat. Canada maintains its competitiveness in exporting both HRS and durum wheat. The EU would lose market share significantly in both the soft and durum wheat markets. Both Australian and Argentinean exports increase substantially, mainly because these two countries become more competitive under the free trade model.

Regional and International Production Competitiveness

Overall competitiveness can be measured, using shadow prices associated with the upper limit of wheat production in a region or a country. The shadow price is interpreted as a marginal value of production in a region or country. Thus, the higher the shadow price in absolute terms, the more competitive the region is in producing wheat.

Shadow prices in the base model solution indicate that the United States is most competitive in producing all classes of wheat, followed by the EU (see Table 5.6). This is because domestic and export subsidies in the United States and the EU are higher than those of other exporting countries. However, the United States is less competitive under the free trade scenario. It is still competitive in producing SRW wheat, but Australia and Argentina become more competitive in producing HRW wheat. Canada is more competitive than the United States in growing HRS wheat, while the United States is more competitive in producing durum wheat.

TABLE 5.6. Shadow Prices Associated with Production Constraints in the Base and Free Trade Model

Country/Class	Base Model	Free Trade Model
	\$/metric ton	
United States		
Soft Red Winter	49.0	39.0
Hard Red Winter	17.9	11.5
Hard Red Spring	20.5	12.0
Durum	14.38	12.68
Canada		
Hard Red Spring	4.23	12.18
Durum	4.23	12.18
European Union		
Soft Red Winter	35.3	17.0
Durum	0	0
Argentina	0	37.3
Australia	0	17.2

CONCLUSIONS

Wheat is produced throughout most of the world, and the total trade volume is about 17 percent of wheat produced worldwide. The world wheat market has changed dramatically in the past decade, mainly because of the UR agreement, regional free trade agreements, including NAFTA, and the recent financial crisis in Asia. The objective of this chapter is to assess U.S. competitiveness in exporting wheat to major importing countries.

This study shows that trade policies and internal support programs used by major exporting countries affect each class of wheat differently in terms of production and exports. Under the existing programs and trade policies, the United States and the EU have a competitive advantage over the other countries in producing all classes of wheat. However, when the subsidies are eliminated in the free trade scenarios, the United States and the EU lose their competitiveness. Australia and Argentina become more competitive than the United States in producing HRW wheat. Canada is more competitive than the United States in producing HRS wheat, while the United States is more competitive in producing durum wheat and is still competitive in producing SRW wheat.

Approaching WTO negotiations are expected to focus on further reductions of trade barriers and subsidies. This may not necessarily be beneficial for all classes of wheat produced in the United States. Further liberalization would not benefit HRW wheat producers in the United States, while providing some benefits to other wheat producers. Furthermore, successful completion of the Free Trade Area of the Americas (FTAA), now underway, could substantially alter the competitive situation in major wheat-producing countries of the hemisphere, Argentina, Canada, and the United States.

REFERENCES

Canadian Wheat Board. (1996-1998). *Statistical Tables.* Winnipeg, Manitoba.
FAO. (1999). *FAOSTAT Database.* Food and Agriculture Organization, United Nations. <http://apps.fao.org>.
Golz, Joel and Won W. Koo. (1993). *Spatial Equilibrium Analysis of the World Wheat Market under Alternative Trade Policies,* Agricultural Economics Report No. 297, Department of Agricultural Economics, North Dakota State University, Fargo.
Haley, Stephen L. (1989). *Evaluation of Export Enhancement, Dollar Depreciation, and Loan Rate Reduction for Wheat.* Staff Report No. AGES 89-6. USDA, Agriculture and Trade Analysis Division, ERS, Washington, DC.
Harwood, Joy L. and Kenneth W. Bailey. (1990). *The World Wheat Market: Government Intervention and Multilateral Policy Reform.* Staff Report No. AGES9007. USDA, CED, ERS, Washington, DC.

International Grains Council. (1997). *World Grain Statistics.* London, England.

Koo, Won W., Joel Golz, and Seung-Ryoung Yang. (1993). "Competitiveness of the World Durum Wheat and Durum Milling Industries under Alternative Trade Policies." *Agribusiness* 9(1):1-14.

McElroy, Robert G. (1987). *State Level Costs of Production*, ERS Report No. AGES870113, USDA, Economic Research Service, Washington, DC.

Ortmann, Gerald, F., Norman Rask, and Valter J. Stulp. (1989). *Comparative Costs in Corn, Wheat, and Soybeans Among Major Exporting Countries.* Research Bulletin, Agricultural Research and Development Center, Wooster, OH.

Stanton, Bernard F. (1986). *Production Costs for Cereals in the European Community: Comparison with United States*, Agricultural Experiment Station Report No. 86-2, Cornell University, Ithaca, NY.

U.S. International Trade Commission. (1990). *Durum Wheat: Conditions of Competition Between the U.S. and Canadian Industries.* Washington, DC.

Chapter 6

Rice

Gail L. Cramer
James M. Hansen
Kenneth B. Young

INTRODUCTION

Rice is produced in more than 100 countries with a wide range of growing conditions, from monsoon-dependent crops in the subtropics to upland rice in mountain highlands. About 95 percent of the world's rice is produced in developing countries and 92 percent of it in Asia (Juliano, 1993). In 1998, China was the principal rice producer (34.9 percent), followed by India (21.9 percent), Indonesia (8.3 percent), Bangladesh (4.9 percent), Vietnam (5.0 percent), and Thailand (3.9 percent); the United States produced only about 2.1 percent of world rice. More than 80 percent of the world's rice is milled and consumed within five miles of where it is grown. About 15 percent is sent to nearby cities, and less than 5 percent enters into world trade.

Six states in the United States produce rice, with a concentration in the Arkansas Grand Prairie, Northeastern Arkansas, the bootheel of Missouri, the Mississippi River delta (in Arkansas, Mississippi, and upper Louisiana), southwestern Louisiana, the coastal prairie of Texas, and the Sacramento Valley in California. The area devoted to rice averages less than 1 percent of the total cropland harvested in the United States. Total U.S. rice production was 6.17 million metric tons (t.) in 1998 (rough basis) with 47 percent in Arkansas, 15 percent in Louisiana, 8.5 percent in Texas, 4 percent in Missouri, 8.3 percent in Mississippi, and 18 percent in California. Long-grain

(LG) rice comprised 75 percent and medium-grain (MG) 25 percent. Most of the medium-grain is produced in California.

Season average farm rough rice prices in 1998 were US$185 per t for LG and US$223 per t for MG. Export prices for milled rice in 1998 averaged US$369 per t FOB Houston (U.S. No. 2 LG) and US$470 per t FOB California (U.S. No. 2 MG). U.S. rice typically sells for a premium price compared to exports from other major rice-exporting countries because of the high quality and high import demand in the North and Central American market.

Rice is an especially important farm crop in Arkansas and Louisiana, where rice farms made up 11.6 percent and 8.3 percent, respectively, of all farms in 1992 (U.S. Dept. of Commerce, 1992). Rice farms constituted only 0.6 percent of all U.S. farms in 1992 compared with grain farms, which comprised 22 percent. Rice provides higher income per acre than other grain crops because of the high yield and relatively high price per t. For example, the average long-grain rough yield in Arkansas was 5,630 pounds per acre in 1997, providing a return of $545 per acre at an average market price of $214 per t.

UNITED STATES RELATIVE COST OF PRODUCTION AND MILLING

Production costs for rough rice in Arkansas, Texas/Louisiana, Thailand, Vietnam, and Guyana are compared in Table 6.1. Thailand has the lowest reported cost per acre, followed by Guyana and Vietnam. Production cost per acre is $110.71 in Thailand, $199.05 in Guyana, and $220.60 in Vietnam. Production cost per acre in the United States is nearly six times the level in Thailand because of the high land, water, and mechanization costs compared to other countries. However, on a per-t basis the costs are more comparable between countries ranging from a low of $110 per t in Vietnam to a high of $248 per t in Texas and Louisiana. Production costs in Arkansas are less than those of Texas and Louisiana by $42 per t, but transport costs to ship rice to the export gulf port from Arkansas are higher.

TABLE 6.1. Rough Rice Production Costs in Arkansas, Texas/Louisiana, Thailand, Vietnam, and Guyana

Cost Item (US$)	Arkansas	Texas/LA	Thailand	Vietnam	Guyana
Variable cash expenses:					
Labor	49.09	69.73	69.10	57.00	26.48
Seed	20.36	22.92	5.11	na[a]	17.49
Fertilizer	50.14	58.43	10.65	na[a]	39.18
Other	216.18	250.94	3.99	110.05	95.09
Subtotal	335.77	402.02	88.85	167.05	178.24
Fixed costs:					
Land tax/rent	132.53	115.11	16.61	10.99	8.93
Depreciation	51.34	58.99	0.57	na[a]	na[a]
Other	42.65	50.50	4.68	42.56	11.88
Subtotal	226.52	224.60	21.86	53.55	20.81
Total per acre	562.29	626.62	110.71	220.60	199.05
Average yield/acre	2.74	2.53	0.81	2.00	1.38
Average cost/t	$205.22	$247.68	$136.68	$110.30	$144.24

Sources: USDA, 1998; Isvilanonda, 1999; Khiem, 1998; Cramer, 1998.

[a] Cost included in other categories.

Estimated FOB port supply costs for milled rice range from a low of $200 per t for 10 percent brokens from Vietnam to a high of $464 per t for U.S. No. 2 with 4 percent brokens (Table 6.2). Guyana has the second highest milled rice supply cost, with 20 to 25 percent brokens because of the low milling yield, requiring 2.21 t of rough rice to produce 1 t of milled rice. Land taxes and rent make up nearly one-quarter of the rice supply cost in the United States. Other countries have only about 15 percent or less of the land cost per acre for rice production.

UNITED STATES AGRICULTURE POLICY

The 1996 FAIR Act significantly changed the price and income mechanisms for rice and other grains. It established a seven-year payment contract for eligible farmers and ranchers. Under this system,

TABLE 6.2. Estimated FOB Supply Cost for Milled U.S., Thailand, Vietnam, and Guyana Rice Exports

Cost Item	United States	Thailand	Vietnam	Guyana
Rough rice cost	$205	$137	$110	$144
Milled rice standard (% broken)	4%	5%	10%	25%
Requirement/t milled rice	1.52 t	1.73 t	1.64 t	2.21 t
Rough rice cost/t milled rice	$312	$237	$180	$318
Milling cost/t milled rice	67	19	25	47
Other net marketing cost		7	20	80
Less value by products			(25)	(51)
FOB port cost	464	263	200	394

Sources: USDA, 1998; Isvilanonda, 1999; Khiem, 1998; Cramer, 1998.

rice producers are provided nearly complete flexibility in planting decisions. They receive a rice contract payment whether they produce rice or not. Thus, production decisions are determined primarily by relative market returns. Nonrecourse loans continue to be available to rice producers at a maximum rate of $6.50 per hundredweight (cwt). The act also retains export assistance programs for rice and other grains. These programs include Export Credit Guarantee (GSM), Market Access Programs (MAP, promotion), P.L. 480 food aid, and the Export Enhancement Program (EEP). EEP subsidizes exports into markets as a countervailing policy against unfair export competition. Export programs have been traditionally important for the U.S. rice industry, but have been reduced substantially in recent years.

RICE TRADE AGREEMENTS, RULES, AND REGULATIONS

World rice trade has accounted for less than 5 percent of production since the 1960s. This thin global market combined with variable production causes greater price volatility for rice compared to other grains traded internationally. High variability in annual rice production occurs, in part, because of Asian dependence on the annual monsoon rains; outside of Asia, rice is mostly irrigated. A further destabilizing factor in the global market is that rice is highly segregated by type and quality, mainly to satisfy different ethnic markets (Bierlen, Wailes, and Cramer, 1997; Cramer, Wailes, and Shui, 1993; Wailes, Cramer, Hansen, and Chavez, 1998; Young, Cramer, and Wailes, 1998).

Government intervention in Asian countries has included both domestic rice price stabilization and stringent rice import controls. Notable examples of extreme price support and stabilization are found in the more developed Asian countries (Japan, Korea, and Taiwan) where the domestic price is supported at several times the world market price (Wailes, Young, and Cramer, 1993, 1994). Rice imports were previously banned in these countries except in emergency shortage situations. The Uruguay Round of the GATT agreement in 1995 partially opened the Japanese and Korean rice markets to international competition with an import quota system to cover a small percentage (4 to 8 percent) of consumption requirements by 2004. The Uruguay Round agreement proposes that the initial small import quotas of these countries be replaced with an import tariff in the future. Taiwan and China are expected to enact minimum access import requirements as a condition to becoming members of the World Trade Organization (WTO). China has maintained strict state control over the domestic economy and international trade, with a variety of different trade restrictions, including trading rights assigned only to government and selected private agents, state trading, nontariff measures, import and export licensing, and sanitary and phytosanitary measures that have been practiced for decades. The aim of the WTO in accession negotiations is to create transparency by requiring all WTO members to replace all current nontariff trade barriers with tariffs and tariff rate quotas that can be uniformly applied and more easily liberalized over time.

Examples of the current base rice import tariff rates and rates bound by the Uruguay Round GATT agreement are shown in Table 6.3 for selected countries. It may be noted that the dominant rice exporter, Thailand, has a base ad valorem rice import tariff of 58 percent compared to a specific import tariff of only US$0.022 per kg in the United States. The United States currently imports nearly 10 percent of its rice, mostly from Thailand.

Implementation of the Uruguay Round agreements and NAFTA contributed significantly to recent expansion of U.S. agricultural exports, including rice. U.S. rice exports reached 2.86 million t in 1998, representing a 23 percent increase above 1990. The value of

TABLE 6.3. Current Base and Bound Import Tariff Rates for Milled Rice Stipulated in the 1993 GATT Agreement (Selected Countries)

Country	Base Rate		Bound Rate	
	Specific	Ad Valorem (%)	Specific	Ad Valorem (%)
Argentina		38		38
Australia		2		2
Brazil		55		55
Canada	5.51 Can$/t		3.53 Can$/t	
China		150		114
Colombia		210		189
Costa Rica		56		36
Cuba		40		40
EU 15	650 ECU/t		416 ECU/t	
Guyana		140		140
Haiti		66		66
Indonesia		180		160
Jamaica		115		115
Kenya		100		100
Malaysia		45		40
Mexico		50		45
Nigeria		230		230
Pakistan		100		100
Peru		185		68
Thailand		58		52
Turkey		50		45
United States	2.2 c/kg		1.4 c/kg	
Uruguay		25		55
Venezuela		135		122

Source: WTO, 1999.

rice exports to Mexico increased from $25 million to $97 million and those to Japan increased from $0.62 million to $127 million from 1990 to 1998. Specific accomplishments of the Uruguay Round for rice include:

- new rice market access to Japan and South Korea, and
- new trade concessions negotiated with the EU allowing U.S. high-quality rice exporters to receive a tariff rebate.

However, significant trade barriers still remain for U.S. rice. Upcoming WTO agricultural trade negotiations after 1999 are expected to continue the reform process, including monitoring the compliance of current WTO members, eliminating use of non-scientifically based sanitary and phytosanitary (SPS) standards that unfairly restrict U.S. access to foreign markets, requiring full compliance of new applicant nations with existing WTO rules, and formulating improved WTO rules such as the further reduction of tariffs and elimination of export subsidies.

Major U.S. concerns on rice trade currently include the following:

- MERCOSUR partners Argentina and Uruguay have preferential tariffs that make U.S. rice uncompetitive in Brazil (a $500 million rice market)
- State trading enterprises in several major importing countries (e.g., Indonesia and the Philippines) monopolize purchases, restricting market access
- Trade activities of some single-desk exporters (e.g., Australia) undermine market transparency
- Nigeria has excessive import tariff rates for parboiled rice (see Table 6.3)

The United States continues to be active in expanding rice trade opportunities, including seeking new fast-track negotiating authority, participating in more regional liberalization through the Asian Pacific Economic Cooperation (APEC) forum and the Free Trade Area of the Americas initiative (FTAA). APEC has the goal of attaining free trade for all developed countries in the Asia Pacific region by 2010 (2020 for developing countries). The FTAA process has the goal of achieving free trade in the Western Hemisphere by 2005.

IMPACT OF EL NIÑO ON GLOBAL RICE PRODUCTION

Weather accounts for the largest variability in production, consumption, and trade in the world rice economy. Weather anomalies associated with El Niño affect some major Asian rice-producing countries as well as some Caribbean and Central and South American countries (Hansen, Wailes, and Cramer, 1998). The climates of these countries are typically influenced by oceanic conditions. In the Pacific region, the Philippines, Indonesia, eastern Malaysia, and Australia have been affected most, with Vietnam affected to a lesser degree. In the Indian Ocean region, all of India and Sri Lanka are affected. El Niño usually results in abnormally dry conditions and warmer climates for these countries. The drought of 1982-1983 was the result of El Niño climatic conditions.

Rice production, consumption, and trade in regions affected by El Niño have a significant impact on the world rice market. Asian and Central American countries have droughts and South America has a variety of weather conditions—northern Brazil has droughts, while southern Brazil, northern Argentina, and Uruguay have floods and cooler weather. The 1997-1998 El Niño strongly affected India, Malaysia, the Philippines, Indonesia, and Australia; countries which account for about one-third of world rice production and consumption. Indonesia, the Philippines, and Malaysia have imported 18 percent of world rice trade, while India and Australia shipped 16 percent of world rice exports on average during 1994 through 1998. These countries also account for one-third of the world's rice inventory.

In recent years Central America has accounted for approximately 1 percent of world rice trade, but less than 0.14 percent of total world production, while South America imported 8 percent, exported 7 percent, and produced 3 percent of the world's rice during 1994-1998. South America's rice production decreased by 9.5 percent in 1997, mostly in countries affected by El Niño, Argentina, Uruguay, Brazil, Ecuador, and Guyana. South American rice exports decreased 23 percent and imports increased 26 percent from 1996 to 1997.

THE ARKANSAS GLOBAL RICE MODEL

The Arkansas Global Rice Model (AGRM), a representation of the world rice economy, is used to simulate rice trade. The model consists of 22 econometric submodels (one for each country or region), including the United States, Thailand, Pakistan, China, India, Burma, Vietnam, Australia, Japan, South Korea, Taiwan, Indonesia, the EU, Spain, Italy, Egypt, Iran, Iraq, Saudi Arabia, Brazil, Argentina, Uruguay, and the rest of the world (ROW). The international rice market is heavily influenced by government policies, and the model attempts to capture this by explicitly incorporating policy variables for individual countries in its supply, demand, export (or import), stocks, and price transmission equations, which are, thus, implicitly reflected in the model solution.

Major components of each submodel include supply, demand, trade, and price linkage equations. Computationally, the simulation model solves for the set of farm, wholesale, and export (import) prices that simultaneously clears all markets in a given year for given exogenous factors. Due to the dynamics of supply and demand, market clearing prices must be obtained recursively for each future year simulated (see Wailes, Cramer, Hansen, and Chavez, 1998).

It is assumed that the supply of rice is determined by profit-maximizing producers. The industry equation describing planted acreage is a function of expected output price and input prices. Yield is specified as a function of expected output price, input prices, and technological change. It is assumed that the demand for rice is determined by utility-maximizing consumers. The trade sector is a function of domestic production, consumption, and ending stocks. For an exporting country, farm price, P_t, is modeled as a function of the retail price. Retail price is a function of the deflated FOB price and a time trend that captures improvements in marketing efficiency. Export price is modeled as a function of the Thai (5 percent broken) export price. The international price, Thai (5 percent broken), is solved to close the model such that world imports and exports are equal in each year.

An Analysis of a 10 Percent Increase in World Price

The scenario for analyzing the international rice market using the Arkansas Global Rice Model is an increase in the world rice market price by 10 percent. It is developed to represent actual conditions that could occur in the world rice economy. In this case, it is assumed that weather conditions cause Indonesia's harvested rice area to decrease by 374,000 hectares (-3.26 percent) in the initial year (1999); this corresponds to a 10 percent increase in the Thai FOB price on the international market. All countries are included in solving for equilibrium, but only Indonesia's harvested rice area has been shocked in the first year.

The initial effect is a decrease in Indonesia's rice production by 1,081,000 t (-3.26 percent from the base), which leads to increased imports for Indonesia by 1,077,000 t (42 percent). Indonesia's increased rice imports affects world rice trade, international prices, domestic prices, stock levels, and world consumption. The Thai FOB price is solved for equilibrium so that world net trade balances, resulting in a 10 percent increase, from US$283 to US$312 per metric ton. The subsequent equilibrium price affects Indonesia's current consumption but not production. In 1999, the first year of the shock, production in other countries is not affected, because all countries are assumed to have one planting season and production decisions are based on the previous year's price.

Major exporting countries' responses in the first year to an increase in imports by Indonesia is to raise exports as international prices increase: Thailand by 33,139 t (0.53 percent), the United States by 29,062 t (1.17 percent), India by 32,548 t (1.7 percent), Italy and other EU countries by 16,024 t (1.49 percent), Myanmar by 8,143 t (3.2 percent), Vietnam by 15,292 t (0.4 percent), China by 7,413 t (0.49 percent), Australia by 4,733 t (0.65 percent), Argentina by 2,491 t (0.42 percent), and Egypt by 1,040 t (0.39 percent).

Most importing countries decreased imports in the first year because of higher international prices. China decreased imports by 34,426 t (-7.39 percent), the European Union by 9,584 t (-0.64 percent), Saudi Arabia by 6,118 t (-0.78 percent), the United States by 2,922 t (-0.91 percent), Iran by 1,215 t (-0.11 percent), Iraq by 1,045 t (-0.15 percent), Brazil by 748 t (-0.05 percent), and ROW by 870,209 t (-7.41 percent). Importing countries reduce their stocks and decrease consumption due to the increased prices. Domestic rice consumption decreases in almost all countries. In the first year, consumption decreased in Thailand by 5,980 t (-0.07 percent), China by 506,104 t (-0.37 percent), the United States by 4,432 t (-0.12 percent), Myanmar by 22,580 t (-0.23 percent), Vietnam by 15,292 t (-0.10 percent), Egypt by 6,441 t (-0.23 percent) Saudi Arabia by 6,118 t (-0.78 percent), Indonesia by 4,588 t (-0.01 percent), and the ROW by 808,744 t (-1.31 percent).

In the second year, world rice production increases as producers respond to the higher price in the first year. World rice area harvested and production are greater by 294,459 hectares (0.2 percent) and 1,307,065 t (0.33 percent), respectively, from the base. World exports increase by 215,731 t (0.98 percent) because of increased production. The world rice price is lower by US$15.41 per t because of the increased production and exports in the world market. World rice consumption is higher by 517,872 t (0.13 percent) and world imports are greater by 215,731 t (0.98 percent) from the base because of the lower world price. By the third year, the model begins to converge back toward the initial equilibrium. The price is only $3.23 (1.08 percent) greater than the base, US$299.01 per t.

By the sixth year and thereafter the effect of decreased production in Indonesia in 1999 on the world rice market is to lower world trade by 0.08 percent, or about 15,000 t. World production is slightly greater on average for the final eight years of the simulation. World consumption is reduced by 10,000 to 12,000 t and the world rice price is approximately 0.07 percent higher than the base (only about 20 to 40 cents per t).

United States Projections from the Arkansas Global Rice Model

Total rice production projections for the United States are presented in Table 6.4 (separate models were run for long- and medium-grain rice, but these results are not shown in the table). Acreage is generally determined by net returns to producers, while changes in yields over time are driven by research expenditures. Total U.S. rice area planted decreased from 3.32 million acres in 1994 to 3.09 million acres in 1995. Under the FAIR Act, rice acreage declined by 10 percent, resulting in only 2.80 million acres in 1996, but increased to 3.10 million acres in 1997 and 3.32 million acres in 1998 due to attractive prices; it is estimated to have increased to 3.60 million acres in 1999 mainly due to a higher rice price relative to competing crops. Over the longer run, area harvested is projected to decline gradually, to 3.16 million acres due to expected higher returns to other crops and stiffer competition from other major rice producing countries (Table 6.4). Long-grain harvested acreage increased to 2.61 million acres in 1998 and is estimated to increase to 2.71 million in

TABLE 6.4. Detailed U.S. Rice Supply and Utilization

Variable	Units/Year	1996	1997	1998	1999	2000	2001	2002	2003	2004	2005	2006	2007	2008	2009	2010
Actual yield	(lb/ac)	6121	5911	5980	6130	6154	6193	6250	6290	6334	6376	6432	6488	6545	6604	6663
Program yield	(lb/ac)	4827	4827	4827	4827	4827	4827	4827	4827	4827	4827	4827	4827	4827	4827	4827
Program area/control area	(1000 ac)	4158	4157	4157	4157	4157	4157	4157	4157	4157	4157	4157	4157	4157	4157	4157
Total harvested area	(1000 ac)	2799	3034	3192	2925	2893	2878	2812	2838	2829	2828	2820	2826	2817	2832	2831
Supply (rough basis)	(mil. cwt)	206.3	216.3	226.4	219.4	215.6	214.3	211.0	212.3	213.3	214.7	216.0	218.2	219.8	222.9	225.7
Production	(mil. cwt)	171.3	179.3	190.9	179.3	178.0	178.2	175.8	178.5	179.2	180.3	181.4	183.4	184.4	187.0	188.7
Beginning stocks	(mil. cwt)	25.0	27.2	25.8	30.3	27.3	25.3	24.0	22.1	22.0	21.8	21.6	21.3	21.5	21.4	22.2
Imports	(mil. cwt)	10.0	9.7	9.8	9.9	10.2	10.7	11.2	11.6	12.1	12.6	13.0	13.5	14.0	14.5	14.9
Domestic use (rough basis)	(mil. cwt)	100.7	107.0	110.3	112.6	114.5	116.5	118.4	120.4	122.5	124.6	126.6	128.7	130.8	133.0	134.6
Food	(mil. cwt)	80.0	82.1	84.8	87.2	89.2	91.2	93.0	94.9	96.8	98.7	100.6	102.5	104.5	106.4	108.4
Seed	(mil. cwt)	4.0	4.0	4.2	4.0	3.9	3.7	3.8	3.7	3.7	3.7	3.7	3.6	3.6	3.6	3.0
Brewing	(mil. cwt)	15.4	15.4	15.3	15.4	15.5	15.6	15.7	15.8	16.0	16.2	16.4	16.6	16.8	17.0	17.1
Residual	(mil. cwt)	1.3	5.5	6.0	6.0	6.0	6.0	6.0	6.0	6.0	6.0	6.0	6.0	6.0	6.0	6.0
Exports	(mil. cwt)	78.4	83.5	85.8	79.5	75.8	73.7	70.4	69.8	69.0	68.5	68.1	68.0	67.6	67.8	68.2
Total use	(mil. cwt)	179.1	190.5	196.1	192.1	190.3	190.2	188.8	190.2	191.5	193.1	194.7	196.7	198.4	200.8	202.7
Ending stocks	(mil. cwt)	27.2	25.8	30.3	27.3	25.3	24.0	22.1	22.0	21.8	21.6	21.3	21.5	21.4	22.2	23.0

Variable	Units/Year	1996	1997	1998	1999	2000	2001	2002	2003	2004	2005	2006	2007	2008	2009	2010
Prices																
Loan rate	(US$/cwt)	6.50	6.50	6.50	6.50	6.50	6.50	6.50	6.50	6.50	6.50	6.50	6.50	6.50	6.50	6.50
Season avg. farm price	(US$/cwt)	9.96	9.88	9.54	9.51	9.59	9.41	9.62	9.70	9.82	9.95	10.11	10.20	10.29	10.31	10.36
Long-grain farm price	(US$/cwt)	10.32	10.40	9.69	9.73	9.84	9.49	9.80	9.86	10.02	10.15	10.33	10.41	10.51	10.50	10.54
Medium-grain farm price																
LG-MG margin	(US$/cwt)	9.25	8.79	9.15	9.07	9.11	9.24	9.30	9.40	9.47	9.58	9.71	9.82	9.91	9.97	10.04
	(US$/cwt)	1.06	1.61	0.54	0.67	0.73	0.25	0.50	0.45	0.55	0.57	0.61	0.59	0.60	0.53	0.50
Exp. price, FOB Hou., US #2	(US$/cwt)	20.48	18.96	8.73	18.57	18.78	19.23	19.40	19.72	19.85	20.08	20.43	20.68	20.83	20.94	21.15
Med. grain, FOB CA, US #2	(US$/cwt)	18.79	17.94	8.54	18.43	18.44	18.66	18.70	18.85	18.92	19.07	19.28	19.43	19.53	19.57	19.64
Deficiency/CLD/contract rate	(US$/cwt)	2.77	2.71	2.85	2.75	2.53	2.04	1.98	1.98	1.98	1.98	1.98	1.98	1.98	1.98	1.98
World price (US$/cwt)	(US$/cwt)	7.51	7.83	6.21	6.31	6.31	6.55	6.59	6.77	6.82	6.94	7.14	7.27	7.31	7.32	7.39
EXPP-SAFP margin	(US$/cwt)	6.26	5.23	5.48	5.37	5.47	6.17	6.03	6.25	6.21	6.27	6.39	6.51	6.53	6.62	6.76
Income factors																
Production market value	(mil. US$)	1706	1773	1821	1704	1707	1677	1692	1731	1760	1793	1834	1871	1898	1928	1955
Deficiency/contract payments	(mil. US$)	472	465	500	483	443	358	348	348	348	348	348	348	348	348	348
Marketing loan/ certificates	(mil. US$)	0	0	53	36	36	0	0	0	0	0	0	0	0	0	0
Total income	(mil. US$)	2178	2238	2374	2223	2186	2035	2040	2079	2108	2141	2182	2219	2246	2276	2303
Returns above variable cost	(US$/ac)	228	204	209	208	210	184	195	198	202	207	215	219	231	237	247

1998 before gradually declining to 2.34 million in 2010. Medium-grain acreage, on the other hand, decreased to 709,000 acres in 1998, but is expected to increase steadily, reaching 819,000 acres by 2010.

Arkansas's total rice area increased from 1.39 million in 1997 to 1.53 million acres in 1998, and is estimated to have increased to 1.65 million acres in 1999, before decreasing to 1.47 million acres over the forecast period. Louisiana's total rice area increased by nearly 38,000 acres in 1998, with all the gains coming from long-grain area. Texas' area increased to 283,000 acres in 1998. Missouri's rice area increased from 117,000 acres in 1997 to 143,000 acres in 1998, and is estimated to have increased by more than 12,000 acres in 1999. Mississippi's harvested acreage increased to 268,000 acres in 1998, and is estimated to have increased to 298,000 acres in 1999. California's acreage decreased to 478,000 acres in 1998 and is estimated to have increased by nearly 100,000 acres in 1999. Acreage declines are expected to be offset partially by yield gains resulting from varietal research. Long-grain yields are projected to grow at 1 percent per year while medium-grain rice yields are projected to grow about 0.5 percent per year. The average U.S. rice yield decreased from 58.97 cwt per acre in 1997 to 56.59 cwt in 1998, due to unfavorable weather. Yields are projected to increase to 65.53 cwt by 2010.

In 1998, the higher acreage offset the decline in yield, resulting in a 2.8 percent increase in production. Production is estimated to have increased substantially in 1999 (to 212.7 million cwt), due to increases in both acreage and yields. Thereafter, total rice production is projected to decrease to 205.7 million cwt in 2002 before increasing to 207 million cwt by 2010. On the average, long-grain production would decline slightly, while medium-grain production is expected to remain relatively flat over the projection period.

Total U.S. rice supply increased to 226 million cwt in 1998 from 219.4 million in 1997, and is expected to range between 246 and 263 million cwt during the rest of the projection period. Imports are projected to grow at 3.3 percent per year, driven by the decline in real Thai (5 percent broken) FOB price and the growth in domestic U.S. rice consumption. Domestic use of rice increased to 115.9 million cwt in 1998 from 106.5 million in 1997. It is projected to increase steadily to 131.6 million cwt by 2010. With a stable population growth of less than 1 percent over the forecast period, the expansion in rice consumption is a result of increased per capita direct and processed rice consumption. The main processed uses of rice are cereals, pet food, and package mixes. Pet food is the fastest-growing sector in the processed category. The increase in food consumption is driven by growth in income and declining real retail prices, assuming low levels of inflation over the period.

Exports increased from 85.2 million cwt in 1997 to 87.6 million in 1998, and are estimated to have increased to 91.2 million in 1999. However, because of increased foreign competition, exports are projected to remain relatively flat throughout the projection period to 2010. There has been a significant shift in U.S. exports from milled to rough rice, especially over the last four years. During the four-year period 1990-1993, rough rice accounted for less than 7 percent of total rice exports. The share of rough rice exports started to increase dramatically in 1994, accounting for 18 percent of total rice exports, up from 5 percent in 1993. Over the last two years (1997-1998), rough rice averaged about 29 percent of total rice exports. The main reason for this shift is the increased demand for rough rice from a number of Latin American countries, notably Mexico, Brazil, Costa Rica, and Venezuela, but also including Colombia, Ecuador, Panama, El Salvador, Honduras, Guatemala, and Nicaragua. These countries prefer to import rough rice to improve utilization of their milling capacity and encourage this by setting lower tariffs for rough rice compared to milled rice imports. U.S. rough rice is well-positioned to maintain its competitive edge in this market segment, not only geographically but because there are very few countries that allow rough rice exports. No other major rice supplier exports significant volumes of rough rice.

U.S. long-grain exports are projected to decrease from 72.1 million in 1998 to 64.2 million cwt in 2010, as both the real Thai FOB price and U.S. export supplies decline. Medium-grain exports, on the other hand, could increase from 15.5 million in 1998 to 25.8 million cwt in 2010, mainly due to the increase in exportable supply. The WTO minimum access requirements for export markets in Japan and South Korea also support the growth of medium-grain exports.

The nominal season average farm price decreased slightly to $9.70 per cwt (rough basis) in 1997 (from $9.96 in 1996) and declined to $8.81 in 1998 due to larger U.S. production and ending stocks and weaker international prices. Farm prices were projected to decline further to $7.38 per cwt in 1999 and to $7.08 per cwt by 2000, but to then gradually increase to $8.46 per cwt by 2010. The average medium-grain farm price decreased from $9.26 per cwt in 1996 to $8.79 in 1997 due to weaker export demand for high-quality japonica rice, but increased to $10.13 in 1998. The average medium-grain price was projected to decrease in 1999 to $9.40 and then decrease steadily to $8.60 by 2001, before increasing to $8.93 by 2010. The long-grain export price (FOB Houston) decreased to $16.74 per cwt (milled basis) in 1998 from $18.82 in 1997, and was predicted to decrease to $14.28 in 1999 before increasing to $17.87 by 2010. The medium-grain export price (FOB California) increased to $21.32 per cwt (milled basis) in 1998 and was expected to decrease to $19.56 in 1999. In real terms, both U.S. farm and export prices decline steadily over the projection period.

At the global level, the growth in world rice production necessary to satisfy the projected consumption levels over the next 12 years (1999-2010) will mainly come from yield increases, as it has over the past two decades. Most of the wet land suitable for rice production has already been developed. World harvested area was projected to increase only slightly from to 151.3 million ha in 1998 to 152.0 million ha in 1999. Average world yield is projected to increase by 0.93 percent per year from 2.53 t per ha in 1998 to 2.9 t by 2010. World net rice trade (exports minus imports) is projected to increase from 21.1 million t in 1998 to 23.8 million t in 2010. Current major world exporters are Thailand, Vietnam, the United States, India, and Pakistan. Nearly 90 percent of total trade is long-grain and aromatic types, such as jasmine and basmati. Aromatic types constitute the bulk of current U.S. rice imports. The United States is projected to lose market share and increase imports over time because of the limited growth in production.

Projected exports by major rice-exporting countries in 2010 are 193,000 t for Thailand; 413,000 t for Vietnam; 4,040,000 t for India; 2,939,000 t for the United States; and 1,875,000 t for Pakistan. Projections for 2010 for imports are 1,061,000 t for Japan;

2,717,000 t for Indonesia; 1,349,000 t for Iran; 1,009,000 t for Iraq; 1,149,000 t for Saudi Arabia; 205,000 t for South Korea; 132,000 t for Taiwan; 1,142,000 t for Brazil; and 16,754,000 t for the remaining importing countries.

The United States produces only 2 percent of the world's rice, but accounts for 15 percent of the world's rice exports. The U.S. rice export market is quite varied with respect to types of markets, which have different criteria for their rice consumption. Four major export markets exist for U.S. rice: Mexico and potentially Latin America; Japan and potentially South Korea and Taiwan; middle eastern countries; and Europe.

Mexico has been the largest importer of U.S. rice in the past three years, followed by the Netherlands and Japan (Hansen et al., 1997). Mexico imports long-grain rough rice from the United States, which benefits Mexican rice millers. Importing rough rice also eliminates competition from Asia's major exporting countries, Thailand and Vietnam, because rough rice exports are not permitted by these governments. Japan has been one of the major importers for U.S. rice. Japan imports milled japonica rice. One of the significant achievements of the Uruguay Round agreement for agricultural trade liberalization was the minimum access requirement for rice imports into Japan and South Korea. Rice imports were previously banned due to the high levels of domestic rice price support in both countries. Initial minimum access requirements were 4 percent for Japan increasing to 8 percent by 2000, and 2 percent for South Korea increasing to 4 percent by 2004. The next multilateral trade negotiations are approaching and there will be strong pressure by the export countries to push for tariffication of the minimum access import requirements and large tariff reductions in line with other grains and oilseeds.

Article 5 of Annex 5 of the WTO agreement includes a voluntary clause that permits the special MA markets to develop tariffication before the grace period ends. On December 21, 1998, the Japanese government notified the WTO that it proposed to introduce a tariffication system for rice imports for the 1999 and 2000 fiscal years. Without serious objection from other WTO members, Japan initiated rice tariffication on April 1, 1999. Japan was able to reduce its rice imports in 1999 and 2000 by implementing tariffication.

The United States maintains a stable export market for about 70 percent of Canada's rice imports. The remaining 30 percent is aromatic rice imported mostly from Thailand and India. Canada imports high-quality milled and semimilled long-grain rice from the United States. Canadian rice imports have continued to expand at about 4.6 percent per year since 1980. This market is quite stable and expected to continue to increase at 4.5 percent per year.

The Netherlands has consistently been one of the major importers of U.S. rice over the past decade. The Netherlands does not produce rice but processes imported rice and exports about 33 percent to other European countries. Most imports from the United States are brown long-grain rice. The United States has maintained an average market share of 31 percent over the past five years and this trend is expected to continue as European markets liberalize.

The Middle East has been an important export market for the United States over the past decade, taking approximately 19 percent of U.S. rice exports. Saudi Arabia has accounted for 5 to 6 percent of total U.S. exports to this region. This market has been quite competitive, with exports of basmati fragrant rice from Pakistan and India.

CONCLUSIONS

Rice production costs vary from $248 per t in Texas to $110 per t in Vietnam. Developed countries have higher costs of production because of higher variable and fixed costs. Countries that plant high-yield varieties on irrigated land also have higher production costs per acre because of the additional costs of fertilizer, herbicides, pesticides, and seeds, but higher yields tend to offset the higher per-acre costs. Among countries that export long-grain rice, the United States has the highest cost of production per metric ton, while Vietnam is among the lowest. Total cost per acre is lowest in Thailand at $111, because traditional varieties, which require low inputs, are planted. Vietnam's per-acre costs are twice as high as those of Thailand at $221, but higher yields more than offset the cost of production on a per-ton basis. Labor costs contribute little to cost differences among countries because the United States utilizes little labor at a high cost while Asia has high labor inputs at relatively low per unit costs.

The quality of rice is highest in the United States with 4 percent broken as the milled rice standard. Vietnam has 10 percent broken, and Guyana has the highest brokens at 25 percent. The rough rice cost per metric ton of milled rice is highest for Guyana and the United States, followed by Thailand and Vietnam. Milling, marketing, and transportation costs are highest for the United States at $152 per t and lowest for Thailand at $26 per t.

Based on the Arkansas rice model, U.S. rice exports could decline from 2.98 million t in 1999 to 2.94 million t by 2010, due to declines in long-grain exports. Medium-grain rice exports, however, are expected to continue to expand as the United States will continue to compete in medium-grain markets in Southeast Asian countries. The other major competitors for these markets are Australia, Italy, and China. Northern China currently exports high-quality medium-grain rice to South Korea and Japan.

The United States may lose some market share in long-grain rice because of increasing competition by both low- and high-quality long-grain producers. Vietnam, Myanmar, and China will continue to provide low-quality long-grain rice for countries such as Indonesia, the Philippines, Bangladesh, and some African countries. High-quality rice from Thailand and India will continue to compete in Asian countries, Middle East markets, and in the aromatic-rice markets of North America and Europe. Argentina and Uruguay are expected to continue to be major competitors in South America, especially for the Brazilian market. The United States should continue to obtain a premium price because of high quality. Exports of rough rice to Mexico and Latin America will probably continue in the short run as these countries liberalize and impose restrictions against milled rice.

REFERENCES

Bierlen, Ralph, Eric J. Wailes, and Gail L. Cramer. (1997). *The Mercosur Rice Economy.* Arkansas Agricultural Experiment Station Bulletin No. 954. University of Arkansas, Fayetteville.

Cramer, Gail L. (1998). *Rice Pricing at Farm, Mill, and Export Levels in Guyana.* Report to the Guyana Rice Development Board and the Inter-American Development Bank, Washington, DC, February.

Cramer, Gail L., Eric J. Wailes, and Shangnan Shui. (1993). "The Impacts of Liberalizing Trade in the World Rice Market." *American Journal of Agricultural Economics* 75(February):219-226.

Hansen, James, Ralph Bierlen, Eric W. Wailes, and Gail L. Cramer. (1997). "The Impacts of Regional Integration on International Markets: The World Rice Market and Expanded MERCOSUR Trade." Paper presented at the American Agricultural Economics Meeting, Toronto.

Hansen, James M., Eric J. Wailes, and Gail L. Cramer. (1998). "The Impacts of El Niño on World Rice Production, Consumption and Trade." Selected Paper, Southern Agricultural Economics Association, Annual Meeting. Little Rock, Arkansas.

Isvilanonda, Sampornn (1999). Department of Agricultural Economics, Kasetsart University, Bangkok, Thailand, personal correspondence, September.

Juliano, Bienvenido O., ed. (1993). *Rice: Chemistry and Technology.* St. Paul, Minnesota: The American Association of Cereal Chemists, Inc.

Khiem, Nguyen Tri (1998). Can Tho University, Vietnam, personal correspondence, July.

U.S. Dept. of Commerce. (1992). *Census of Agriculture.* Washington, DC.

USDA. (1998). *Production, Supply, and Demand Data.* United States Department of Agriculture, Economic Research Service, Washington DC.

USDA. (1997). *Rice Situation and Outlook Yearbook.* RCS-1997, United States Department of Agriculture, Economic Research Service, Washington, DC.

Wailes, Eric J., Kenneth B. Young, and Gail L. Cramer. (1993). "Rice and Food Security in Japan: An American Perspective." In L. Tweeten, C.L. Dishon, W.S. Chern, N. Amamura, and M. Morishima (Eds.), *Japan and American Agriculture: Tradition and Progress in Conflict.* Boulder, CO: Westview Press, Inc., pp. 377-393.

Wailes, Eric, Kenneth Young, and Gail Cramer. (1994). "The East Asian Rice Economy After GATT." In L. Tweeten and H. Hsu (Eds.), *Changing Trade Environment After GATT: A Case Study of Taiwan.* Council of Agriculture, Taipei, Republic of China, pp. 13-42.

Wailes, Eric J., Gail L. Cramer, Eddie C. Chavez, and James M. Hansen. (1998). *Arkansas Global Rice Model: International Baseline Projections for 1998-2010.* Special Report 177, Arkansas Agricultural Experiment Station, University of Arkansas, Fayetteville.

Wailes, Eric J., Gail L. Cramer, James M. Hansen, and Eddie Chavez. (1998). *Structure of the Arkansas Global Rice Model.* Department of Agricultural Economics and Agribusiness, University of Arkansas, Fayetteville, Arkansas.

WTO. (1999). World Trade Organization Web site. <http:/www.wto.org/>.

Young, Kenneth B., Gail L. Cramer, and Eric J. Wailes. (1998). *An Economic Assessment of the Myanmar Rice Sector: Current Developments and Prospects.* Research Bulletin 958, Arkansas Agricultural Experiment Station, Fayetteville, Arkansas.

Chapter 7

Soybeans

Donald W. Larson

INTRODUCTION

Competitiveness and access to world markets are central negotiating issues in the General Agreement on Tariffs and Trade/World Trade Organization (GATT/WTO), North American Free Trade Agreement (NAFTA), and other trade agreements among countries and regions. Agricultural trade issues are a major sticking point in many of these negotiations. For the United States and many other countries, agricultural exports are an important component of this new focus on the ability of countries to compete in world markets. In the United States, soybeans and their products of oil and meal are very important export commodities, accounting for about one-third of soybean sales and about one-fifth of agricultural exports.

Slow growth of world markets, large crops, and rapidly accumulating stocks in the late 1980s and early 1990s placed increased pressure on competitiveness and export market shares among countries. Market shares cannot be considered as static and constant. Some countries have gained market shares through time, and others have lost market share. In some cases, countries have resorted to various anticompetitive domestic and trade policies such as tariff and nontariff barriers as well as subsidies to increase market share and competitiveness. World and regional trade liberalization became the preferred vehicle to restore competitiveness and dismantle the trade barriers among countries. This combination of factors explains much of the current interest in gaining market share and increasing competitiveness among producing countries. Among soybean-producing countries, these changing market shares include a loss of market share in soybeans for the United States and market share gains for Argentina and Brazil.

The objectives of the present chapter are: (1) to describe and analyze the changing world market shares for soybeans, soymeal, and soyoil among major producing countries including Argentina, Brazil, the European Union, and the United States; (2) to examine soybean competitiveness for Argentina, Brazil, and the United States; and (3) to evaluate the impact of trade agreements and policy changes on competitiveness and opportunities to enhance income of U.S. soybean producers.

Soybeans and soybean products have enjoyed more rapid growth than food grains in world markets, largely because of a strong consumer demand for more income-elastic products such as edible oils and higher value meat products that require protein for livestock feeding. World soybean production has increased rapidly, from slightly over 40 million metric tons in 1970 to almost 140 million metric tons in 1994-1995. The demand pull of a world hungry for more protein has fueled this rapid growth of production. The leading countries in soybean production and their mid-1990s share of world production are the United States (50 percent), Brazil (19 percent), Argentina (9 percent), and China (12 percent) (see Table 7.1). The European Union (EU-15), Paraguay, and other countries also produce some soybeans, but the amounts are not important in world trade. China is a major producer but not a major exporter. Only three countries, the United States, Brazil, and Argentina, accounted for over 80 percent of world soybean production and over 90 percent of world trade in the mid-1990s. Their shares of world trade in soybeans are 71 percent, 11 percent, and 8 percent, respectively (see Table 7.1). While still a small player, Paraguay became a significant exporter in the 1990s, with a 5 percent market share. The importance of Paraguay today is not unlike that of Argentina and Brazil 25 years ago when those two countries were beginning to produce soybeans on a commercial scale and were beginning to compete with the United States in world markets. In the early 1970s, the United States had over 90 percent of the soybean export market, and Brazil and Argentina had about 5 percent of the export market.

TABLE 7.1. Market Shares for Soybean Production and Exports

Countries	Quantity (Million Metric Tons)	Market Share (Percent)
Soybean Production		
United States	68.5	49.8
Brazil	25.9	18.8
Argentina	12.5	9.1
China	16.0	11.6
Other	14.8	10.7
Total	137.7	100.0
Soybean Exports		
United States	22.8	70.8
Brazil	3.6	11.1
Argentina	2.5	7.8
Paraguay	1.5	4.7
China	0.4	1.2
Other	1.4	4.4
Total	32.2	100.0
Soymeal Exports		
United States	6.1	19.7
Brazil	10.5	34.0
Argentina	6.7	21.6
European Union	3.7	11.9
China	1.3	4.2
Other	2.7	8.6
Total	30.9	100.0
Soyoil Exports		
United States	1.2	20.4
Brazil	1.5	25.4
Argentina	1.4	23.7
European Union	1.3	22.0
Other	0.5	8.5
Total	5.9	100.0

Sources: USDA, 1989; 1998.

When world market conditions changed dramatically in the early 1970s due to a worldwide drought (a severe El Niño of 1972-1973), devaluation of the U.S. dollar, and strong demand, soybean prices increased rapidly to reach record highs of "beans in the teens." These high prices attracted more competitors, especially Argentina and Brazil, to produce for world markets. After the crisis, the world's huge appetite for protein and edible oils demanded more and more production. The world market for soybeans and soybean products has been very dynamic, increasing at a 3.3 percent annual compound growth rate, compared to 2.3 percent for corn, 2.5 percent for wheat, and 2.6 percent for rice from 1973 to 1993 (USDA, 1989, 1998). Brazilian soybean production grew at an impressive average compound annual rate of 12 percent from 1970 to 1995 (Warnken, 1999). Today, competitiveness in world soybean markets depends primarily on what happens to production and productivity in the United States, Brazil, and Argentina.

Market shares in soymeal and soyoil exports differ markedly from those for soybeans where the United States has a dominant position. The soymeal and soyoil market shares are more evenly divided among the United States, Brazil, Argentina, and the EU (see Table 7.1). Brazil has the largest market share in soymeal (34 percent) and soyoil (25 percent). Argentina has second place in soymeal (22 percent) and soyoil (24 percent) followed by the United States in third place at a 20 percent market share for each product. Just as in soybeans, the soyoil and soymeal market shares for Brazil and Argentina have increased steadily over the last 25 years while the United States and EU have lost market share. Although the EU-15 does not produce large quantities of soybeans, the EU is an important exporter of soyoil and soymeal because they import large quantities of beans from world markets for processing and reexport some of these as soymeal and soyoil. Argentina and Brazil play a much greater role as exporters of soymeal and soyoil compared to soybeans, because their domestic tax polices favor local processing of soybeans to gain the value added from the processing activity (Warnken, 1999). However, Brazil and Argentina are becoming increasingly large consumers of soyoil and soymeal. As their per capita income increases, consumers change their diet to consume more high-value foods, especially poultry and pork, that increase the local demand for soymeal as a livestock feed.

Soyoil faces strong competition for world market share from palm, sunflower, rapeseed (canola), cottonseed, peanut, coconut, olive, and fish oil. Palm oil ranks first in export market share (35 percent), followed by soyoil (20 percent), sunflower seed (12 percent), and rapeseed (10 percent) among major vegetable and marine oils in the mid-1990s (USDA, 1998). Palm oil and more recently rapeseed have gained market shares at the expense of other oils in the past few years. Palm oil has gained market largely because of rapidly growing world production and very competitive prices. Soyoil is price competitive with most other oils, except palm oil, which usually sells at a discount to soyoil. When the price discount approaches 10 cents per pound, palm oil imports tend to increase rapidly in the U.S. market (Houston, 1998). Rapeseed production and exports increased rapidly in the 1980s and early 1990s because of EU support prices that were two to three times the world price. The United States' granting of Generally Recognized as Safe (GRAS) status for low-erucic acid rapeseed/canola oil in 1985 opened up the food-use market in the United States (McCormick and Hoskin, 1991).

Many health-conscious consumers prefer canola oil, corn oil, and, more recently, olive oil to many other oils because of their lower saturated fat levels. Among major vegetable oils, the saturated fat content is highest for coconut (87 percent), followed by palm (49 percent), cottonseed (26 percent), peanut (17 percent), soybean (15 percent), corn (13 percent), sunflower (10 percent), and canola (6 percent) (McCormick and Hoskin, 1991). Changing consumer tastes and preferences and market prices will determine which oils will gain market share in the future.

Soymeal tends to dominate the world protein market shares with no close substitutes such as are present in the vegetable oil market. Soymeal accounts for about 65 percent of the world protein market, followed by fish meal (10 percent), rapeseed meal (8 percent), sunflower meal (5 percent), peanut meal, and others in the mid-1990s. Soymeal has gained market share and is likely to gain more as the world supplies of other proteins such as fish meal become increasingly scarce and uncertain.

SOYBEAN TRADE RESEARCH

Many models, ranging from global trade and competitiveness models to regional trade models, to individual country models, and others have been developed in the soybean research literature to examine issues of trade barriers, trade agreements, exchange rates, government policy, transport, quality, cost, new seed technology, and many other issues. The studies reviewed here will highlight the results of these models in terms of the estimated impact on soybean trade and competitiveness.

The market and welfare effects of a joint yield increase and change in soybean composition was recently studied (Gallagher, 1998). A simulation of the development of a new soybean seed that yields more and has a lower protein content (from 44 percent to 40 percent) and higher oil content demonstrates that grower revenue might increase because of the more elastic demand for oil than for meal. The yield increase simulation benefits consumers, but producers lose from the technology change. The composition change benefits the livestock producers who pay lower soymeal prices, and soyoil becomes more competitive in world markets. Results indicate a net benefit for the United States.

Loss of U.S. soybean market share has sometimes been attributed to the higher quality of soybeans from Argentina and Brazil (Hill et al., 1996). Results of a study of soybean quality from Argentina, Brazil, and the United States indicate that Brazilian soybeans are higher in oil, meal, and foreign material than U.S. soybeans. U.S. soybeans are superior in terms of damage, free fatty acids, moisture content, and test weight. Argentine soybeans are lower in oil content than those of the United States or Brazil but equal to those of the United States in protein content. An estimated processing value based on FOB Rotterdam prices for oil and meal show that Brazilian soybeans had a $6.61/ton advantage over U.S. soybeans and a $1.10 advantage over Argentina for the period 1988 to 1992. The Brazil advantage is reduced to $4.15 per ton when the estimated value of other quality factors is added to the model (Hill et al., 1996).

The impact of different environmental regulations and standards among countries regarding competitiveness on world markets has been widely debated in GATT Uruguay Round and NAFTA trade negotiations. Based on the theory of comparative advantage, a grain trade model was developed to study environmental regulation effects on trade (Krishna Valluru and Peterson, 1997). A regression model with a dependent variable of net grain trade for 40 countries as a function of six proxy variables for environmental regulations found that environmental regulations have little impact on comparative advantage in grains. Grain trade patterns are well explained by factor endowments.

Japan is the world's largest importer of soybeans and rapeseed. The increasing market share of rapeseed imports compared to soybean imports during the last twenty years is of concern to U.S. soybean growers. A recent study identified the variables affecting imports of soybeans and rapeseed in the Japanese market (Persaud, 1998). A six-equation import demand model was estimated for Japan's imports of soybeans, rapeseed, soybean meal, rapeseed meal, and crude soybean oil for the period 1972 to 1994. Results indicate that soybean and rapeseed competition is a function of Japan's meat imports, Japan's meat production, and the import prices of soybeans, rapeseed, soybean meal, and corn. Higher meat imports and lower meat production associated with Japan's liberalization of beef imports tend to lower soybean imports and increase rapeseed imports. U.S. beef producers benefitted from more beef exports to Japan, and soybean producers lost market share in Japan but sold more soymeal to U.S. beef producers. Canada, a major exporter of rapeseed to Japan, benefitted from U.S. and Australian efforts to open Japan's market to beef imports.

Monetary policy may also influence competitiveness and market shares among exporters through its impact on exchange rates. A study of the impact of U.S. monetary policy on market shares for aggregate U.S. soybean exports to the Japan/EC-12 countries found that a weak dollar increased imports of U.S. soybeans and soybean meal significantly (Hwang, 1989). However, expansionary monetary policy did not significantly raise U.S. market shares. Overvalued exchange rates have been identified as a major tax on Brazilian soybean production and exports for most of the last 25 years. It is argued that Brazilian soybean production could expand rapidly in the future if a market equilibrium exchange rate regime were to be implemented (Warnken, 1999).

The impact of selected policy changes on Brazilian soybean production and trade identified policies to improve competitiveness (Abooki Bahiigwa, 1997). An econometric model of supply and demand for the soybean industry in four regions of Brazil was specified. The model, containing 37 equations, was estimated using two-stage least squares (2SLS) for the period 1970 to 1994. Two main policy options were analyzed: (1) elimination of the ICMS taxes (differential export taxes introduced in late 1996) on soybeans, soymeal, and soyoil exports; and (2) a reduction of the transportation cost of moving soybeans from production points to export points. Elimination of the ICMS taxes could increase domestic prices, as well as production and raw soybean exports, by as much as 47 percent per year in the next ten years. A 50 percent reduction in transport cost in the central west and northeast could increase producer prices by about 7 percent and exports by about 5 percent per year in the next ten years.

Perhaps the most dramatic way to liberalize trade is the zero-for-zero approach discussed but not adopted in the Uruguay Round of GATT negotiations. This approach involves an agreement to eliminate export subsidies, import tariffs, and export taxes in a number of sectors. An examination of the zero-for-zero approach for the oilseeds sector was recently completed using a modified version of the Organization for Economic Cooperation and Development (OECD) AGLINK model (Meilke and Wensley, 1998). The five major oilseeds studied are soybeans, cottonseed, rapeseed (canola), sunflower seed, and palm oil, and their products. The main simulation sets all oilseed and oilseed product tariffs, export taxes, and palm oil tariffs to zero for 1996 and then simulates the results over a five-year period to 2001.

World vegetable oil prices would increase by about 6 percent and meal prices by almost 2 percent. U.S. prices of oilseed would increase 2.3 percent, oilseed oil 5.0 percent, and oilseed meal 1.4 percent. World oilseed production increases by 0.2 percent, crush by 0.2 percent; oil consumption decreases by -0.1 percent, and meal increases by 0.3 percent. U.S. production would increase 0.5 percent and crush 3.6 percent; oil consumption would decrease by 1.0 percent and meal by 0.2 percent. Oil prices and crush would fall the most in China and Japan. Oilseed producers gain in all regions (Argentina, Brazil, Canada, China, the European Union, Japan, and the United States) except the rest of the world. Oilseed producers generally gain from the zero-for-zero proposal, while livestock producers are largely unaffected by it.

COMPETITIVE POSITION OF THE U.S. SOYBEAN INDUSTRY

Technology, input costs, infrastructure, natural resource endowments, and geographic location influence the competitiveness of countries in world markets, although government policies can enhance or, in some instances, retard the expression of these economic factors. The above factors have all affected production, productivity, and costs of production among Argentina, Brazil, and United States.

Production and Productivity

Total factor productivity in U.S. soybeans increased 1.6 percent annually from 1974 to 1983, compared to 2.4 percent per year for Brazil and Argentina (Cooke and Sundquist, 1993). Total economic surplus in the world increased between $1.7 and $3.2 billion.

Soybean area harvested among major producing countries reveals some very dramatic increases from 1975-1976 to 1995-1996 (see Table 7.2). Areas harvested in Argentina and Brazil increased very rapidly in this period, compared to relatively slow growth in the United States because of limited land area available for expansion. Brazil expanded rapidly because of the availability of large unexploited land areas (*cerrados*—a vast tropical savannah scrub land) in the central and southern regions of Brazil. This *cerrado* region is located in the states of Minas Gerais, Mato Grosso do Sul, Mato Grosso, Goiás, Tocantins, Distrito Federal, Bahia, and Maranhao. In addition, the Brazilian government invested in research to develop the technology (tropical soybean seed varieties) to produce in tropical conditions. Areas harvested in Brazil increased at an aver-

TABLE 7.2. Soybean Area Harvested in Major Producing and Exporting Countries

Country	1975-1976	1985-1986	1995-1996	% Increase 1975-1996
	Millions of Hectares			
United States	20.8	24.9	24.9	20
Brazil	5.8	9.5	11.7	100
Argentina	0.4	3.4	5.7	13,250

Sources: USDA, 1982; 1986; 1996.

age annual compound rate of 9.1 percent from 1970 to 1995 and is now about half that of the United States (Warnken, 1999).

Area harvested in the United States reached a peak of 69.4 million acres in 1982-1983 and did not return to that level until 1997-1998 because of lower market prices and because government programs that idled feed grain area harvested also indirectly reduced soybean area harvested. The 1996 farm bill, the FAIR Act, eliminated the acreage set-asides and price supports to increase competitiveness on world markets. The loan rate now provides income support through the loan deficiency payment but no longer supports market prices directly. Better financial returns from alternative crops (especially in the Southeast) have also slowed the rate of growth of soybean acreage in recent years.

Area harvested in Argentina has increased faster than for any other country from 1975-1976 to 1995-1996 (see Table 7.2). The 5.7 million hectares harvested in Argentina are about 40 percent of the harvested area in Brazil and 20 percent of the U.S. area. Favorable world prices that caused conversion of pasture land to crops in Argentina explain most of this rapid expansion in area.

Soybean yields in the United States, Argentina, and Brazil increased significantly in this period (Table 7.3). Brazilian soybean yields have not increased as fast as those of Argentina during this period, which indicates that Brazil may have expanded to less fertile lands than Argentina. Even so, Brazilian yields have increased at a high average annual compound rate of 2.7 percent from 1970 to 1995. U.S. yields have increased more slowly at about 0.33 bushels per acre annually (0.022 metric tons per hectare) since the 1950s. Improved fertilization, management practices, harvesting methods, and new varieties will likely lead to continued growth in U.S. yields as well as those of Argentina and Brazil.

TABLE 7.3. Soybean Yields in Major Exporting Countries

Country	1975-1976	1985-1986	1995-1996	% Increase 1975-1996
	Metric Tons per Hectare			
United States	1.94	2.29	2.37	22
Brazil	1.72	1.49	2.22	29
Argentina	1.60	2.20	2.19	37

Sources: USDA, 1982; 1986; 1996.

Costs of Production

Natural resources, technology, management, and government policy all interact to determine costs of production among soybean-producing countries. Variable and fixed costs of production differed widely among Argentina, Brazil, and the United States in 1986 (see Table 7.4). For example, subsidization or taxation of inputs will result in different kinds of production technology and/or different levels of input use. Subsidization of agriculture in general or of particular competitive crops can raise land rents. For example, corn and wheat subsidies in the United States in that period led to higher land values and rents that must be paid by soybean farmers if they are to compete with corn and wheat for available land.

TABLE 7.4. Soybean Production Costs in Selected Countries

Cost Item	Argentina	Brazil	United States
Variable Costs: Dollars/Acre			
Seed	$13.86	$10.61	$10.16
Fertilizer	—	40.09	10.29
Chemicals	8.01	10.80	19.36
Custom Operations	23.52	—	4.01
Fuel and Lube	11.27	15.19	12.83
Repair	8.87	4.79	8.07
Hired Labor	—	—	1.52
Miscellaneous	—	4.44	.29
Interest	2.29	2.92	3.20
Total Variable Costs	67.82	88.84	69.73
Fixed Costs:			
Overhead	—	1.89	11.53
Taxes and Insurance	11.75	3.40	12.59
Capital Replacement	9.31	9.83	26.10
Labor	11.79	4.72	13.16
Interest	6.88	4.72	9.08
Land Charge	18.99	31.13	49.68
Total Fixed Costs	58.72	55.69	122.14
Total Production Costs	$126.54	$144.53	$191.87
YIELD (Bu/Ac)	31.2	26.8	29.0

Source: Larson and Rask, 1992.

Note: Numbers shown are in U.S. dollars, 1986 prices, and 1986 exchange rates.

Conversely, taxing of export crops in Argentina leads to lower returns for crop farmers and hence much lower farm rents (40 percent of U.S. levels) and lower capital inputs. This significant difference in agricultural policy results in a much lower overall farm-level cost of soybean production for Argentine farmers ($126.54/acre compared to $191.87/acre in the United States). Yet, natural resource endowments in the two countries are very similar. It is likely that in the absence of market-distorting polices in each country, the costs of production would be similar.

The taxing of crop exports also results in a very low level of land use intensity in Argentina. For example, a majority of the land area in the Pampas region of Argentina (comparable to the U.S. Corn Belt) is in pasture and forage crops. Animal and green manure are principal sources of fertility, as low farm prices and taxed fertilizer make commercial fertilizer use uneconomical. Clearly, a change in price policy in Argentina could result in a significant increase in input use and cropping intensity, leading to much stronger competition for world soybean market share.

Brazil has a more neutral government policy toward agriculture and experiences land rent levels intermediate between those of Argentina and the United States. However, the natural resource situation is more tropical, necessitating greater levels of fertilizer inputs.

The United States is competitive on a variable cost basis, but has significantly higher fixed costs, principally land rents, capital replacement, and overhead (see Table 7.4). The higher fixed cost reflects a more prosperous agriculture derived primarily from a higher level of government policy support. Government transfers to U.S. crop farmers lead to a high level of capital acquisition, excessive bidding for land, and ultimately to a high fixed cost of production. Lower prices and/or reduced levels of policy support may put downward pressure on land prices and rents.

Beyond the farm gate, marketing and transport cost advantages accrue to the United States (see Table 7.5). Marketing costs in the United States are about half of the Brazilian costs and about 67 percent of the Argentine costs. The United States is fortunate to have a well-defined Ohio-Mississippi water transport system from its major producing regions to the Gulf Coast. In Brazil, the Paraná-Paraguay waterways flow through some of the world's richest farm land (similar to

TABLE 7.5. Production and Marketing Costs for Soybeans Landed in Rotterdam and Japan for Various Exporting Countries (US$/bu)

Cost Item	Argentina	Brazil	United States
Variable Costs	$2.17	$3.32	$2.41
Fixed Costs	1.88	2.08	4.21
Total Production Costs	4.05	5.40	6.62
Marketing Cost	0.99	1.18	0.67
Price FOB Port	5.04	6.58	7.29
Ocean Freight	0.50	0.45	0.34
Rotterdam Landed Cost	5.54	7.03	7.63
Ocean Freight	0.88	0.93	0.70
Japan Landed Cost	5.92	7.51	7.99
Land Rent	0.61	1.16	1.72
Yield (Bu/Ac)	31.2	26.8	29.0

Source: Larson and Rask, 1992.

Note: Numbers shown are in U.S. dollars, 1986 prices, and 1986 exchange rates.

the Ohio-Mississippi system) from near the coast inland to central South America and then exit the continent through the Rio de la Plata system in Argentina. This is a long but not well-developed system; therefore, Brazilian soybeans must be trucked long distances (especially from the new frontier production regions in the interior of the country) to export ports. Brazil does not have an adequate railroad system to serve as an alternative to trucking. Similarly, in Argentina, though producing regions are relatively closer to the sea, transportation inadequacies lead to higher marketing costs. However, the governments of Brazil and Argentina, and many private firms (including Archer Daniels Midland, Cargill, and Louis Dreyfus) are investing large sums of money to improve the efficiency of the waterways, ports, and other infrastructure system (Friedland, 1997).

With both Brazil and Argentina located in the Southern Hemisphere, the United States also has a slight ocean freight advantage for the principal soybean markets in Europe and Japan; however, the landed cost is still higher for the United States. Since Brazil and Argentina harvest soybeans just prior to our planting season, they enjoy an off-market price advantage.

TRADE AGREEMENTS

Trade agreements such as GATT/WTO and NAFTA have already had profound impacts on world trade in many products, including soybeans and soybean products. Additional gains from trade can be expected from the Uruguay Round of GATT, because developing countries have until 2004 to implement agreed commitments on market access, tariff reductions, and domestic support. Even though the world market for oilseeds and oilseed products is considered to be one of the least protected and distorted among those of agricultural commodities, trade agreements have achieved even lower trade barriers (OECD, 1994, see Table 7.6). Other trade agreements such as the European Union 15, the common market for Brazil, Argentina, Uruguay, and Paraguay (Mercado Comun del Cono Sur, MERCOSUR), Association of Southeast Asian Nations (ASEAN), and the Central American Common Market (CACM) have also increased trade within member country groups and among countries. Discussions under way to extend NAFTA to include all nations of the Western Hemisphere in a Free Trade Area of the Americas (FTAA) agreement have the potential to affect future trade in soybeans and soybean products. Formation of regional groups such as MERCOSUR and NAFTA probably will facilitate and, indeed may accelerate, formation of the FTAA so that all countries in the Western Hemisphere will benefit from increased trade liberalization.

NAFTA, implemented on January 1, 1994, removes most barriers to trade among Canada, the United States, and Mexico. NAFTA eliminates all quantitative restrictions on agricultural trade between the United States and Mexico. In addition, the agreement immediately eliminated many tariffs, with others to be phased out over periods of five to ten years. All agricultural provisions will be fully implemented by the year 2008. Mexico's bound rate tariff (the official tariff legally committed to under the WTO) on soyoil is 45 percent, soymeal 22.5 percent, and soybeans 9 percent (see Table 7.6). The agricultural provisions of the Canada-United States Trade Agreement (CUSTA), dated January 1, 1989, were incorporated into NAFTA. In the case of Canada and the United States, all tariffs on agricultural trade were to be eliminated on January 1, 1998, except for a few items covered by the Uruguay Round tariff-

TABLE 7.6. WTO Base and Bound Rate Tariffs for Selected Countries, 1998

Country/Region	Product	Net Oilseed Trade (1,000 tons)	Base Rate Tariff 1995 (%)	Bound Rate Tariff (%)*
Argentina	Soyoil	2,517	35.0	35.0
	Soymeal	8,108	35.0	35.0
	Soybeans	3,247	35.0	35.0
Brazil	Soyoil	987	55.0	35.0
	Soymeal	8,510	35.0	0.4
	Soybeans	3,585	35.0	35.0
Canada	Soyoil	401	7.5	4.8
	Soymeal	299	0.0	0.0
	Soybeans	2,665	0.0	0.0
China	Soyoil	-939	19.0	19.0
	Soymeal	1,352	5.0	0.0
	Soybeans	426	7.0	0.0
EU-15	Soyoil	1,256	5.0	3.2
	Soymeal	-13,448	0.0	0.0
	Soybeans	-16,124	0.0	0.0
Former USSR	Soyoil	-427	15.0	15.0
	Soymeal	-1,076	5.0	5.0
	Soybeans	848	9.0	0.0
Indonesia	Soyoil	-60	40.0	35.0
	Soymeal	-1,015	50.0	30.0
	Soybeans	-715	30.0	27.0
Japan	Soyoil	-11	0.0	0.0
	Soymeal	-1,052	0.0	0.0
	Soybeans	-6,671	0.0	0.0
South Korea	Soyoil	-71	30.0	5.4
	Soymeal	-1,559	20.0	1.8
	Soybeans	-1,437	541.0	487.0
Mexico	Soyoil	-322	50.0	45.0
	Soymeal	-260	25.0	22.5
	Soybeans	-2,827	50.0	45.0
United States	Soyoil	643	22.5	19.1
	Soymeal	4,771	0.7¢/kg	0.45¢s/kg
	Soybeans	20,644	0.0	0.0

Sources: USDA, 1998; Meilke and Wensley, 1998, for net oil seed trade data.

*The bound rate tariff is the rate committed to under the WTO.

rate quotas (dairy, poultry, eggs, and sugar products). The Canadian bound rate tariff for soybeans and soymeal is zero, but a 4.8 percent bound rate tariff continues for soyoil (see Table 7.6). The Canadian-Mexican agreement immediately eliminated most tariffs, but some are eliminated during longer periods, five, ten, or fifteen years. The tariff rate quotas between the two countries for dairy, poultry, eggs, and sugar are maintained.

U.S. agricultural exports to NAFTA countries have increased 44 percent and imports 57 percent since 1993 (USDA, 1998). Soybean exports to Mexico have increased at double digit rates every year since 1994, except for 1995 because of the Mexican peso crisis. U.S. soybean exports to Mexico totaled 3 million tons in 1997 ($794 million), which accounted for 97 percent of Mexico's soybean imports. U.S. soybean and soybean product exports to Canada are substantial, but not as large as those to Mexico. In contrast to Mexico, Canada is a net exporter of oilseeds and oilseed products to world markets, which reduces their import demand. In fact, the United States has become a major importer of canola oil from Canada because of strong consumer preference for canola oil among health-conscious consumers.

But trade barriers for soybeans still remain that may be reduced or eliminated in the future. Tariffs remain high in some countries of Eastern Europe and Asia, which may be reduced as those countries join the EU-15, WTO, ASEAN, or other trade groups (see Table 7.6). Phytosanitary regulations, health standards, quarantines, licensing requirements, and other regulations are used to protect domestic consumers, but may also be used as nontariff barriers by countries when pressures arise to protect local interest groups. In addition, barriers exist in terms of consumer and market acceptance of new technologies applied to seeds.

Biotechnology creates new dynamics in world markets. Canadian investments in rapeseed technology created a new variety of rapeseed (canola, also defined as the "double low" and "double 00" variety) that is low in erucic acid and glucosinolates, elements that are potentially harmful to humans and that reduce the palatability and nutritional value of meal for livestock feed (McCormick and Hoskin, 1991). After obtaining Food and Drug Administration approval to market canola in the United States, sales of both the oil and meal have grown rapidly since 1985. Canola created opportunities for U.S. producers to grow a new oilseed and new products for consumers as well as more competition for all oil and meal products such as soyoil and soymeal.

Large Brazilian investments in research to develop high-yielding soybean varieties that were suitable for tropical conditions and low-fertility soils gave birth to a major new agricultural industry in the country (Warnken, 1999). The soybean miracle occurred because the right combination of natural resources, technology, policy, and management was put in place by the public and private sectors in Brazil. No one could have imagined in the early 1970s that the world's appetite for protein would become so large or what the soybean industry would become in Brazil, Argentina, and the United States.

Biotechnology will likely change the soybean seed (genetically modified organisms) in many unexpected ways in the future. The biotechnology advances will likely include higher yielding varieties that are also resistant to diseases, insects, and chemicals (e.g., Round-Up Ready soybeans). In addition, new seed varieties will be custom-made for specific end-use characteristics. Soybean quality traits in development include high oleic acid, improved protein, high stearic acid, and phyto-manufacturing (Kalaitzandonakes and Maltsbarger, 1998). Problems of consumer acceptance of these biotechnology advances in terms of food health, safety, and the environment will most likely be major issues affecting commercial success of such products.

CONCLUSIONS

The world market for soybeans (the miracle crop) has been truly dynamic during the past 25 years, increasing much faster than most anyone would have expected and probably faster than most other products studied in the present book. Because of the world's huge appetite for food, soybean production has been able to increase from about 40 million metric tons in 1970 to nearly 140 million metric tons in 1994-1995. A rapidly growing population plus increasing per capita income permitted many consumers to improve the quality and variety of their diet by consuming more high-value products such as livestock and dairy products, especially poultry, pork, and vegetable oils. This increased demand for high-value products increased the demand for crops such as soybeans to supply edible oil for consumers and protein meal for livestock production.

These market dynamics attracted new suppliers and competitors to world soybean markets. Countries invested in new technologies, opened new land to soybeans, built new infrastructure, and changed policies to promote soybean production. Production and market shares changed as Argentina and Brazil emerged as major producers and exporters. Today, the United States (71 percent), Brazil (11 percent), and Argentina (8 percent) supply over 90 percent of the soybeans to world markets. Although soybeans continue to be the dominant oilseed on world markets, rapeseed (canola), palm oil, and others have become significant competitive products in both the oil and meal markets. When world market growth slowed in the late 1980s and early 1990s, while at the same time the world harvested bumper crops and large stocks began to accumulate, countries began to address seriously the issues of market access and competitiveness. Market access and competitiveness in agricultural products became central issues in the GATT/WTO, NAFTA, and other trade negotiations.

Recent trade agreements including GATT/WTO, NAFTA, MERCOSUR, and others have reduced tariffs and reduced or eliminated quotas and other barriers to promote increased trade. Usually, these agreements, GATT/WTO for example, allow countries to phase in tariff reductions or phase out other barriers during periods as long as 15 years. Globally, agricultural product tariffs are still at about 40 percent compared to industrial products at about 4 percent (Davis, 1999). Even though tariff and trade barriers in oilseeds tend to be lower than for most agricultural products, additional gains from trade liberalization are expected.

Agricultural policies in Argentina, Brazil, and the United States have changed to become more competitive on world markets. U.S. farm policy has become more market oriented through elimination of acreage set-asides and target prices and reduction of loan rates on wheat and feed grains that also adversely affected soybean production and competitiveness. U.S. monetary policy and lower interest rates have also favored exports in the 1990s. Export taxes have been reduced or eliminated in Argentina and Brazil on soybeans and soybean products. Brazil and Argentina have improved their competitiveness and have reduced the indirect tax on exports of an overvalued exchange rate by attempting to maintain a market clearing equilibrium exchange rate.

The future looks bright for the miracle crop. With appropriate domestic policies and trade liberalization, many countries will achieve faster economic growth, which means higher incomes for consumers who will want more and higher value foods. Technological advances in soybeans will be needed not only to increase yields to maintain competitiveness in current markets, but also to supply new markets with specific characteristics and qualities for particular end users.

REFERENCES

Abooki Bahiigwa, Godfrey B. (1997). "The Brazilian Soybean Industry: An Econometrics Framework for Policy Impact Analysis." Unpublished PhD dissertation, Faculty of the Graduate School, University of Missouri-Columbia.

Cooke, Stephen C. and W. Burt Sundquist. (1993). "The Incidence of Benefits from U.S. Soybean Productivity Gains in a Context of World Trade." *American Journal of Agricultural Economics* 75(1): 169-180.

Davis, Bob. (1999). "Gore to Urge Sharp Global Cuts in Farm Aid, Tariffs." *The Wall Street Journal,* January 29, p. A16.

Friedland, Jonathan. (1997). "South America Reaps a Harvest of Reforms." *The Wall Street Journal,* August 14, p. A10.

Gallagher, Paul W. (1998). "Some Productivity-Increasing and Quality-Changing Technology for the Soybean Complex: Market and Welfare Effects." *American Journal of Agricultural Economics* 80(1):165-174.

Hill, Lowell, Karen Bender, Glen Bode, Kyle Beachy, and Jennifer Dueringer. (1996). "Quality Choices in International Soybean Markets." *Agribusiness: An International Journal* 12(3): 231-246.

Houston, Jack E. (1998). "Structural Change in the U.S. Edible Vegetable Oils Processing Industry." Chapter 11 in Donald W. Larson, Paul W. Gallagher, and Reynold P. Dahl (Eds.), *Structural Change and Performance of the U.S. Grain Marketing System.* Columbus: The Ohio State University, pp. 217-232.

Hwang, Tsorng-Chyi. (1989). "A Monetary Approach to the Competition of U.S. Soybean Exports in the World Market." Unpublished PhD dissertation, The Ohio State University, Columbus, Ohio.

Kalaitzandonakes, Nicholas and Richard Maltsbarger. (1998). "Biotechnology and Identity-Preserved Supply Chains." *Choices* Fourth Quarter: 15-18. American Agricultural Economics Association.

Krishna Valluru, S. and E. Wesley Peterson. (1997). "The Impact of Environmental Regulations on World Grain Trade." *Agribusiness: An International Journal* 13(3): 261-272.

Larson, Donald W. and Norman Rask. (1992). "Changing Competitiveness in World Soybean Markets." *Agribusiness: An International Journal* 8(1): 79-91.
McCormick, Ian and Roger Hoskin. (1991). "Canola: Prospects for an Emerging Market." *Oil Crops: Situation and Outlook Report,* OCS-31. Washington, DC: U.S. Department of Agriculture, Economic Research Service, October.
Meilke, Karl D. and Mitch Wensley. (1998). *Trade Liberalization of the International Oilseed Complex.* Economic and Policy Analysis Directorate, Policy Branch, Agriculture and Agri-Food Canada, Ottawa, Ontario.
OECD. (1994). *The World Oilseed Market: Policy Impacts and Market Outlook.* Organization for Economic Cooperation and Development, Paris.
Persaud, Suresh C. (1998). "Japan's Import Demand for Soybeans, Rapeseed, and Their Respective Meals and Oils." Unpublished Master's thesis, The Ohio State University, Columbus, Ohio.
USDA (1982, 1986, 1996). Foreign Agricultural Service circulars. U.S. Department of Agriculture, Foreign Agricultural Service, Washington, DC.
USDA (1989). *Oil Crops: Situation and Outlook.* OCS-23, U.S. Department of Agriculture, Economic Research Service, Washington, DC.
USDA (1998). Web site: <http://www.fas.usda.gov/>. U.S. Department of Agriculture, Foreign Agricultural Service.
Warnken, Philip F. (1999). *The Development and Growth of the Soybean Industry in Brazil.* Ames, IA: Iowa State University Press.

Chapter 8

Cotton

Darren Hudson
Don Ethridge

INTRODUCTION

Cotton is an important crop to the United States, especially the southern tier of states from California to North Carolina. The United States has consistently been in the top three countries in terms of cotton production, averaging 18.2 million bales per year over the 1970-1997 period (USDA, 1985-1998). At a farm price of 65 cents per pound, this represents a farm value of production of $5.7 billion per year. U.S. production has represented an average of 25 percent of world cotton production from 1970 to 1997, with a low share of 18 percent in 1984 and a high of 29 percent in 1980.

IMPORTS AND EXPORTS

Cotton exports are an important part of the overall cotton sector, with an average export of 6 million bales per year (USDA, 1985-1998). Exports have ranged from 11 percent of production in 1985 to 50 percent in 1979, with an average of 33 percent of production over the 1970-1997 period. The United States has also been an important part of the world market. Figure 8.1 shows the relationship between U.S. exports and total world trade. On average, U.S. exports represented 22 percent of world trade over the 1970-1997 period. However, U.S. exports as a percentage of world trade have increased in recent years, to an average of 25 percent from 1987 to

FIGURE 8.1. Relationship Between U.S. Exports and Total World Trade

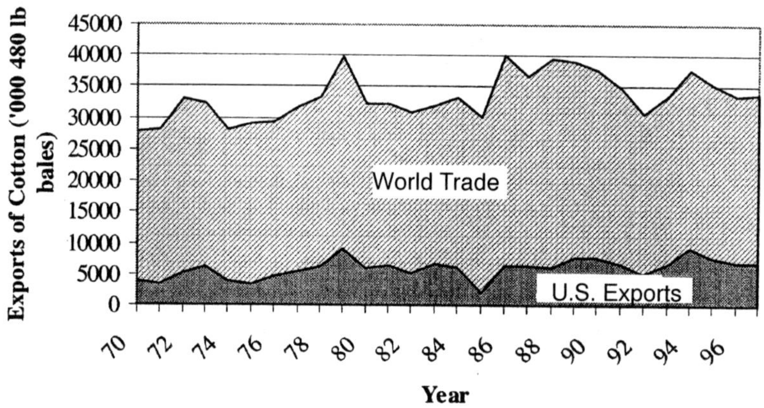

Source: USDA, 1985-1998.

1997. Thus, it is apparent that U.S. cotton production is important to both the domestic and international markets. Additionally, the interrelationships among cotton markets around the world implies that domestic policies in other cotton producing, exporting, and importing countries, as well as international trade policies, have significant implications for U.S. cotton producers and consumers.

Figure 8.2 shows the average distribution of cotton exports from the United States by region of destination. Asia has been the primary importer of U.S. cotton in recent years, with the majority of those shipments going to China, Japan, and South Korea. Countries of the Western Hemisphere are the next largest export customers, having increased consumption of U.S. cotton by about 43 percent since the 1995-1996 marketing year, with Mexico being the primary source of that growth (FAS, 1998). NAFTA has had a beneficial impact on cotton exports to that region. In contrast, exports of cotton to Canada have remained essentially unchanged since 1995. The surge in textile production in Mexico and U.S. imports of textiles from Mexico, accompanied by U.S. direct investment in the textile industry in Mexico, likely account for the differential growth rates in U.S. exports of cotton to Mexico as compared to Canada.

FIGURE 8.2. Average Distribution of U.S. Cotton Exports, 1995-1998

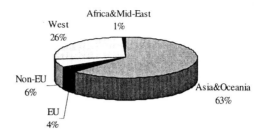

Source: Compiled from FAS, 1998.

The top eleven markets for U.S. cotton in the past three years have been (1) China, (2) Mexico, (3) Japan, (4) South Korea, (5) Indonesia, (6) other Western Hemisphere countries as a group, (7) Turkey, (8) Taiwan, (9) Canada, (10) Thailand, and (11) Brazil. With the exception of Turkey, the other top countries are all in the Western Hemisphere or Asia. The primary difference is that exports to most Asian countries tend to be on a downward trend, while exports to Central and South American countries tend to be on an upward trend.

Table 8.1 shows the average total imports by major importers and the percentage of total imports made up of U.S. exports to those countries over the 1995-1997 period. What can be seen is that for those countries that are large markets for the United States, the United States makes up a large percentage of their total imports. There is a predominance of Asian countries in the major importer list. U.S. exports to these areas are mainly cotton grown in the Western (mostly California) portion of the United States. The combination of quality characteristics of this cotton is particularly suited to spinning technologies employed by Asian mills and end products that are produced in that region as well.

The primary competitors for the United States in cotton export markets have been (1) Uzbekistan, (2) the African continent, (3) Australia, (4) Pakistan, and (5) Paraguay. Uzbekistan has become a prominent component of export markets since the breakup of the Soviet

TABLE 8.1. Major Global Importers of Cotton and U.S. Market Share in Those Countries

Country	Total Imports (1,000 480-lb Bales)	U.S. Percent Market Share
European Union	4,658	6.8
Russia	1,017	1.4
Japan	1,371	58.7
Indonesia	2,146	28.8
South Korea	1,522	44.9
Thailand	1,320	18.9
Taiwan	1,342	22.0
Hong Kong	669	25.1
China	2,953	49.0

Source: Compiled from USDA, 1985-1998.

Union (for a full discussion of the cotton industry in Uzbekistan, see Isengildina, Cleveland, and Herndon, 1998). As the primary producer in the former Soviet Union, Uzbekistan has been left with a large cotton production base, but only a minor domestic textile industry. Thus, much of its production is placed on export markets. A primary concern with Uzbekistan is foreign market share and not domestic consumption. The proximity of Uzbekistan to the major Asian importers may explain some of the loss of market share the United States has experienced in Asia. Changes in trade policy in Pakistan in 1995 and continued problems with disease have raised questions about Pakistan's ability to be an active participant in the cotton exporting in the near future (Hudson and Ethridge, 1998). China has been a net importer of cotton since 1995 (USDA, 1985-1998). China has shifted its support policies to address concerns about food production and imports. At the same time, additional polyester production is offsetting some domestic cotton fiber consumption, thus reducing Chinese and world demand for cotton fiber. Nevertheless, China remains an enigma and has a large impact on world market outcomes.

Cotton fiber is an industrial raw product; its primary use is in textile products. A full understanding of cotton markets requires some perspective on textile markets as well. Many of the major cotton-producing countries discussed previously are also major cotton-consuming countries. Countries such as India and China, for example, consume most, if not all, of their large domestic production. The United States consumes about 67 percent of its annual production, or about 12 million bales per year. There are exceptions to this, however. Uzbekistan and Australia, for example, have small textile industries and, thus, consume a small percentage of domestic production. At the same time, countries of the EU and Japan are large cotton consumers with little or no cotton production. Historically, very few countries have been able to sustain large domestic textile industries for extended periods of time without also having large cotton production industries. Likewise, few countries have historically been able to maintain large cotton industries without having large domestic textile industries. This likely stems from not having a sufficient and stable (a) raw material supply and/or (b) product market. However, policy (especially textile policy) may have some impact on the ability of some of these countries to sustain either a cotton or a textile industry without having the other industry present. Thus, it is apparent that some discussion of textile policy is also relevant to understanding the future of cotton trade.

Another issue of importance in understanding U.S. cotton trade is some perspective on the "net" imports of cotton and cotton products. That is, the relationship between U.S. exports of cotton lint and the cotton lint equivalent of processed products versus the imports of lint and cotton products is relevant to understanding the U.S. position in the world market. Table 8.2 shows the exports of cotton lint and the cotton lint equivalent content (the cotton content of products) as well as cotton lint imports and the cotton lint equivalent imports over the 1970-1996 period. The percentages of total exports and imports that are in the unprocessed (lint) form and domestic production are also shown.

In the early part of the period illustrated, total cotton exports (cotton fiber and the cotton content of manufactured products) were much larger than total cotton imports, although manufactured product exports were smaller than imports. Exports of manufactured goods have increased through time, as have fiber exports. However, imports of cotton manufactured goods have increased at a much

TABLE 8.2. Cotton Fiber and Fiber Equivalent Exports and Imports in the United States, 1970-1996 (1,000 Pounds)

Year	Exports				Mill Consumption	Imports		
	Cotton Equiv. Products	Fiber	Total	Percent Unprocessed		Cotton Equiv. Products	Fiber	Total
1970	199,186	1,870,560	2,069,746	90.4	3,890,400	329,258	17,760	347,018
1971	226,311	1,624,800	1,851,111	87.8	3,876,480	451,072	34,560	485,632
1972	290,444	2,549,280	2,839,724	89.8	3,684,000	480,453	16,320	496,773
1973	325,197	2,939,040	3,264,237	90.0	3,455,320	465,319	23,040	488,359
1974	392,493	1,884,480	2,276,973	82.8	2,782,560	371,252	16,320	387,572
1975	353,663	1,589,280	1,942,943	81.8	3,436,800	400,376	44,160	444,536
1976	413,154	2,296,320	2,709,474	84.8	3,165,600	479,487	18,240	497,727
1977	369,462	2,632,320	3,001,782	87.7	3,079,680	530,715	2,400	533,115
1978	355,745	2,966,400	3,322,145	89.3	3,017,280	642,587	1,920	644,507
1979	477,962	4,429,920	4,907,882	90.3	3,090,720	524,973	2,400	527,373
1980	528,233	2,844,480	3,372,713	84.3	2,796,960	540,644	12,960	553,604
1981	376,300	3,152,160	3,519,460	89.6	2,503,680	639,076	12,480	651,556
1982	253,342	2,499,360	2,752,702	90.8	2,619,360	807,096	9,600	816,696
1983	219,614	3,257,280	3,476,894	93.7	2,809,440	1,069,490	5,760	1,075,250
1984	206,081	3,000,000	3,206,081	93.6	2,635,200	1,342,569	12,000	1,354,569
1985	213,224	940,800	1,154,024	81.5	3,048,960	1,629,166	15,840	1,645,006
1986	274,828	3,208,320	3,483,148	92.1	3,544,800	1,910,477	1,440	1,911,917
1987	298,004	3,159,360	3,457,364	91.4	3,631,200	2,335,692	960	2,336,652
1988	330,266	2,951,040	3,281,306	89.9	3,701,280	2,118,775	2,400	2,121,175
1989	507,422	3,476,160	3,983,582	87.3	4,169,280	2,353,918	960	2,354,878
1990	664,752	3,541,440	4,206,192	84.2	4,124,160	2,416,410	1,920	2,418,330
1991	722,885	3,047,040	3,769,925	80.8	4,583,040	2,592,913	6,240	2,599,153
1992	844,928	2,337,120	3,182,048	73.4	4,891,200	3,193,165	480	3,193,645
1993	958,309	3,146,400	4,104,709	76.7	4,966,080	3,574,383	2,880	3,577,263
1994	1,107,443	4,512,960	5,620,403	80.3	5,332,320	3,795,917	9,600	3,805,517
1995	1,321,663	3,684,000	5,005,663	73.6	5,040,960	4,048,667	195,840	4,244,507
1996	1,501,097	3,304,800	4,805,897	68.8	5,289,600	4,171,554	193,440	4,364,994
Avg.	508,260	2,846,116	3,354,375	85.4	3,672,828	1,600,571	24,516	1,625,086

Source: USDA, 1985-1998.

faster pace. In fact, the United States has moved from being a large net exporter of cotton and cotton products to being a small net exporter. Excluding exports of cotton fiber, the United States is a large *importer* of cotton. Additionally, the proportion of total cotton exports in unprocessed form is large, ranging from 69 to 91 percent, suggesting that much of the potential value-added has been exported with this cotton fiber. Much of the exported cotton manufactured products was in the form of yarn and fabrics; again, several steps (value-added) removed from the final consumer (USDA, 1985-1998). This point is further emphasized by the fact that the cotton content of imports is approaching domestic consumption of fiber. These facts indicate that much of the value-added embodied in cotton products consumed within the United States is captured in other textile-producing countries.

This discussion has outlined the current situation and some important issues related to cotton trade. The remainder of this chapter will be devoted to summarizing the important trade literature as it relates to U.S. cotton exports as well as the competitive position of U.S. cotton in the world marketplace, including a discussion of relevant information regarding textiles.

TRADE RESEARCH IN COTTON

There are several relevant analytical studies that relate to cotton. Pick and Park (1991) analyzed the competitive structure of U.S. agricultural exports, including cotton. Using imperfect competition models, these authors attempted to ascertain whether U.S. exporting firms practiced price discrimination in exports of different commodities. For cotton, they found no evidence that (1) U.S. firms were discriminating between markets or (2) large importers had an overall influence on the price of cotton. Monke, Cory, and Heckerman (1987) analyzed the issue of cotton storage versus disposal on the world market. Their findings suggest that the elasticity of demand has a great influence on the optimal decision of storage versus disposal. When cotton is price inelastic, permanent storage or destruction of the surplus is preferable to disposal on world markets. This could be important as countries decide how to dispose of surplus stocks of cotton, which could ultimately affect global cotton prices and trade flows.

Ayuk and Ruppel (1988) analyzed the relationship between cotton export sales and shipments. These authors contend that export *sales* is an economic variable, whereas export *shipments* is a logistical variable. Based on this premise, the authors performed an econometric analysis and found that export sales of cotton do respond to economic variables such as price more readily than export shipments do. This finding has implications for trade modeling, suggesting that export sales is the more meaningful variable when elasticity estimate derivation is the goal of the analysis. Also, consideration of export sales appears to be the relevant variable for assessing the impacts of trade policy on cotton trade.

Estimates of the elasticity of export demand are important pieces of information when making policy decisions. Past estimates of the elasticity of export demand have ranged from highly inelastic (Blakely, 1962; Cathcart and Donald, 1966) to highly elastic (Alipoe, 1984; Johnson, 1977; Liu and Ronigan, 1985; Wohlgenant, 1986). Duffy, Wohlgenant, and Richardson (1990) more recently conducted an analysis of export demand elasticity for cotton, using both more recent data and a more complete analytical framework. This analysis was based on the premise that imports by a country are differentiated on the basis of region of origin. The authors extended this framework to estimate elasticities that accounted for the impacts of changes in the U.S. price of cotton on the price of cotton from other regions in the vein of Buse (1958). The estimation of market shares by these authors is also similar to previous work by Sirhan and Johnson (1971). Duffy, Wohlgenant, and Richardson (1990) found that the direct elasticity of export demand for U.S. cotton was highly elastic for exports to Europe and centrally planned economies (CPEs) (elasticities of -7.143 and -14.448, respectively), less elastic for Asia and Japan (-1.238 and -1.892, respectively), and inelastic for Canada (-0.696). The share-weighted direct elasticity of export demand was found to be -1.50 without consideration of CPEs, and -3.97 with consideration of CPEs. Zhang (1991) found similar direct elasticities of export demand.

These numbers represent direct elasticities of export demand. Duffy, Wohlgenant, and Richardson (1990) also estimated total elasticities of export demand. They estimated the total share-weighted demand elasticity to be -0.959 when CPEs are not considered price responsive, but -2.46 when CPEs are considered price responsive. This study highlights the importance of the assumption about the price responsiveness of CPEs in the calculation of export demand elasticities.

Interpretations based on these analyses for the current market situation are limited due to several factors. First, the data used in all these studies do not cover periods under the current trading regimes of the Uruguay Round (UR) of GATT or NAFTA. Second, the regions to which U.S. cotton is exported are not defined as NAFTA, MERCOSUR, or the European Union countries. Recent shifts in exports of cotton from Asian to NAFTA countries suggest that the elasticities of export demand have changed.

Analyses of cotton sectors and trade policies in other countries have also been conducted. Hudson and Ethridge (1997), for example, modeled the cotton and cotton yarn sectors in Pakistan. The objective of this analysis was to determine the economic impacts on the cotton and textile (yarn) sectors of an export tax on the cotton fiber sector. The findings of this analysis suggest that the export tax significantly decreased exports of cotton and increased domestic consumption. As a result of the indirect subsidy to the domestic yarn spinning industry, cotton yarn production and exports significantly increased. This indicates that the export tax policy in Pakistan reduced cotton export competition for countries such as the United States, but increased competition for U.S. textile exports. These results show that domestic policies in other countries have a significant impact on export markets and global cotton trade flows, including U.S. cotton exports and cotton textile imports.

The relative scarcity of analytical literature on cotton trade is supplemented by a body of descriptive literature. For example, Hudson (1997) analyzed the production system in Turkey. Isengildina, Cleveland, and Herndon (1998) examined production and marketing in Uzbekistan. Regmi and Roberson (1997) discussed production in Pakistan. Hudson and Ethridge (1996) described patterns in production, consumption, and exports in Pakistan and India. MacDonald and Whitton (1994) and Glade and Meyer (1994) outlined issues related to cotton trade and U.S. competitiveness in export markets. Finally, MacDonald (1997) has examined global end-use demand for cotton. All of these studies provide some perspective on the issues surrounding cotton exports, both domestically and internationally.

COMPETITIVENESS OF U.S. COTTON

Figure 8.3 shows the comparative costs of production, excluding land rent and the value of cotton seed. Among major producers, the United States appears to have the highest cost of production. Direct interpretation of competitive positions based on these data is cautioned against, however. Several countries such as the United States and Argentina have positive rates of protection for cotton production, while others such as Turkey and Pakistan have negative rates of protection. The effects of these protection rates are not accounted for in these data. Additionally, differences in cultural practices, input subsidies, and other factors make cross-country comparisons

FIGURE 8.3. Comparative Costs of Cotton Production in Major Cotton-Producing Countries Excluding Land Rent and Value of Cotton Seed

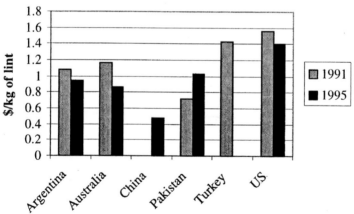

Sources: ICAC, 1992; 1996.

difficult. These data are collected directly from ICAC member countries and there are no checks on data accuracy or standardization of accounting procedures. Nevertheless, some perspective on relative costs of production is relevant.

Three general factors affect the cost of cotton in dollars per kilogram of lint (exclusive of exchange rates)—cost of production per hectare, yield of seed (not ginned) cotton per hectare, and the lint percentage of seed cotton. For example, a country with higher lint percentage of seed cotton can offset some higher costs per hectare and result in a similar cost per kilogram of lint as a country with lower lint yields and lower costs. This highlights the difficulty in making cross-country comparisons. For example, increasing costs in Pakistan combined with steady or declining yields have resulted in increasing costs per kilogram. In contrast, the United States has seen moderate decreases in costs per hectare and increasing yields, resulting in lower costs per kilogram of lint.

Domestic Policy Affecting Exports

Domestic agricultural policy has changed dramatically since the implementation of the Federal Agriculture Improvement and Reform (FAIR) Act of 1996. Direct subsidization of cotton production has been replaced by Production Flexibility Contract Payments (PFCPs), which are not tied to the production of any commodity. This change in policy has introduced planting flexibility and eliminated government supply controls, which has had some impact on U.S. cotton production, especially in the Mid-South (Mississippi, Louisiana, Arkansas, Missouri, and Tennessee) and Southeast (Alabama, Georgia, North Carolina, and South Carolina) regions. U.S. policy has retained the nonrecourse loan price support program and the "three-step competitiveness" program (USDA, 1991).

The competitiveness program centers on two primary elements. First is the "Step 2" provision, which provides payments to domestic users and exporters equal to the difference between the U.S. and world price when several conditions are met.[1] The U.S. transportation cost-adjusted price must exceed the transportation cost-adjusted world price by 1.25 cents per pound for four consecutive weeks to trigger these payments.[2] Additionally, the adjusted world price (AWP) cannot exceed the current crop year loan rate by 130 percent for payments to be made. Thus, this marketing loan program is unique in that it promotes consumption as well as exports. It has been estimated that the Step 2 program adds 100,000 bales to U.S. exports annually (USDA, 1998), but this is likely a lower bound estimate.

Import policy is also covered in the competitiveness program. The "Step 3" provision of the program adds flexibility to the currently established tariff rate quotas (TRQs). That is, when the U.S. price of cotton exceeds world prices by 1.25 cents per pound (adjusted for Step 2 payments) for ten consecutive weeks, the Step 3 provision triggers additions to the TRQs that already exist by an amount equal to one week's domestic mill consumption. However, these TRQs have source and quality restrictions, but the effect of the Step 3 provision is to allow U.S. textile manufacturers to remain cost competitive with manufacturers in other countries. The established TRQs are only occasionally exhausted and U.S. prices do not tend to persist above world prices for ten consecutive weeks, so the Step 3 provision is rarely triggered.

The operation of this policy is somewhat confusing because the trigger mechanisms are very similar. The Step 2 provision can be thought of as maintaining competitiveness of U.S. fiber exports and domestic mills when U.S. prices are temporarily above world prices (periods of about four weeks). That is, payments made under Step 2 are intended to offset temporary differences between U.S. and world prices. Step 3 is intended to rectify persistent differences between U.S. and world prices (periods longer than ten consecutive weeks) by allowing additional imports into the United States. These additional imports put downward pressure on U.S. prices, thus bringing them back in line with world prices.

The money budgeted to last through 2002 for the Step 2 program was exhausted in 1999, but new money was appropriated to continue the program. Continued funding of this program should help maintain the competitiveness of U.S. cotton exports, at least in the short run.

Trade Agreements

The implementation of the GATT agreement and NAFTA has had implications for global trade in cotton. In addition to these, the potential of the Free Trade Area of the Americas (FTAA) also needs to be considered. First, the implementation of the UR of GATT is not expected to have any substantial *direct* impact on cotton production either in the United States or in foreign countries (Varangis and Thigpen, 1995). The UR deals directly with levels of subsidization. However, subsidization levels in the United States have been below levels required by the UR agreements since their implementation, making subsidization level reductions moot for the United States. Additionally, other types of subsidization such as conservation reserve programs and PL-480 shipments are not covered by the UR agreement. Levels of subsidization in many foreign cotton-producing countries are often negative and are therefore not subject to reductions called for in the UR of GATT.

The effect of the UR agreement on cotton comes indirectly through reduction in textile tariffs and the gradual phase-out of the Multi-Fiber Arrangements (MFAs).[3] As Varangis and Thigpen (1995) point out, this phase-out of the MFAs will allow expanded exports of textiles from lesser-developed countries (LDCs) due to (1) their proximity to cotton production, (2) lower labor costs, and (3) preferential treatment of LDCs in the GATT agreement. Much of the world's cotton is produced in countries that possess LDC status. At the same time, many of these countries (e.g., Pakistan and India) are also large cotton consumers and textile producers. This suggests a shift in textile production to these areas and away from the developed economies such as the United States. However, the textile industries of the United States and EU have developed specialty/brand loyalty markets and adopted efficient technology that will likely allow them to flourish, even after the full phaseout of the MFAs (Varangis and Thigpen, 1995).

The second major trade agreement affecting cotton exports from the United States is NAFTA. According to the USDA (1998), NAFTA has led to a surge in direct investment and exports of cotton from the United States to Mexico. At present, the United States accounts for virtually all of the cotton imports by both Mexico and Canada. The earlier free trade agreement between the United States and Canada had virtually no impact on U.S. cotton exports (USDA, 1998). The addition of Mexico and the formation of NAFTA, however, altered textile competitiveness and investment, which contributed to increases in exports to both Mexico and Canada.

South America (especially Brazil) has been an important market for the United States as well. Exports to Brazil increased during the early 1990s as import tariffs were eliminated and domestic production waned in the face of dwindling farm credit (USDA, 1998). However, MERCOSUR's (of which Brazil is a member) Common External Tariff (CET) has increased to 6 percent and domestic support of cotton production has increased, leading to decreased imports of cotton from the United States. Argentina, a large cotton producer, has benefited from MERCOSUR because of (1) preferential tariff treatment with other MERCOSUR members and (2) its proximity to major cotton consumers such as Brazil.

Economic liberalization in Central and South America has likely had as much impact on U.S. exports to the region as any trade agreement. That is, liberalization in countries such as Colombia and Guatemala has allowed resources to shift away from cotton production to other uses, allowing U.S. exports to the region to increase from negligible levels to 200,000 bales per year (USDA, 1998). Thus, it appears that MERCOSUR has had only marginal impacts on exports of U.S. cotton to Central and South America.

This information has been used by USDA (1998) in projections on the potential impact of U.S. membership in the FTAA. Based on previous estimates of elasticities and a static computable general equilibrium model, the findings of this analysis suggest that membership in the FTAA will not substantially alter U.S. cotton exports. The upper-bound impact was estimated to be an increase in U.S. exports of 100,000 bales.

CONCLUSIONS

Several conclusions can be offered based on the above discussion. First, the United States continues to be a dominant player in world cotton fiber markets. The United States has consistently exported about 25 percent of world trade. The future of U.S. cotton trade depends on many factors. These include the outcome of trade agreements, the ability to financially sustain domestic trade policy such as the three-step competitiveness provisions, and the impact of changes in domestic policy on cotton production. The importance of the cotton and textile sectors to the overall agricultural complex suggests careful consideration of these factors is warranted.

First, the available evidence suggests that GATT will have no direct impact on cotton exports from the United States. However, GATT provisions on textile production and trade could have a substantial impact on the countries producing textiles, significantly altering the location of textile production. If this result holds, it could significantly alter global trade flows of cotton because GATT is expected to benefit textile producers in developing countries, many of which are currently large cotton producers as well. Shifting the location of textile production could alter where cotton is consumed, and thus, where cotton is traded. The fact that the United States has been a large supplier of cotton in the past and the acceptance of U.S. cotton in foreign markets due to reliability of supply and quality suggest that the United States will continue to be a large supplier of cotton in the future. However, the United States is facing increased cotton export competition from countries such as Uzbekistan and Australia that have no real incentive to keep their domestic production in their countries (no large textile industry), which may mean increased price competition. The nonprice advantage of U.S. cotton through the reliability of cotton quality is slowly eroding as Australia and other countries adopt grading standards similar to those used in the United States.

Second, budgetary constraints have forced a decrease in government outlays for programs such as the three-step competitiveness provisions. In fact, some estimate that the budgeted amount for this program will be exhausted before the end of the current farm legislation in 2002 (USDA, 1998). Given the apparent success of the program in stimulating both exports and domestic consumption of cotton, some consideration of additional appropriations for this program appears warranted.

Finally, changes in U.S. farm policy have changed the environment for production agriculture, affording producers more flexibility in making cropping decisions. In some areas such as the Southeast and Mid-South regions of the United States, this has led to more variability in cotton acreage. At the same time, global demand for cotton is not expected to increase in coming years (ICAC, 1999). Thus, export markets for U.S. cotton appear to hinge greatly on the production situation in other major producing countries as well as the impact of the FAIR Act on U.S. production.

A second important issue is the value added from the manufacturing of cotton-containing products and where that value-added is captured. Increasingly, it appears that the value-added embodied in the products that U.S. consumers purchase is captured by other textile-producing countries. Some of the value-added processes are labor intensive, putting the United States at a competitive disadvantage. However, some processes are not. It would appear advantageous to the United States to employ strategies and policies that would maximize the amount of value-added production that can occur within the United States, while not significantly hindering free trade. For example, a large capital gains tax may inhibit the development and adoption of new cost-saving technology that would allow the United States to perform some of these value-added processes. A reduction in the tax would stimulate the development and adoption of that technology while not interfering with free trade. These issues need careful consideration because unnecessary importation of value-added only serves to worsen trade balances.

More generally, the process of globalization and the negotiations of trade agreements presents many opportunities for the United States and the rest of the global community. At the same time, as is the case with cotton and textiles, this globalization and the implementation of trade agreements creates many adjustment problems. The rapid movement of cotton consumption alluded to by Varangis and Thigpen (1995) may cause "growing pains" in the emerging textile producers while creating displacement of workers and capital in existing textile-producing countries. Addressing these adjustment problems is a major part of the trade negotiations. No matter the apparent equity of a trade agreement, enforcement is the only mechanism through which the equity is achieved. Trade *agreements* that benefit the United States may require stringent enforcement of trading partners' adherence for benefits to be realized.

NOTES

1. "Step 1" simply incorporates into law the discretionary changes to the "adjusted" world price adjustment that USDA implemented on October 3, 1989.

2. This threshold has been increased to 3 cents per pound for new program participants in an effort to decrease government expenditure. The threshold remains at 1.25 cents per pound for existing participants.

3. The Multi-Fiber Arrangements are bilateral trade agreements between textile producing/consuming countries controlling the quantity of textiles coming from one country that can enter another (Erzan and Holmes, 1990).

REFERENCES

Alipoe, D. 1984. *Econometric Analysis of the Structural Relationships of the U.S. Cotton Economy.* Unpublished PhD dissertation, Department of Agricultural Economics, Texas Tech University, Lubbock, TX.

Ayuk, E. and F. Ruppel. 1988. "Cotton Exports: An Analysis of the Relationships Between Sales and Shipments." *Southern Journal of Agricultural Economics,* 20: 159-169.

Blakely, L. 1962. *Quantitative Relationship in the Cotton Economy with Implications for Economic Policy.* Oklahoma State University Technical Bulletin Number T-93, Stillwater, OK.

Buse, R. 1958. "Total Elasticities—A Predictive Device." *Journal of Farm Economics,* 40: 881-891.

Cathcart, W. and J. Donald. 1966. "Analysis of Factors Affecting U.S. Cotton Exports." Agricultural Economic Report Number 70, U.S. Department of Agriculture, Economic Research Service, Washington, DC, May.

Duffy, P., M. Wohlgenant, and J. Richardson. 1990. "The Elasticity of Export Demand for U.S. Cotton." *American Journal of Agricultural Economics,* 72(2): 468-474.

Erzan, R. and P. Holmes. 1990. "Phasing Out the Multi-Fibre Arrangement." *The World Economy,* 13: 191-211.

FAS. 1998. "Cotton: World Markets and Trade." U.S. Department of Agriculture, Foreign Agricultural Service, Washington, DC, October.

Glade, E. Jr. and L. Meyer. 1994. "U.S. Cotton Competitiveness, Price Relationships, and Certificate Values." *Beltwide Cotton Conferences, Proceedings,* P. Dugger and D. Richter, eds. Cotton Economics and Marketing Conference, National Cotton Council, Memphis, TN, pp. 432-435.

Hudson, D. 1997. "The Turkish Cotton Industry: Structure and Operation." *Beltwide Cotton Conferences, Proceedings,* P. Dugger and D. Richter, eds. Cotton Economics and Marketing Conference, National Cotton Council, Memphis, TN, pp. 285-287.

Hudson, D. and D. Ethridge. 1996. "Patterns in Exports of Cotton Fiber and Yarn from Southwest Asian Countries." *1996 Beltwide Cotton Conferences, Proceedings,* P. Dugger and D. Richter, eds. Cotton Economics and Marketing Conference, National Cotton Council, Memphis, TN, pp. 442-446.

Hudson, D. and D. Ethridge. 1997. *The Effects of an Export Tax on Cotton Lint on the Domestic Cotton and Textile Sectors, Trade, and Sectoral Economic Growth in Pakistan.* Cotton Economics Research Report Number CER-97-13,

Department of Agricultural Economics, Texas Tech University, Lubbock, TX, July.

Hudson, D. and D. Ethridge. 1998. "The Pakistani Cotton Industry: Impacts of Policy Changes." *Beltwide Cotton Conferences, Proceedings,* P. Dugger and D. Richter, eds. Cotton Economics and Marketing Conference, National Cotton Council, Memphis, TN, pp. 294-297.

ICAC. 1992, 1996. "Survey of the Costs of Production." International Cotton Advisory Committee, Washington, DC.

ICAC. 1999. "ICAC Press Release." Monthly press release, International Cotton Advisory Committee, Washington, DC.

Isengildina, O., O.A. Cleveland, and C. Herndon. 1998. "Cotton Industry in Uzbekistan: Structure and Current Developments." *Beltwide Cotton Conferences, Proceedings,* P. Dugger and D. Richter, eds. Cotton Economics and Marketing Conference, National Cotton Council, Memphis, TN, pp. 297-301.

Johnson, P. 1977. "The Elasticity of Foreign Demand for U.S. Agricultural Products." *American Journal of Agricultural Economics,* 59(4): 735-736.

Liu, K. and V. Ronigan. 1985. "The World Grain-Oilseed-Livestock (GOL) Model, a Simplified Version." Staff Report Number AGES850128, U.S. Department of Agriculture, Economic Research Service, Washington, DC, February.

MacDonald, S. 1997. "Global End-Use Demand for Cotton: A Time-Varying Parameter Model." *Beltwide Cotton Conferences, Proceedings,* P. Dugger and D. Richter, eds. Cotton Economics and Marketing Conference, National Cotton Council, Memphis, TN, pp. 278-282.

MacDonald, S. and C. Whitton. 1994. "World Cotton Trade: Prospects and Issues for the Next Decade." *Beltwide Cotton Conferences, Proceedings,* P. Dugger and D. Richter, eds. Cotton Economics and Marketing Conference, National Cotton Council, Memphis, TN, pp. 428-431.

Monke, E., D. Cory, and D. Heckerman. 1987. "Surplus Disposal in World Markets: An Application to Egyptian Cotton." *American Journal of Agricultural Economics,* 69: 570-579.

Pick, D. and T. Park. 1991. "The Competitive Structure of U.S. Agricultural Exports." *American Journal of Agricultural Economics,* 73: 133-141.

Regmi, A. and R. Roberson. 1997. "Cotton Production in India and Pakistan." *Beltwide Cotton Conferences, Proceedings,* Cotton Economics and Marketing Conference, National Cotton Council, Memphis, TN, pp. 285-287.

Sirhan, G. and P. Johnson. 1971. "A Market-Share Approach to Foreign Demand for U.S. Cotton." *American Journal of Agricultural Economics* 53(4): 593-599.

USDA. 1985-1998. *Cotton and Wool Situation and Outlook: Yearbook.* U.S. Department of Agriculture, Economic Research Service, Washington, DC.

USDA. 1998. *Free Trade in the Americas: Situation and Outlook Series.* International Agriculture and Trade Reports, U.S. Department of Agriculture, Economic Research Service, Washington, DC, November.

Varangis, P. and E. Thigpen. 1995. "The Impacts of the Uruguay Round Agreement on Cotton, Textiles and Clothing." *Beltwide Cotton Conferences, Pro-*

ceedings, P. Dugger and D. Richter, eds. Cotton Economics and Marketing Conference, National Cotton Council, Memphis, TN, pp. 370-373.

Wohlgenant, M. 1986. *Impact of and Export Subsidy on the Domestic Cotton Industry.* Texas Agricultural Experiment Station Bulletin B-1529, Lubbock, TX, April.

Zhang, P. 1991. *Projecting U.S. Cotton Prices in an International Market.* Unpublished PhD dissertation, Department of Agricultural Economics, Texas Tech University, Lubbock, TX.

Chapter 9

Peanuts

Stanley M. Fletcher

The peanut is an annual soil-enriching legume native to South America. The crop was carried to Africa and China by early explorers and missionaries. During the United States' early colonial days, the peanut was introduced from Africa. Presently, peanut production can be found on six continents, although four continents account for the majority. Over 95 percent of the world's peanut area and about 92 percent of production are found in developing countries (Freeman et al., 1999). Production is concentrated in Asia and Africa. Peanuts are one of the world's principal oilseeds. Until the mid-1980s, peanuts ranked number three behind soybeans and cottonseed. Now, rapeseed (canola) has passed peanuts in terms of world production, with sunflower seed close behind.

Peanuts are a regional crop in the United States and are very important to Southern agriculture. Seven states (Virginia, North Carolina, Georgia, Florida, Alabama, Texas, and Oklahoma) produce about 98 percent of the U.S. peanut crop, with an average annual value of over $1 billion in cash receipts for producers. Based on the 1992 *Census of Agriculture* for these seven states, 36 percent of the peanut-producing counties had 35 percent or more of their total crop income from peanuts. Twenty-four percent of the counties had 50 percent or more of their crop income from peanuts. From a state perspective, 70 percent of the crop income in Alabama's peanut-producing counties was generated from peanuts. For Virginia, the percentage was 48 percent, and in Oklahoma, 21 percent of the peanut-producing counties received 50 percent or more of their crop income from peanuts. In some Oklahoma counties, peanuts accounted for over 80 percent of the crop income. Thus, one can see readily how the peanut industry and the Southern rural economies are deeply intertwined.

PRODUCTION

World peanut production has shown a continual increase, with major growth in the 1980s and 1990s (see Table 9.1, Carley and Fletcher, 1995; Fletcher, Zhang, and Carley, 1992). The average production for 1966-1968 was 16,435,000 metric tons (t). The average production for the 1970s increased by only about 3 percent, to 17,005,400 t. In contrast, average production in the 1980s was 19,806,800 t, approximately a 17 percent increase over the 1970s average production. The average peanut production in the 1990s has increased approximately 43 percent, to an average level of 28,233,000 t.

Although world peanut production has increased, there have been different growth patterns for individual countries. China, which has experienced the largest growth, increased production by approximately 120 percent from the 1970s to the 1980s and by about 44 percent from the 1980s to the 1990s. A major reason for this growth was rural economic reform that boosted China's peanut industry. Argentina had the next largest rate of growth in the 1990s. Average production increased about 39 percent over the 1980s. This growth occurred after 1995 in response to the trade benefits Argentina received in the Uruguay Round of GATT. India increased production by 26 percent in the 1990s over the 1980s. The United States had one of the lowest percentage increases in production, due to the supply management program, which restricts domestic production to equal domestic consumption.

World peanut production is relatively concentrated, although more than 50 countries grow the crop. China and India produced almost 60 percent of the peanut crop in the 1990s (see Table 9.1). While the United States is the third largest peanut producer, it is a distant third in terms of share produced. In the 1970s, the United States accounted for about 10 percent of the world production. Now, the U.S. share has fallen to approximately 6 percent.

TABLE 9.1. Peanut Production of Selected Countries, 1993-1998

Country	1993	1994	1995	1996	1997	1998	Average 1991-1998	Share 1991-1998
	\multicolumn{7}{c}{1,000 metric tons}	%
World	26,275	28,832	29,548	31,531	30,160	30,972	28,233	100.00
Argentina	333	297	340	660	403	929	465	1.63
China	8,496	9,763	10,327	10,218	9,728	9,650	8,825	31.02
India	7,830	8,062	8,055	8,800	8,000	8,300	8,088	28.84
Mexico	83	80	91	112	96	96	96	0.34
Nicaragua	20	56	24	40	37	31	29	0.10
S. Africa	151	175	117	216	158	108	149	0.53
USA	1,539	1,927	1,570	1,661	1,604	1,783	1,783	6.43

Source: FAO, 1999.

TRADE

Peanuts are utilized several ways (Freeman et al., 1999). The peanut can be crushed for oil and meal. The edible oil is an important source for human consumption, especially in Asia and Africa. The meal is used for livestock feed. It also is used directly for food (i.e., edible use), which is the primary form consumed in the industrial countries (e.g., United States, Canada, and EU). Finally, a small proportion of peanut production is used for feeding animals directly, industrial purposes, and seed.

The peanut trade has changed character since the 1960s (Carley and Fletcher, 1995; Fletcher, Zhang, and Carley, 1992). In the 1960s and early 1970s, peanut oil was the major peanut item traded, and the edible peanut trade was negligible. Since that period, the edible peanut has become the major driving force in trade, while peanut oil plays a minor role in the world market. Although overall world peanut trade increased by approximately 4 percent in the 1980s over the 1970s, the 1990s have seen slower growth, about 3 percent. Furthermore, peanut trade accounts for only a small proportion of world production. Approximately 4 percent of world production was traded in the 1990s, which is down from 6 percent in the 1980s. While this share is small, a larger variation exists among the individual countries (see Tables 9.1 and 9.2). For example, Nicaragua exported approximately 43 percent of its production while India exported only about 1 percent of its output during the 1990s. Argentina exported approximately 31 percent of its crop while China exported only about 4 percent. The U.S. share of production exported dropped from about 20 percent in the 1980s to approximately 13 percent in the 1990s.

World edible peanut exports are concentrated among few exporters, although many countries export peanuts (see Table 9.2). The United States, China, India, and Argentina shared about 47 percent of world exports during the 1970s. The export market became even more concentrated in the 1980s, when those four countries accounted for 66 percent of the trade. In the 1990s their share increased to about 67 percent. Although these four countries dominate world trade, their individual shares have changed significantly over the last three decades. Argentina has seen an increase from about 4 percent in the 1970s to almost 12 percent in the 1990s. China has seen an even larger shift from about 4 percent in the 1970s to almost 28 percent in the 1990s. In contrast, the United States has seen a reversal, as its share declined from approximately 32 percent in the 1970s to about 20 percent in the 1990s. One of the reasons for the decline in U.S. share has been increased weather variability. Given the relative difference in price between the U.S. domestic market and the export market, the domestic market will be given priority over exports when domestic production is relatively low.

TABLE 9.2. Edible Peanut Exports of Selected Countries, 1993-1997

Country	1993	1994	1995	1996	1997	Average 1991-1997	Share 1991-1997
			metric tons				%
World	1,148,853	1,278,111	1,377,041	1,283,703	1,265,311	1,211,452	100.00
Argentina	119,314	104,432	165,117	197,967	179,610	143,960	11.84
China	304,327	460,612	377,898	335,922	157,563	334,371	27.81
India	91,350	50,455	117,987	147,358	258,500	96,211	7.59
Mexico	20	1,232	4,106	7,084	6,090	2,653	0.20
Nicaragua	6,772	13,799	20,457	19,822	22,903	12,640	0.99
S. Africa	21,040	48,071	24,515	32,936	49,321	30,480	2.48
USA	231,236	163,654	278,010	201,736	227,112	236,946	19.90

Source: FAO, 1999.

In contrast to the export market, the import market is less concentrated (see Table 9.3). Although Netherlands had the largest share (approximately 17 percent), a significant amount is reexported to other EU countries. After the Netherlands, Indonesia had the largest share, about 10 percent during the 1990s. All other countries' shares were in single-digit percentages. The EU has experienced a declining share since the 1970s (67 percent for the 1970s, 47 percent for the 1980s, and 45 percent for the 1990s). France and Italy are the major contributors to the decline. Since 1995, the United States has become a noticeable importer (approximately 3 percent of total trade). This is due to GATT and NAFTA provisions, which eliminated Section 22 protection for the domestic peanut industry, converting it into tariff rate quotas (TRQs).

TABLE 9.3. Edible Peanut Imports of Selected Countries, 1993-1997

Country	1993	1994	1995	1996	1997	Average 1991-1997	Share 1991-1997
			metric tons				%
Japan	45,424	40,158	41,989	43,094	42,450	43,308	3.77
Mexico	30,759	44,019	35,013	57,391	51,590	37,961	3.00
USA	12	1,921	31,373	47,210	49,660	19,539	1.46
Canada	88,268	89,146	93,369	91,417	93,461	87,870	7.11
Indonesia	108,097	150,902	148,854	162,799	170,789	127,488	10.03
Singapore	75,328	104,766	56,423	36,811	24,839	60,185	4.92
Hong Kong	32,913	18,289	14,104	10,716	10,916	22,799	1.94
EU	509,282	608,030	588,691	546,712	560,736	552,969	44.71
France	31,961	56,755	69,444	52,312	36,227	49,833	4.03
Germany	100,124	92,164	73,684	77,606	96,690	94,226	7.73
Italy	18,618	23,121	27,216	22,492	27,870	23,166	1.86
Netherlands	189,424	263,437	221,895	203,164	209,420	205,884	16.54
UK	112,096	105,253	101,482	111,868	122,143	111,810	9.11

Source: FAO, 1999.

COMPETITIVE POSITION OF THE U.S. PEANUT INDUSTRY

The 1996 Federal Agriculture Improvement and Reform Act (FAIR) is a sign that the U.S. peanut industry is entering an era characterized by less or no government financial support. In addition, movement toward a global marketplace in the world economy continues, as characterized by the implementation of NAFTA and the Uruguay Round of GATT. This is further supported by the upcoming negotiations of the Millennium Round of WTO and the Free Trade Area of the Americas (FTAA). Thus, the U.S. domestic peanut industry has to meet the competition from the rest of the world.

A significant amount of literature exists concerning the analysis of competitiveness of major agricultural commodities (e.g., Ahearn, Culver, and Schoney, 1990; Van Duren, Martin, and Westgren, 1991; Polopolus, 1986; Sharples, 1990; Sumner, 1986). However, little is known about the competitive position of U.S. peanuts in the world market because of the difficulty of obtaining information from foreign countries. The only known study that examined the competitiveness of U.S. peanuts is by Chen et al. (1996). This study examines the competitiveness of U.S. peanuts relative to China for the period 1988-1993. The intent of the study was not to comprehensively determine the competitive position of American peanuts but, rather, to investigate peanut competitiveness in terms of economic costs of production and net returns to management and risks both for the United States and China.

Table 9.4 summarizes the findings of Chen et al. (1996). There is a statistically significant difference in peanut production value per acre ($348 per acre) between China and United States, due primarily to the domestic price received by producers. Little difference existed between the average yields (i.e., 2,300 lb per acre for the United States versus 2,283 lb per acre for China). Then and now, U.S. producers receive a significantly higher domestic price due to the U.S. peanut support program. This aspect has been a major point of contention in the domestic policy arena. China has a statistically significant lower cost of production relative to the United States. If all economic costs are considered, U.S. costs are over three and one-half times higher than China's. A major contributor to the cost difference is quota rent and land value. These two main costs do not exist in China. In addition, quota rents do not exist for any peanut-producing country except the United States. This rent is derived from the supply management program for peanuts in the United States.

TABLE 9.4. Comparison of Economic Costs and Returns in Peanut Production Between the United States and China, 1988-1993

	Average		
Item	U.S.	China	Difference[a]
Production value			
Production value ($/acre)	682.02	334.09	347.93*
Production value (¢/lb)	29.85	14.74	15.11*
Economic costs (¢/lb)			
All[b]	30.03	8.77	21.30*
No quota (with peanut land)[c]	24.98	8.77	16.21*
No quota (with cotton land)[d]	23.99	8.77	15.22*
No quota and land[e]	21.48	8.77	12.70*
Net returns (¢/lb)			
All[b]	-0.18	5.97	-6.15*
No quota (with peanut land)[c]	4.87	5.97	-1.10
No quota (with cotton land)[d]	5.86	5.97	0.11
No quota and land[e]	8.37	5.97	2.40**
Net returns ($/acre)			
All[b]	-0.02	135.55	-135.57*
No quota (with peanut land)[c]	114.38	135.55	-21.17
No quota (with cotton land)[d]	137.23	135.55	1.68
No quota and land[e]	193.80	135.55	58.25***

Sources: USDA, 1988-1994; *China Agricultural Statistical Yearbook,* 1988-1991; China State Statistical Bureau, 1994.

[a] The Mann-Whitney test for the mean difference between the U.S. and China.
[b] Both quota rent and peanut land value were included.
[c] Quota rent was excluded, but peanut land value was included.
[d] Quota rent was excluded, but cotton (not peanut) land value was included.
[e] Both quota rent and peanut land value were excluded.

* Indicates significance at 1 percent level; ** indicates significance at 5 percent level; *** indicates significance at 10 percent level.

Even if quota rent and land value are eliminated in the United States, the U.S. total economic cost would still be over two and one-half times higher than China's (see Table 9.4). All cost components (i.e., seed, fertilizer, chemicals, and other expenses) except labor, were statistically significantly higher for the United States than for China on a per acre basis. In particular, the United States' other expenses were almost eight and one-half times more than China's ($209.74 per acre versus $24.77 per acre). The large other expenses category results from the cost of using and maintaining peanut-farming equipment such as the costs of fuel, lubricants, electricity, repairs, and capital replacement.

Although the net returns (excluding quota rent) look comparable between the two countries (see Table 9.4), those values are based on the respective domestic prices received in each country. If the U.S. domestic price is reduced to the world trade price (approximately $350 per ton of farmer stock peanuts), the U.S. net returns would be negative. Although this study was conducted for the period of 1988-1993, the findings would hold true today. U.S. peanut production costs have continued to increase due to chemical costs and other expenses. Thus, U.S. peanut producers will have to drastically change their production costs if they are to compete in the world market without a domestic support program.

AGRICULTURE AND TRADE POLICY

The U.S. domestic peanut program was initiated in the early 1930s (Jurenas, 1994). The domestic program influenced where peanut production would exist and price trends. The focus of the program was to control production and, thus, support the incomes of peanut producers. Until 1978 the program used acreage controls similar to those used for other crops. However, in the 1970s a new peanut variety (Florunner) was introduced that significantly increased yields. In 1977, peanut program costs escalated, with projections of further cost increases. Thus, in the 1978 Farm Bill the peanut program was modified to be a supply management program. Instead of acreage controls, production limits were set that would match domestic market demand. Producers were offered a guaranteed price (i.e., price support) for quantities up to the specified limit (referred to as quota peanuts). This price support was significantly above the world market price and was adjusted annually to reflect any increases in the cost of producing peanuts. To maintain the integrity of the peanut program, Section 22 was utilized with the import quota set at 775 metric tons.

Any producer was allowed to produce peanuts above this limit (referred to as additionals), but that production had to be exported or crushed for oil and meal. The price received for the additionals was determined by the market. A price support was provided for the additionals based on competing oil markets, which have historically been below the market price ($149-$179 per ton versus $250-$350 per ton). Furthermore, any farmer was allowed to produce peanuts even if they had never produced peanuts in the past. However, these peanuts also were classified as additionals.

The 1996 FAIR Act further modified the peanut program (Jurenas, 1997). This act demonstrates that the U.S. peanut industry is entering an era characterized by reductions in government program costs and a shifting of U.S. peanuts to an increasingly competitive arena. Under the new program, the quota support price was reduced from $678 to $610 per ton and the price escalator was eliminated. Thus, the support price was frozen at this new level for seven years (until 2002). In addition, the revised program enabled the movement of quotas within a state to allow the most efficient production areas to benefit from their competitive advantage. Such movements have been most prevalent in Texas, where peanut production has been shifting out of central to west Texas.

Trade policy has had a major impact on U.S. peanut producers in the 1990s (Carley and Fletcher, 1993; Chen and Fletcher, 1998; Fletcher and Carley, 1994). The Uruguay Round of GATT and NAFTA have also forced the U.S. peanut industry into an increasingly competitive arena, the key reason being the elimination of Section 22, which was converted into a TRQ, also known as minimum access import levels. The U.S. TRQs are shown in Tables 9.5 and 9.6. A TRQ permits imports up to a specified amount to enter at a bound tariff rate or on a duty-free basis. Imports above the TRQ may also enter but at a very high tariff (see Table 9.7). The bound tariff for TRQ under GATT is less than 3 cents a pound for shelled and prepared peanuts and duty-free for peanut butter. Under NAFTA, there is no bound tariff for TRQ (i.e., duty-free).

The minimum access level for peanut imports into the United States was significantly increased under GATT. The original level under Section 22 was 775 t, while the minimum access under GATT started at 33,871 t in 1995, expanding to 56,938 t in 2000 (see Table 9.5). Due to the price difference (domestic versus imports), these additional imports displaced domestic quota production at a cost to

TABLE 9.5. Minimum Access Import Levels for Edible Peanuts under NAFTA and GATT

Year	Argentina	Mexico[a]	Other	Total
	(metric tons)			
1995	26,341	3,478	4,052	33,871
1996	29,853	3,582	5,043	38,478
1997	33,364	3,690	6,034	43,088
1998	36,877	3,801	7,024	47,702
1999	40,388	3,915	8,015	52,318
2000	43,901	4,032	9,005	56,938
2001	43,901	4,153	9,005	57,059
2002	43,901	4,278	9,005	57,184

Sources: USTR, 1993; 1994.

[a] The import year starts April 1 for Mexico.

TABLE 9.6. Minimum Access Import Levels for Peanut Butter under GATT

Year	Canada	Argentina	GSP[a] Countries	Others	Total
			(metric tons)		
1995	14,500	3,650	750	250	19,150
1996	14,500	3,650	920	250	19,320
1997	14,500	3,650	1,090	250	19,490
1998	14,500	3,650	1,260	250	19,660
1999	14,500	3,650	1,430	250	19,830
2000	14,500	3,650	1,600	250	20,000

Sources: USTR, 1993; 1994.

[a] GSP countries are developing countries that are given a generalized system of preferences (GSP), which provide nonreciprocal tariff preferences to those countries to aid their economic development and to diversify and expand their production and exports.

TABLE 9.7. U.S. Ad Valorem Tariffs for Shelled and Prepared Peanuts and Peanut Butter Under NAFTA and GATT

NAFTA		GATT	
Year	Shelled and Prepared Peanuts	Year	Shelled and Prepared Peanuts and Peanut Butter
Base	123.1	Base	155.0
1994	120.0		
1995	116.9	1995	151.1
1996	113.9	1996	147.3
1997	110.8	1997	143.4
1998	107.7	1998	139.5
1999	104.6	1999	135.7
2000	93.0	2000	131.8
2001	81.4	2001	131.8
2002	69.8	2002	131.8
2003	58.1		
2004	46.5		
2005	34.9		
2006	23.3		
2007	11.6		
2008	0.0		

Sources: USTR, 1993; 1994.

the U.S. government prior to the FAIR Act. Under the revised peanut program, imports are taken into account when determining the domestic quota level.

Mexico's minimum access levels were defined within NAFTA, which was implemented in 1994. By 2008 peanut trade between Mexico and the United States will be totally free. Mexico's initial TRQ level was 3,377 t in 1994 (starting year for NAFTA) and increases by 3 percent each succeeding year. After 2008, Mexican-origin peanuts will be allowed to enter the United States freely and with no quantity restrictions.

Another TRQ established under GATT deals with peanut paste/butter (see Table 9.6). Initially, there was no Section 22 provision for peanut paste/butter. The Canada-U.S. Trade Agreement established minimum tariff rates for exporting Canadian-produced peanut paste/butter to the United States. This laid the groundwork for some entrepreneurs to develop an industry, which was done soon after the trade agreement was ratified. Even though Canada was not a peanut-producing country, firms would import the edible peanut and process it into peanut paste/butter for export to the United States. U.S. additional peanuts could not be used in this endeavor based on domestic regulations., i.e., U.S. additional peanuts could not be exported and then reimported in any form for domestic consumption. Severe penalties existed for violators. Thus, the Canadian flow of peanut paste/butter into the United States was capped at slightly above the 1993 import levels. Potential flows of peanut paste/butter from other countries were also given TRQs. However, NAFTA was already finalized and did not cover the peanut paste/butter issue. Mexico's peanut paste/butter, made from Mexican-origin peanuts, was exempt from any TRQ. While Mexico had not produced peanut paste/butter in the past, it began producing some in 1998, exporting it to the United States. Although Mexico does not produce enough peanuts to meet its own domestic consumption needs, the relative price difference encourages Mexico to export what quantities it can into the higher value U.S. market and import a cheaper world peanut.

Table 9.7 presents the U.S. ad valorem tariffs for shelled and prepared peanuts and peanut butter under GATT and NAFTA. The GATT tariff rates were set such that even after the 15 percent reduction over the six years of the agreement, foreign-origin peanuts are not expected to be price competitive in the U.S. market. However, NAFTA tariff rates were set at lower levels and are to be reduced to zero by 2008. If Mexican producers could grow peanuts at a price comparable to Argentineans and Nicaraguans (in the $250-300 per ton range), then Mexican peanuts could be price competitive in the U.S. market starting in the year 2000.

Whereas GATT and NAFTA have had a significant influence on the U.S. domestic peanut industry, the potential Free Trade Area of the Americas (FTAA) agreement would permanently alter the landscape of the U.S. domestic peanut industry. Argentina and Nicaragua would be part of this agreement. As discussed in the sections on production and trade and seen in the respective tables (see Tables 9.1 and 9.2), Argentina has expanded production and trade shares in the 1990s. Nicaragua is rebuilding its peanut infrastructure and is increasing its exports to capture part of the U.S. minimum access levels as well as exporting to Canada for use in the production of peanut paste/butter. Based on conversations with U.S. government trade personnel, the FTAA tariff schedule will not exceed the NAFTA tariff schedule. Countries such as Argentina that are currently covered by the GATT tariff schedule will probably be provided an accelerated tariff schedule that would bring them in line with the NAFTA schedule or the potentially lower FTAA tariff schedule. If this occurs, the new FTAA tariff schedule will be ineffective in controlling peanut imports. Even with the tariff, Argentinean and Nicaraguan peanuts will be very price competitive in the U.S. market. This will force a major revamping of the U.S. domestic peanut supply management program. For example, one option could be the significant lowering of the domestic quota price support as well as domestic quota levels. Based on the U.S. economic costs shown in Table 9.4, U.S. peanut production as known today would be permanently altered.

The Millennium Round of WTO provides a potentially different picture for the U.S. peanut industry. Part of the outcome depends on whether China is granted WTO status. The second key area deals with the sanitary/phytosanitary rules and regulations. GATT addressed this issue to a degree, but this area has become a major nontariff trade barrier issue. For peanuts, the key point is aflatoxin. The United States does not allow aflatoxin to exceed 15 ppb in edible peanuts for human consumption. The EU has started implementing rules with significantly lower aflatoxin levels without approval of the Codex Alimentarius Commission of the Food and Agriculture Organization and the World Health Organization of the United Nations. Another issue is genetically modified organisms, also known as GMOs. In contrast to corn and soybeans, peanuts do not currently have any GMO varieties. However, research is being conducted to develop such varieties.

CONCLUSIONS

The peanut is one of the world's principal oilseeds. In the United States, it is a regional crop that is an integral part of Southern agriculture and its rural economy. Peanuts provide a crop value averaging over $1 billion each year in cash receipts for producers. This translates into at least $2 to $3 billion in economic activity in the rural South. Since the 1930s, there have been farm program supports for peanuts. In 1978, the peanut program was changed to a supply management program that utilized production controls rather than the acreage controls used previously. Furthermore, a price escalator was implemented to ensure that the price reflects any increases in the cost of producing peanuts. In 1996, the FAIR Act significantly changed government policy toward agriculture generally, as well as for peanuts. The domestic support price for peanuts was reduced by 10 percent and the price escalator was eliminated. Production quotas were allowed partial movement within a state to find the most competitive areas for production. In addition, imports under TRQs are taken into account in setting domestic quotas.

With this type of domestic peanut program coupled with the changes, the U.S. peanut industry has been losing world production share, export market share, and increasing imports of peanuts. The economic cost of producing peanuts in the United States relative to other countries, such as China, is significantly higher. Even if the quota rent were eliminated in the United States, economic costs would still be higher. Thus, the U.S. land grant system has a challenge in providing relevant research and disseminating appropriate information to producers if a viable and competitive peanut industry is to be maintained in the United States.

Trade agreements (i.e., WTO and NAFTA) have altered the landscape of the U.S. peanut industry. Potential future agreements (i.e., FTAA) will significantly impact the domestic peanut industry. These forces will create economic difficulties for the U.S. peanut industry. The potential decline in farm income raises concerns about peanut producers' survival in the highly concentrated peanut production regions and those businesses that rely on peanut income. Decreased peanut production would inevitably affect the peanut shelling and manufacturing sectors. In spite of this potential scenario, the U.S. technology advantage and the land grant system can help ensure that the U.S. peanut industry remains viable and competitive.

REFERENCES

Ahearn, M., D. Culver, and R. Schoney. 1990. "Usefulness and Limitations of COP Estimates for Evaluating International Competitiveness of Canadian and U.S. Wheat." *American Journal of Agricultural Economics* 72(5):1283-1291.

Carley, D.H. and S.M. Fletcher. 1993. "NAFTA—Potential Impact on U.S. Peanut Producers." Faculty Series: FS-93-14, Department of Agricultural and Applied Economics, University of Georgia, Athens, GA.

Carley, D.H. and S.M. Fletcher. 1995. "An Overview of World Peanut Markets." In H.E. Pattee and H.T. Stalker (Eds.), *Advances in Peanut Science*. American Peanut Research and Education Society, Inc., Stillwater, OK, pp. 554-577.

Chen, C. and S.M. Fletcher. 1998. "Policy, Trade, and Competition: Forces Shaping the U.S. Peanut Industry." In C.M. Jolly, S.M. Fletcher, P.L. Kennedy, and W. Amponsah (Eds.), *Trade, Policy and Competition: Forces Shaping American Agriculture Proceedings*. Southern Cooperative Series Bulletin 390. Auburn University, Auburn, AL, pp. 13-29.

Chen, C., S.M. Fletcher, P. Zhang, and D.H. Carley. 1996. "The Competitive Position of Peanuts: A Comparison Between the U.S. and China." *Peanut Science* 23(2):104-110.

China Agricultural Statistical Yearbook. 1988-1991. *Zhongguo Nongye Nianjian* (China Agricultural Statistical Yearbook). Government of China, Beijing, China.

China State Statistical Bureau. 1994. *Zhongguo Nongcun Tongji Nianjian* (China Rural Statistics Yearbook, 1992-1993). Rural Social Economic Statistics Division, Government of China, Beijing, China.

FAO. 1999. *FAOSTAT Database*. <http://apps.fao.org/cgi-bin/nph-db.pl?subset=agriculture>.

Fletcher, S.M. and D.H. Carley. 1994. "GATT—What Does It Mean to the U.S. Peanut Industry." Faculty Series: FS-94-01, Department of Agricultural and Applied Economics, University of Georgia, Athens, GA.

Fletcher, S.M., P. Zhang, and D.H. Carley. 1992. "Groundnuts: Production, Utilization and Trade in the 1980s." In S.N. Nigam (Ed.), *Groundnut—A Global Perspective*. ICRISAT, Patancheru, Andhra Pradesh, India, pp. 17-32.

Freeman, H.A., S.N. Nigam, T.G. Kelley, B.R. Ntare, P. Sbrahmanyam, and D. Boughton. 1999. *The World Groundnut Economy: Facts, Trends, and Outlook*. ICRISAT, Patancheru, Andhra Pradesh, India.

Jurenas, R. 1994. *The U.S. Peanut Program*. CRS Report for Congress, 94-475 ENR. Congressional Research Service, Washington, DC.

Jurenas, R. 1997. *Farm Commodity Programs: Peanuts*. CRS Report for Congress, 97-245 ENR. Congressional Research Service, Washington, DC.

Polopolus, L.C. 1986. "The Competitive Position of Southern Commodities in International Market: A Synopsis." *Southern Journal of Agricultural Economics* 18(1):75-78.

Sharples, J.A. 1990. "Cost of Production and Productivity in Analyzing Trade and Competitiveness." *American Journal of Agricultural Economics* 72:1278-1281.

Sumner, D.A. 1986. "The Competitive Position of Southern Commodities: Some Trends and Underlying Forces." *Southern Journal of Agricultural Economics* 18:49-59.

USDA. 1988-1994. "Economic Indicators of the Farm Sector: Costs of Production—Major Field Crops, and Livestock and Dairy." U.S. Department of Agriculture, Economic Research Service, Washington, DC.

USTR. 1993. "Revised Country Schedule and North American Free Trade Agreement." Office of the United States Trade Representative, Washington, DC.

USTR. 1994. "Uruguay Round Negotiations, United States of America." Office of the United States Trade Representative, Washington, DC.

Van Duren, E., L. Martin, and R. Westgren. 1991. "Assessing the Competitiveness of Canada's Agrifood Industry." *Canadian Journal of Agricultural Economics* 39(4):727-738.

Chapter 10

Fruits

Dale Colyer

INTRODUCTION

A large number of both deciduous and citrus fruits are grown in the United States, with production occurring in every state, although production for most individual crops tends to be concentrated in a few areas. International trade is an important activity for most fruits, with around 10 percent of commercial fruit production being exported; this was 5.2 percent for oranges and 12 percent for apples in 1997 (USDA, 1998). Fruit and fruit preparation exports amounted to $3.2 billion in fiscal year (FY) 1998 or about 6 percent of the $53.6 billion in agricultural exports in that period. Fruits are exported in both fresh and processed form, with the importance varying among particular products. The products included in this analysis are limited to the major citrus fruits (oranges, grapefruit, lemons/limes, and tangerines) and nontropical tree fruits (apples, apricots, cherries, peaches, pears, and plums), with most of the emphasis being on the dominant crop in each of those categories, oranges and apples, respectively.

FRUIT PRODUCTION

The U.S. aggregate production and value of production for citrus, noncitrus, and all fruits for 1978 through 1997 are shown in Table 10.1. Fruit production has grown considerably during the last two decades, with citrus fruit production increasing by 13.2 percent,

TABLE 10.1. U.S. Fruit Production, Quantity, and Value

Year	Citrus	Other	Total	Citrus	Other	Total
	1,000 Short Tons			Million Dollars		
1977	15,242	12,605	27,847	1,149	2,754	3,903
1978	14,255	12,790	27,045	1,593	3,244	4,837
1979	13,329	14,008	27,337	1,770	3,651	5,421
1980	16,484	15,504	31,988	1,905	3,780	5,685
1981	15,105	13,332	28,437	1,867	3,898	5,765
1982	12,139	14,658	26,797	1,617	3,870	5,487
1983	13,682	14,168	27,850	1,743	3,596	5,339
1984	10,832	14,301	25,133	1,755	3,695	5,450
1985	10,525	14,191	24,716	2,080	3,830	5,910
1986	11,058	13,874	24,932	1,768	4,204	5,972
1987	11,994	16,012	28,006	2,053	4,421	6,474
1988	12,761	15,911	28,672	2,618	5,103	7,721
1989	13,186	16,345	29,531	2,663	5,279	7,942
1990	10,860	15,640	26,500	2,243	5,525	7,768
1991	11,285	15,740	27,025	2,414	6,021	8,435
1992	12,452	17,124	29,576	2,401	6,037	8,438
1993	15,274	16,563	31,837	2,151	6,135	8,286
1994	14,561	17,341	31,902	2,268	6,269	8,537
1995	15,799	16,258	32,057	2,328	6,818	9,146
1996	15,712	16,114	31,826	2,516	7,261	9,777
1997	17,247	18,390	35,637	2,574	8,080	10,654

Source: USDA, 1998.

noncitrus fruit production by 45.9 percent, and total production by 28 percent. The value of production increased more rapidly due to higher prices for most of the fruit. The increased production was due primarily to increased productivity since the harvested acreage changed very little; for citrus crops average acreage declined slightly, from an annual average of 1,170,000 for 1976-1980 to 1,047,000 in 1993-1997. However, the acreage in noncitrus fruits increased during the same period, rising from 1,653,000 to 1,780,000 acres, 7.7 percent (USDA, 1998, p. 11). Average production per acre for citrus crops increased from 12.7 to 15.0 tons, while noncitrus fruit average production per acre increased from 8.1 to 9.5 tons.

The United States is one of the world's larger producers of fruits, ranking in the top five for most citrus and other nontropical fruits; it accounts for about 41 percent of the output of the four major citrus crops (oranges, grapefruit, lemons/limes, and tangerines), but only about 6.3 percent of the major nontropical tree fruits (apples, apricots, cherries, peaches, pears, and plums)—the top five producers of noncitrus fruits account for a much smaller percentage of total world production since their production is much more widely distributed than is citrus fruit production. Other important producing and hence, competitor, countries in fruit production include Brazil, Mexico, Spain, Israel, Argentina, and China for citrus fruit and China, France, Turkey, Italy, and Spain for the other nontropical tree fruits. China has had some of the more rapid growth rates for production of a number of fruits including apples, pears, peaches, and oranges.

CITRUS FRUITS

Oranges are the most important citrus crop produced in the United States, as well as in the rest of the world. In the U.S., grapefruit is second while tangerines and lemons/limes are both more important for the rest of the world combined. Oranges account for about two-thirds of the world's production of citrus fruits, 70 million of the 104 million short tons of the four major citrus fruits produced in 1997 (USDA, 1998, p. 16). In the United States, almost all the oranges are produced in Florida (75 percent) and California (over 24 percent); Texas and Arizona each produce a small amount of oranges. Worldwide, Brazil is the leading producer (about one-third of the world total) followed by the United States (18 percent), Mexico (6 percent), Spain (4 percent), and China (3 percent). Oranges are consumed primarily as fresh fruit or as orange juice. The latter is most commonly processed and sold as frozen concentrate. In the United States, per capita consumption of oranges is around 90 pounds per year, although this tends to vary considerably; in terms of pounds about five to six times more orange juice than fresh oranges is consumed (see Table 10.2).

The United States is the leading country in grapefruit production, accounting for slightly over half of the production in recent years, 2,888,000 of 5,650,000 tons in 1997 (USDA, 1998, p. 16). Other important producers are Israel, Mexico, Cuba, and Argentina. Florida

TABLE 10.2. U.S. Production, Consumption, and Trade of Oranges

Year	Prod. 1,000 ton	Per Capita Use (Pounds)		Fresh Oranges (Metric Tons)		Orange Juice (Metric Tons)	
		Fresh	Juice	Imports	Exports	Imports	Exports
1978	9,546	13.4	78.3	16,171	324,607	205,896	45,357
1979	9,160	11.5	74.6	28,061	281,922	206,447	52,025
1980	11,832	14.3	81.0	10,070	432,941	101,461	82,548
1981	10,487	12.4	82.8	7,061	406,482	359,753	91,871
1982	7,600	11.7	75.0	14,205	327,019	399,659	76,106
1983	9,519	15.0	91.0	10,726	480,457	384,986	85,730
1984	7,245	11.9	80.3	20,217	363,130	807,893	75,374
1985	6,719	11.6	78.4	28,405	398,130	505,362	55,028
1986	7,476	13.4	82.4	25,170	406,558	762,493	50,477
1987	7,697	12.8	78.0	25,006	386,701	784,287	64,268
1988	8,551	13.9	72.4	15,914	340,032	625,171	77,846
1989	8,949	12.2	73.9	7,163	372,263	532,234	88,278
1990	7,745	12.4	72.0	11,673	518,751	713,271	158,894
1991	7,848	8.5	73.7	68,044	231,054	359,004	147,037
1992	8,909	12.9	63.5	9,808	518,547	435,481	158,663
1993	10,992	14.2	74.6	10,952	553,816	353,287	123,464
1994	10,329	13.1	72.5	15,583	564,807	442,895	87,911
1995	11,432	12.0	78.9	18,250	568,821	237,320	91,475
1996	11,427	12.8	76.7	23,336	514,342	321,403	102,761
1997	12,677	14.1	78.1	31,620	615,508	286,588	121,753

Source: USDA, 1998.

is the leading grapefruit producer, accounting for about 80 percent of the total, followed by California with around 10 percent, and Texas with 7 percent. Although total production has been relatively constant since 1978, varying from 1,978,000 tons in 1989-1990 to 2,986,000 in 1979-1980, per capita production has declined; it was 24.4 pounds per person in 1978 but only 16.4 in 1997. Some 35 to 40 percent is consumed in fresh form, with most of the remainder being juice.

International Trade

International trade in citrus fruits has grown, along with the general increase in agricultural product trade. This has been particularly noticeable in the case of orange juice, where international trade increased by about 240 percent during the last 20 years (FAO, 1999). However, trade in fresh oranges, grapefruit, and other products also has grown. United States exports of fresh oranges have nearly doubled, rising from 324,607 metric tons in 1978 to 615,508 in 1997, while exports of fresh grapefruit also increased substantially, growing from 270,939 metric tons in 1978 to 489,595 in 1997. The U.S. shares of exports in these two products also have increased, with the share for oranges rising from less than 10 percent in 1977 to nearly 14 percent in 1997. The U.S. share for grapefruit exports is much higher, over 43 percent in 1995-1997; this represents an increase from about a 33 percent share in the late 1970s (see Figure 10.1).

U.S. exports of frozen, concentrated orange juice also have grown since the late 1970s, but the U.S. share of juice exports has declined slightly, from just over 10 percent in the late 1970s to 7 to

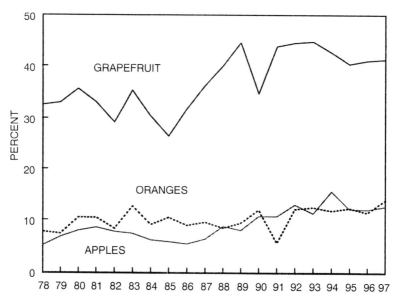

FIGURE 10.1. U.S. Share of World Fruit Trade

Source: USDA, 1999a.

8 percent in recent years. Brazil dominates world trade in orange juice and has supplied nearly 80 percent of exports in recent years, a share that has grown, although it had at least 50 percent of the market share in the late 1970s. The U.S. imports orange juice from Brazil, in part to blend with Florida juice, especially early in the season, to produce a desirable color.

The United States is a net exporter of lemons but a net importer of limes, nearly all of which come from Mexico. In 1997, lemon production in the United States was 935,000 tons, of which 130,600 or 13.9 percent were exported (USDA, 1998). Two-thirds of the exports were to Japan; Canada was the next important destination with 19.8 percent of U.S. lemon exports, while Hong Kong took 8.3 percent (USDA, 1999e, p. 145).

Competitiveness

U.S. competitiveness in citrus fruit trade varies with the product, but tends to be based more on quality than on cost considerations (USDA, 1999c). Brazil is a major competitor for oranges, although it is limited primarily to orange juice, where Brazil dominates the world market due to lower costs. As a result of the North American Free Trade Agreement, Mexico is a potential major competitor, although this could be manifested in competition for the U.S. domestic market rather than replacing the United States in international markets (Gunter, 1993). Cost of production in Mexico, for instance, was estimated to be $1.67 per box of oranges compared to $3.91 in Florida (Gunter, 1993, p. 38). A major difference is in the cost of labor, with Mexican wage rates being reported as 12 percent of those in Florida; labor is particularly important for fruit due to the process of harvesting, which generally is dependent on migrant labor, sometimes imported for that purpose. Trippensee (1996) also reports that orange production costs are lower in Brazil due to lower land and labor costs, a situation that results in a delivered fruit price about one-half of that in Florida; in addition, Brazil benefits from economies of size in processing oranges for juice since most juice is processed by only four firms.

Quality is highly important in citrus fruit markets and production of high-quality products is a major factor in U.S. competitiveness. Andayani and Tilley (1997), for example, report that the high quality of U.S. oranges and other fruits with respect to color, size, and other features has been an important factor in the growth of U.S. export to Indonesia; important competitors in that market are Australia and New Zealand, which benefit from lower transportation costs. High-quality grapefruit also is a factor in the U.S. domination in international trade in that product (USDA, 1999c).

NONCITRUS FRUITS

Apples and grapes are the major noncitrus and nontropical fruits produced in the United States and in the rest of the world (wine grapes are covered in Chapter 12); other important fruits grown in the United States include peaches, cherries, plums, and apricots. Apples and grapes are the more important from the standpoint of international trade. The leading apple-producing countries, in order of importance, are China (32.8 percent of total world output in 1997), the United States (8.3 percent), France (4.4 percent), Turkey (4.2 percent), and Poland (3.7 percent). Production in China has grown very rapidly, increasing from 9.9 million short tons in 1993 (18.2 percent of the total) to 20.3 million in 1997, 32.8 percent of world production (USDA, 1998, p. 16). Apple production in the United States also has grown during the last two decades, with commercial production rising from 3.37 million tons in 1978 to 5.19 million in 1997—a 54.1 percent increase. This occurred on an increase in acreage of only 12.4 percent, implying that yields increased by 37.2 percent (USDA, 1998, pp. 13-14).

Although production occurs in every state, commercial apple production is concentrated in relatively few states; the USDA lists 37 states as producing apples for commercial purposes (USDA, 1999d). Washington dominates, with around half of the commercial apple production (48.8 percent in 1998), followed by New York (10.9 percent), Michigan (9.8 percent), California (9.4 percent), and Pennsylvania (5.1 percent). Thus, these five states produce over 80 percent of the commercial apples.

Apples are consumed fresh, processed for applesauce, juice, and other uses, and dried. Per capita consumption of apples has been increasing in the United States, rising from 31.5 pounds in 1978 to 46.1 pounds in 1997, a 46.3 percent change (USDA, 1998). Most of this change occurred for juice, with apple juice consumption rising from 6.3 pounds in 1978 to 19.1 pounds in 1997—thus, juices accounted for 12.8 of the 14.6 pound increase in per capita consumption.

International Trade

World trade in apples has been rising in recent years, with an 83.4 percent rise in fresh apple exports between 1978 and 1997; apple juice exports, however, rose even more rapidly, with single-strength exports rising by 914 percent and concentrated juice increasing by 1270 percent (FAO, 1999). It should be noted that juice exports were relatively low in 1978, and that the 1997 tonnage of fresh apple exports is still nearly five times that for the two types of juices combined, 5.4 million metric tons of fresh apples compared with 1.1 million for the juices. The U.S. 1997 shares in the international apple trade are 12.6 percent for fresh apples (see Figure 10.1), 5.1 percent for single-strength juices, and 6.2 percent for concentrated juices. Fresh apples dominate in terms of tonnage of U.S. apple exports, 680,249 metric tons in 1997 versus 52,410 for the juices. Between 1978 and 1997, fresh apple exports increased about 4.5 times (453 percent), rising from 150,107 to 680,249 metric tons. Data on juice exports start in 1989, with single-strength exports fluctuating but not increasing, while concentrated juice exports doubled between 1990 and 1997, going from 18,636 to 26,440 metric tons.

The United States both imports and exports apples. It is a large net exporter of fresh apples but is a large net importer of concentrated apple juice. In terms of a total value, the country is a small net exporter of apples and apple juices, with a 1997 net trade balance of $43 million out of a total value of $486 in exports (FAO, 1999). The more important importers of U.S. apples in recent years have been Taiwan, Canada, Mexico, Indonesia, and Hong Kong, which accounted for a combined total of about 63 percent of U.S. exports in 1997; Taiwan and Canada accounted for nearly one-third of the total (USDA, 1998, p. 95). The United States gets the majority of its apple imports from Canada, New Zealand, Chile, South Africa, and Argentina, which, except for Canada, tend to supply fresh apples in the off-season for U.S. producers. Apple juice comes principally

TABLE 10.3. Apples: U.S. Production, Utilization, and Trade

Year	Production and Utilization			Per Capita, lb	Imports	Exports
	Mil lb	Fresh	Process		Metric tons	
1978	7,596.9	4,210.4	3333.6	35.8	60,004	150,107
1979	8,126.1	4,288.6	3812.6	36.0	74,886	214,040
1980	8,818.4	4,934.1	3866.3	39.7	72,073	256,356
1981	7,739.6	4,442.2	3250.7	34.7	76,691	307,835
1982	8,122.0	4,536.7	3573.5	39.7	71,870	264,391
1983	8,378.5	4,620.5	3737.4	41.6	98,196	264,167
1984	8,324.0	4,654.6	3654.5	44.3	103,996	214,967
1985	7,914.5	4,221.7	3605.1	43.2	124,103	191,461
1986	7,859.0	4,463.6	3369.1	43.2	131,631	189,347
1987	10,742.1	5,610.1	4841.2	48.2	133,418	227,795
1988	9,120.0	5,230.0	3840.1	47.3	126,234	306,494
1989	9,916.8	5,822.3	4049.1	48.5	120,267	276,333
1990	9,656.8	5,516.0	4103.2	48.1	109,373	396,930
1991	9,706.7	5,447.0	4189.8	43.9	119,770	431,768
1992	10,568.5	5,767.0	4696.3	46.7	120,409	524,189
1993	10,684.8	6,123.9	4450.0	48.7	113,950	524,940
1994	11,500.5	6,366.2	4966.2	49.7	115,843	738,703
1995	10,585.0	5,843.1	4546.8	45.7	154,423	634,531
1996	10,392.0	6,215.4	4124.6	47.2	182,961	615,476
1997	10,386.1	5,823.1	4496.9	46.1	159,085	680,249

Source: USDA, 1999d.

from Argentina, Germany, Chile, Hungary, and China. Chinese exports jumped significantly in 1997, rising from 4.5 million gallons in 1996 to 20.7 million in 1997 (USDA, 1998, p. 94); this has become a concern for U.S. producers (Naegley, 1999). China, however, exports only a small proportion of its total apple production; in 1997 about 1 percent was exported as fresh fruit and a similar percentage as juice.

Although relatively small compared to apple production and trade, several other deciduous tree fruits have significant shares of production that are exported, including peaches, pears, plums, and cherries. In 1997, 15.1 percent of the pear crop was exported, while for peaches 8.7 percent was exported, for plums 8.8 percent, and for sweet cherries 15.7 percent (USDA, 1998). The United States is the largest producer of sweet cherries, while China is the largest producer of the other fruits in this group. Canada and Mexico tend to be the major destinations for U.S. exports of these fruits.

Competitiveness

Competitiveness for products such as apples is often determined more by factors other than costs, although costs are important especially for trade in juices. Analyses have found that consumers in importing countries are sensitive to prices (Andayani and Tilley, 1997). However, they found that, for Indonesia, apple prices affected imports of other U.S. fruits, oranges and grapes, more than imports of the same fruits from other countries. Often, these same fruit imports were from Southern Hemisphere countries that supplied fruit at a different time of year. Higher land and labor costs in the United States contribute to making it a high-cost producer, but higher yields and efficiency in other aspects of production and processing help to keep the country more competitive.

Another major factor is quality, including variety, since personal preferences differ by variety and other factors (see Johansen, 1999). Quality also was found to be an important factor in U.S. fruit exports to Indonesia (Andayani and Tilley, 1997). The use of controlled atmosphere storage for apples enables producers to maintain higher quality for a longer period of time than is possible without it. The widespread use of this type of storage enables U.S. production to be used yearround, although fresh fruit from the Southern Hemisphere is able to compete in the U.S. consumer market despite relatively high transportation costs.

AGRICULTURAL AND TRADE POLICY IMPACTS

Tariff and nontariff barriers affect the fruit trade. Reductions in protection and tariffs in several Asian countries, however, have contributed to the ability of U.S. fruit producers to increase exports to that region. The creation of the WTO and explicit agricultural provisions of the last round of GATT negotiations is also opening some markets with minimum access provisions. The NAFTA agreement facilitates trade within North America, as most barriers are eliminated over the zero-to-15-year phase-in period. It is expected to be important for Mexican citrus exports to the United States and U.S. noncitrus exports to Mexico and Canada. The United States, for example, agreed to reduce its tariffs on fresh citrus imports by 15 to 20 percent over the first six years, and Mexico is changing its tariffs to correspond to the U.S. rates (Canada has no tariffs on citrus). Similar reductions are being phased in for orange juice: a 15 percent reduction over the first six years and then a phasedown to zero tariffs. For apples, the United States had no tariffs, but Mexico had a 20 percent rate, required import licenses, and had phytosanitary controls that virtually prohibited imports of U.S. apples; the latter are still in effect, but imports are permitted if the apples come from cold storage and are inspected by Mexican officials, paid for by the U.S. industry. Tariff rate quotas with a preferential rate were instituted (55,000 metric tons in 1995, to be increased annually at a 3 percent compounded rate) and tariff reductions are being phased in. Canada had no tariffs on U.S. apples, but restricts trade to a degree through prohibitions on large containers; they also have restricted the sales of Red Delicious apples under Canadian antidumping provisions.

Among the agreements of the Uruguay Round of GATT (WTO) that will affect fruit are the commitment of the EU to cut its export subsidies, which are permitted on a maximum of 906,900 tons of fresh fruits and 158,600 tons of processed products (FAS, 1999a, b). Tariff reductions will include reductions by Japan on citrus, including a cut to 10 percent for grapefruit by the year 2000; a cut by the EU for orange juice from 19 percent to 12.1 percent, a cut from 40 to 30 percent by South Korea for grapefruit and grapefruit juice, a 50 percent reduction by Thailand for oranges and grapefruit (to 30 percent), and elimination of tariffs on grapefruit by Switzerland. While the changes were not as extensive as for citrus fruits, several countries also agreed to a freer trade regime for other fruits. The EU is cutting rates on apples and grapes, Japan on canned peaches and fruit cocktail, Korea on cherries and prunes (Korea also agreed to lift its ban on apples, grapes, and fruit juices), and Norway on apples and pears. The United States agreed to a 15 percent reduction in tariffs for most fruits, to be phased in by equal installments over a six-year period beginning in 1995.

Among the issues in fresh fruit trade are pesticide residues and the transmission of pests such as the Mediterranean fruit fly. Thus, sanitary and phytosanitary provisions are still important factors. Marin, Epperson, and Ames (1997), in an analysis of fruit trade between the United States and Latin America, found that exporters perceive such provisions as a barrier to trade. U.S. exporters found this to be a greater problem than did Latin American exporters of fruit to the United States. Australia, Chile, China, Japan, and Korea also impose phytosanitary rules that affect fruit imports, especially apples, from the United States (Krissoff, Calvin, and Gray, 1997). These regulations are designed to control apple maggots, fire blight, and coddling moths. Australia, Chile, and Korea bar imports of U.S. apples while China and Japan permit limited imports of Red and Yellow Delicious apples from Washington and Oregon (and Idaho for China). Japan requires expensive treatments for the apples to be imported.

CONCLUSIONS

International trade has become an important factor in fruit marketing, resulting in larger production and higher prices than would exist with only the domestic market, although this latter still accounts for most of the production and consumption in the United States. The country both exports and imports most of the major fruits. International trade takes advantage of differences in seasonal production and types of products, enabling U.S. consumers to benefit from a more constant and varied supply of fresh fruits and juices. The nation's fruit producers have remained competitive in international markets for many fresh fruits, based on market shares. The advantage has depended on the large size of U.S. production and the maintenance of a high-quality product. When compared to other countries, especially the developing countries, fruit production and processing costs in the United States tend to be high (although U.S. costs are lower than for most European producers); labor costs are a significant factor in this situation.

The development of freer trade regimes is a major factor in international trade in fruits; a lowing of tariffs and other barriers has, for example, enabled the U.S. fruit industry to expand exports to Asia (although the economic crisis in that region has caused a reduction in imports in the most recent years). NAFTA and the WTO agricultural provisions also are affecting trade patterns and quantities. To date their effects have been relatively minor, although apple exports to Mexico have increased. Many barriers to trade still exist and distort the competitive situation.

REFERENCES

Andayani, Sri R.M., and Daniel S. Tilley. 1997. "Demand and Competition Among Supply Sources: The Indonesian Fruit Import Market." *Journal of Agricultural and Applied Economics* 29(2, December):279-289.

FAO. 1999. *FAOSTAT Database*. Rome: FAO, United Nations, Food and Agriculture Organization. Electronic Database: <http://apps.fao.org/Trade.CropsLivestock Products&Domain=SUA>.

FAS. 1999a. "Citrus." U.S. Department of Agriculture, Foreign Agricultural Service. FAS Online, <http://www.fas-usda.gov:80/itp/Policy/GATT/citrus.htm>.

FAS. 1999b. "Vegetable and Non-Citrus Fruits." U.S. Department of Agriculture, Foreign Agricultural Service. FAS Online, <http://www.fas-usda.gov:80/itp/Policy/GATT/vegetables.htm>.

Gunter, Dan L. 1993. "Trade Issues Affecting Florida's Citrus Industry." In J.E. Reynolds (Ed.), *International Trade and Florida Agriculture*. SP93-23, Food and Resource Economics Department, University of Florida, Gainesville, pp. 34-41.

Johansen, Asle E. 1999. "Norwegian Customers Will Pay for U.S. Quality." U.S. Department of Agriculture, FAS Online, <http://www.fas-usda.gov/info/agexporter/1999/norwegia.htm>.

Krissoff, Barry, Linda Calvin, and Denice Gray. 1997. "Barriers to Trade in Global Apple Markets." *Fruit and Tree Nuts Situation and Outlook,* FTS-280. Economic Research Service, U.S. Department of Agriculture, Washington, DC, pp. 42-51.

Marin, Rebecca, J.E. Epperson, and Glenn C.W. Ames. 1997. "A Probit Analysis of the Characteristics of Firms Engaged in the Fruit and Vegetable Trade Between the United States and Latin America." Paper presented at the annual meeting of the Western Agricultural Economics Association, Reno, NV, July 13-16.

Naegley, Stella K. 1999. "Chinese Orchards Reach New Heights." *American Fruit Grower* 119(3, March):26-28.

Trippensee, Austin. 1996. "The Juice Trade in the Americas." *Journal of Food Distribution Research* 28(1, February):13-14.

USDA. 1998. *Fruit and Tree Nuts: Situation and Outlook Report*, Economic Research Service, U.S. Department of Agriculture. Washington, DC, October.

USDA. 1999a. "Export Share of U.S. Agricultural Production." U.S. Department of Agriculture. ERS Electronic Data Base: <http://www.econ.ag.gov/epubs/pdf/AgTrade/expshare.pdf>, accessed March 22.

USDA 1999b. "Outlook for U.S. Agricultural Trade." ERS-AES-21, Economic Research Service, U.S. Department of Agriculture. Washington, DC, February 22.

USDA. 1999c. *Agricultural Outlook*. Economic Research Service, U.S. Department of Agriculture. Washington, DC, February 22.

USDA. 1999d. *Noncitrus Fruits and Nuts: 1998 Preliminary Summary*. National Agricultural Statistics Service, U.S. Department of Agriculture. Washington, DC, January.

USDA. 1999e. *Foreign Agricultural Trade of the United States: Calendar Year 1997 Supplement*. Economic Research Service, U.S. Department of Agriculture. Washington, DC.

Chapter 11

Vegetables

John J. VanSickle

INTRODUCTION

International trade in vegetables has taken on more significance over the last three decades as demand for fruit and vegetable products has increased and improved technologies have made the products more accessible. Average annual imports of fresh vegetables increased from 1.44 billion pounds in the 1970s to 3.77 billion pounds in the 1990s (USDA, 1998), an increase of 161.8 percent. Average annual imports of frozen vegetables increased from 38.3 million pounds in the 1970s to 647.0 million pounds in the 1990s, an increase of 1,589.3 percent. Imports of canned vegetables increased from 262.5 million pounds in 1970 to 467.7 million pounds in 1997, an increase of 78.2 percent.

Exports have also become more important to U.S. vegetable producers, as average annual exports of fresh vegetables increased from 1.26 billion pounds in the 1970s to 3.14 billion pounds in the 1990s, an increase of 149.2 percent. Average annual exports of canned vegetables rose from 131.5 million pounds in the 1970s to 317.9 million pounds in the 1990s, an increase of 141.7 percent. Imports of canned vegetables rose from 87.3 million pounds in 1970 to 1.210 billion pounds in 1997, an increase of 1,290.8 percent.

Mexico remains the largest supplier of imported fresh and frozen vegetables in the United States. Mexico accounted for 81.5 percent of the total value of fresh vegetable imports over the 1991 to 1997 period and 68.9 percent of the total value of frozen imports. The value of fresh and frozen vegetable imports from Mexico increased from $749.7 million in 1991 to $1.221 billion in 1997, a 63.0 percent increase. The top five fresh vegetable imports from Mexico in 1997 (in terms of value) were tomatoes ($517.0 million), bell peppers ($129.8 million), cucumbers ($89.1 million), squash ($58.9 million), and chile peppers ($56.8 million). The only other countries to account for more than $50 million in import value for a fresh product were Canada for fresh tomatoes ($58.9 million) and the Netherlands, also for fresh tomatoes ($52.9 million). Broccoli was the leading import value item from Mexico for frozen vegetables, totaling $91.2 million in 1997.

Spain and Canada are the largest suppliers of canned vegetables imported by the United States, totaling $51.6 million and $50.3 million, respectively, in 1997. Tomato products account for $31.5 million of the canned vegetable imports from Canada and artichokes account for $38.5 million of the canned imports from Spain.

The top destination for U.S. fresh vegetable exports is Canada, accounting for an average annual value of $596.1 million, 72.3 percent of the total value of fresh vegetable exports, over the 1991 to 1997 period. Lettuce is the leading U.S. export item, with a total export value of $159.1 million in 1997. Tomatoes ($131.1 million), broccoli ($94.7 million), onions ($89.8 million), and melons ($80.2 million) are the next four ranked items in total export value.

China is the world's leading producer of most vegetable products. The United States is the second-ranked producer of lettuce, tomatoes, carrots, and mushrooms. The United States is the third largest producer of potatoes, dry onions, asparagus, watermelon, and spinach.

VEGETABLE TRADE RESEARCH

Until recent years, most of the trade research on vegetables focused on competition between the United States and Mexico in the winter fresh vegetable industry. This focus was driven by the fact that most fresh vegetables consumed in the United States before 1960 were grown in the United States, with relatively little imported. Total fresh vegetable imports never exceeded 1.0 billion pounds prior to 1970, but have done so in every year following 1970. Fresh market tomato imports never exceeded 500 million pounds prior to

1970, but have done so every year since, peaking at 1.636 billion pounds in 1997. These gains have been driven by improved technologies allowing increased quality and longer shelf life, and by increased demand. Per capita consumption of fresh vegetables has increased from 107.0 pounds per person in 1970 to 163.1 pounds per person in 1998. Per capita consumption of processed vegetables (excluding potatoes) increased from 114.1 pounds per person in 1970 to 128.0 pounds person in 1998, a 12.1 percent rise. Per capita consumption of potatoes increased from 121.7 pounds per person in 1970 to 142.7 pounds per person in 1998, a 17.2 percent rise.

Competition in the winter fresh vegetable market has been monitored through the years. Increased competition from Mexico led to several trade disputes between U.S. and Mexican growers of fresh vegetables. U.S. growers filed antidumping petitions against Mexican growers of fresh vegetables in 1970, 1978, 1995, and 1996. These petitions were resolved using data generated by several studies contributing to the knowledge about trade in fresh vegetables. Many of those studies evaluated the cost-competitive position of fresh vegetable producers, including Fliginger et al. (1969), Simmons, Pearson, and Smith (1976), Zepp and Simmons (1979), Buckley et al. (1986), and VanSickle et al. (1994). These studies focused on costs of producing fresh vegetables for the winter market and on the average net revenues received by shippers of these vegetables.

Econometric approaches have also been used in assessing competition in the fresh vegetable industry and the forces affecting market shares. Bredahl, Schmitz, and Hillman (1987) evaluated alternative means for increasing returns to producers in the winter fresh tomato industries of Mexico and the United States. They concluded that a cooperative relationship could be developed that would increase returns to growers in both areas. However, they also suggested that it may be impossible to form an effective coalition between the growers in these regions and that growers must be content with the economic rents they receive from free trade.

Jordan and VanSickle (1995) evaluated market integration in the winter fresh tomato industry using a dynamic model of spatial price adjustment. Their results showed that Florida and Mexico were integrated in the same market and that prices received by shippers in each region were dependent on prices received by shippers in the other area.

Douglas (1997) used cointegration techniques to evaluate the impact of exchange rate adjustments on the U.S. tomato industry. The results of his analysis implied that pass-through of exchange rate adjustments to tomato prices was incomplete in the short run but complete in the long run. His results demonstrated that changes in exchange rates can provide significant short-run comparative advantage to Mexican producers and lead to significant increases in market share for them.

Modeling efforts have also been used to provide an understanding of trade in the fresh vegetable industry. Spreen et al. (1995) developed a model of the winter fresh tomato industry to evaluate the impacts a ban on methyl bromide would have on Florida producers of fresh vegetables. Methyl bromide is a critical soil fumigant used in the production of several fresh vegetables throughout the world. The results of their modeling effort indicated that banning methyl bromide would result in a loss of more than $600 million in shipper revenues if better alternatives were not developed and that most of that loss was absorbed by gains in Mexican production and by losses in consumer welfare. That model was also used to determine the impact of the large devaluation of the Mexican peso in 1994 to 1995 on the Florida vegetable industry (VanSickle, Spreen, and Jordan, 1996). Their results demonstrated that the devaluation was the most significant factor in explaining increased imports of fresh vegetables from Mexico. Weather had minor impacts on increased imports, but was small relative to the devaluation in explaining changes in trade flows.

COMPETITIVE POSITION OF THE U.S. VEGETABLE INDUSTRY

The relative costs of production for fresh and processed vegetables in the United States and its major competitors are important in determining relative shares in the U.S. market and the ability of U.S. producers to export their products. Other factors are also important in determining competitiveness. Tariff and nontariff barriers including sanitary and phytosanitary regulations, seasonality, domestic support, and related factors are important in determining the competitive position of vegetable producers. Technology has played a significant role in determining market shares. Taylor and Wilkowske (1984) demonstrated that productivity growth has been a prime factor in Florida's ability to retain a competitive position in the U.S. domestic winter vegetable market.

Cost Comparisons

Comparative cost data for winter vegetables supplied by Florida and Mexico are shown in Figure 11.1. Mexico has significantly lower preharvest costs, but marketing costs including transportation to the border, crossing fees, tariffs, and selling costs offset that advantage, keeping the market relatively competitive between the two producing regions. Florida producers had a cost-competitive advantage in the 1990-1991 season relative to Mexico for tomatoes, cucumbers, and eggplant. Mexican producers held advantages for bell peppers and squash (VanSickle et al., 1994). These costs have changed because of changes in economic conditions in Mexico. VanSickle and Douglas (1997) estimate that cost advantages provided by the peso devaluation and by investments in technology significantly altered cost competitive position, giving Mexican producers a clear advantage in supplying U.S. markets.

Trade Agreements

The principle trade agreement that affects the vegetable industry is the North American Free Trade Agreement (NAFTA). NAFTA primarily affects trade between the United States, Canada, and Mexico. Most tariffs for vegetables are being phased out over a ten-year period. The significance of these tariffs varies by product, but generally represents a small part of the total cost of producing and marketing these products. VanSickle et al. (1994, p. 65) found tariffs to represent 4.2 percent of the total cost for Mexican squash and 14.4 percent of the total cost of Mexican cucumbers. The primary effect of NAFTA felt by U.S. producers resulted from the change in

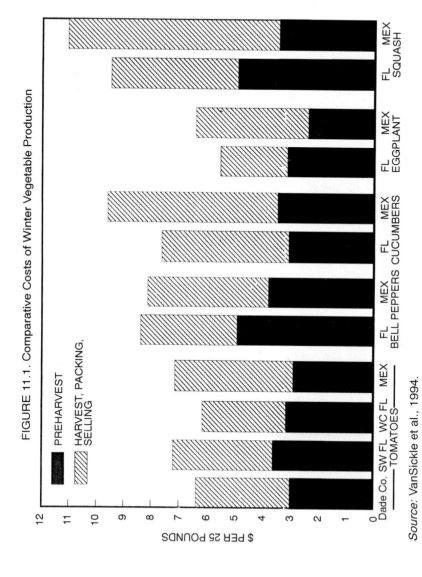

FIGURE 11.1. Comparative Costs of Winter Vegetable Production

Source: VanSickle et al., 1994.

the investment climate. Changes in rules regarding investment and land tenure increased productivity in Mexico relative to U.S. producers and contributed to the surge in imports from Mexico following the implementation of NAFTA. Those increases in technology and the large devaluation of the peso have had far bigger impacts on competitiveness than the lowering of tariffs brought on by NAFTA.

CONCLUSIONS

International trade in vegetables has increased because of improvements in technology and because of increased demand for vegetable products. Vegetables are a labor-intensive crop and consequently rely on technological improvements to remain competitive with imports from emerging economies such as Mexico. Unstable economies also impact competitiveness because of short-run advantages provided by currency devaluations.

The United States will require a constant flow of technical innovation to remain competitive in the vegetable industry. The rapid flow of technology throughout the world has removed some of the advantages provided by technical innovation in earlier years, but those investments remain critical. The relatively large research and development infrastructure relative to its major competitors will continue to help U.S. growers to compete. The United States should maintain a strong competitive position in the fresh and processed industries for vegetables.

REFERENCES

Bredahl, Maury E., Andrew Schmitz, and Jimmye S. Hillman. 1987. "Rent Seeking in International Trade: The Great Tomato War." *American Journal of Agricultural Economics* 69(1): 1-10.

Buckley, Katherine C., John J. VanSickle, Maury E. Bredahl, Emil Belibasis, and Nicholas Gutierrez. 1986. "Florida and Mexico Competition for the Winter Fresh Vegetable Market." AER-556. Washington, DC: Economic Research Service, U.S. Department of Agriculture.

Douglas, Charles D. 1997. "The Impact of Exchange Rate Adjustments on the United States Fresh Tomato Industry: Policy Implications for the United States and Mexico." Unpublished PhD dissertation. University of Florida, Gainesville, FL.

Fliginger, C. John, Earle E. Garett, Joseph C. Podany, and Levi A. Powell Sr. 1969. "Supplying U.S. Markets with Fresh Winter Produce: Capabilities of U.S. and Mexican Production Areas." AER-154. Washington, DC: Economic Research Service, U.S. Department of Agriculture.

Jordan, Kenrick H. and John J. VanSickle. 1995. "Integration and Behavior in the U.S. Winter Market for Fresh Tomatoes." *Journal of Agricultural and Applied Economics* 27(1): 127-137.

Simmons, Glenn L., James L. Pearson, and Ernest B. Smith. 1976. "Mexican Competition for the U.S. Fresh Winter Vegetable Market." AER-348. Washington, DC: Economic Research Service, U.S. Department of Agriculture.

Spreen, Thomas S., John J. VanSickle, Anne Moseley, M.S. Deepak, and Lorne Mathers. 1995. "The Use of Methyl Bromide and the Economic Impact of Its Proposed Ban on the Florida Fresh Fruit and Vegetable Industry." University of Florida Agricultural Experiment Station Bulletin 898, Gainesville, FL.

Taylor, Timothy G. and Gary H. Wilkowske. 1984. "Productivity Growth in the Florida Fresh Winter Vegetable Industry." *Southern Journal of Agricultural Economics* 16(1): 55-61.

USDA. 1998. "Vegetable and Specialties Situation and Outlook." VGS-275. Washington, DC: Economic Research Service, U.S. Department of Agriculture.

VanSickle, John J., Emil Belibasis, Dan Cantliffe, Gary Thompson, and Norm Oebker. 1994. "Competition in the Winter Fresh Vegetable Industry." AER-691. Washington, DC: Economic Research Service, U.S. Department of Agriculture.

VanSickle, John J. and Charles Douglas. 1997. "Monitoring Competition in the Winter Fresh Tomato Market." EN-37, Food and Resource Economics, University of Florida, Gainesville, FL.

VanSickle, John J., Thomas H. Spreen, and Kenrick Jordan. 1996. "An Economic Evaluation of the Impact of Devaluation of the Peso and Adverse Weather in Florida on the North American Winter Fresh Vegetable Market." Paper presented at the Tri-National Research Symposium NAFTA and Agriculture: Is the Experiment Working? San Antonio, TX, November 2.

Zepp, Glenn A. and Richard L. Simmons. 1979. "Producing Fresh Winter Vegetables in Florida and Mexico." ESCS-73. Washingtonm, DC: Economics, Statistics and Cooperative Service, U.S. Department of Agriculture.

Chapter 12

Miscellaneous Crops

Glenn C. W. Ames

Wine, sugar, and tobacco combined represent a significant share of the farm value of U.S. agricultural production. Wine and tobacco, as value-added products, are important export commodities, while the United States is a net sugar importer. Agricultural policy in the United States and abroad determines the competitive position of these commodities in the world market. This chapter is devoted to a discussion of production and trade policy, international trade agreements, and the competitiveness of U.S. wine, sugar, and tobacco on the world market.

WINE

Commercial vineyards are found in more than 40 states and produce the eighth largest agricultural crop in the United States, valued at $2.69 billion in 1997. Nearly 55 percent of the grape crop was used for wine, with an estimated farm value of $1.25 billion compared with $547 million a decade earlier (Love, 1997, p. 12). The United States is fourth among the world's wine producers, after Italy, France, and Spain.

World markets are important for the wine industry to bolster sales, farm-gate prices, and farm income. In 1997, exports accounted for an estimated 12 percent of U.S. production, up from 5.6 percent in 1990 (Love, 1997, p. 13). The value of U.S. wine exports reached $425 million, an increase of 30 percent more than 1996 (see Figure 12.1). Wine export volume grew to 60 million

FIGURE 12.1. U.S. Wine Exports and Market Access Program Funds Expenditures

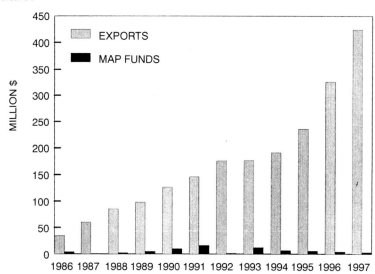

Source: Horiuchi, 1998.

gallons, an increase of 26 percent over 1996. California is the leader in wine exports, with an estimated $383 million exported in 1997.

The United Kingdom, Canada, and Japan are the leading destinations among 164 countries for U.S. wine exports, accounting for 53 percent of the total value—$108, $79, and $40 million in 1997, respectively (see Figure 12.2). In 1997, U.S. wine sales to the European Union (EU) continued an eight-year growth trend as exports reached $224 million, 33 percent more than 1996. In 1997, U.S. wine exports to Canada increased 9 percent in value and 7 percent in volume over the previous year. Wine exports to Japan rose to $40 million, 26 percent more than 1996. Japan agreed to reduce its tariff on U.S. wine from 21.3 percent to 15 percent as part of the Uruguay Round of GATT.

The United States, however, is a net importer of wine. In 1997, 18.8 million gallons with a value of $1.149 billion were imported. This was a 63.8 percent increase in volume in two years. Italy is the largest supplier by volume, accounting for 41 percent, while France

FIGURE 12.2. Top Ten U.S. Wine Export Markets

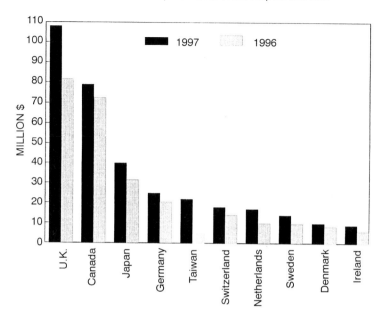

Source: Horiuchi, 1998.

was the largest supplier in value terms, accounting for 47.5 percent of the total value of all wine imports.

Market Access and Trade Agreements

Prospects for U.S. wine trade depend on the industry's ability to expand market access and reduce trade barriers. Export subsidies, tariffs, and protectionist policies inhibit the competitiveness of U.S. wines. Under the Canada-U.S. Trade Agreement, Canadian tariffs on U.S. wines were eliminated on January 1, 1998. Canada, however, applies certain "cost-of-service" markups on wines sold in Canadian stores. In Ontario, this differential was 6 percent in 1998. Under NAFTA, Mexican tariffs on grape wine and grape brandy are being eliminated over ten years. U.S. wine exports to Mexico peaked in 1994 at $6.9 million, then fell to $3.8 million in 1996 (USDA, FAS, 1998c).

Trade agreements have opened some markets and reduced trade-distorting subsidies, but significant barriers still exist. In the Uruguay Round of the GATT negotiations, the EU agreed to cut subsidized wine exports by 21.1 percent to 2.43 million hectoliters (64.2 million gallons) by the year 2000 and to lower the value of export subsidies by 35.9 percent to 41.3 million ECU ($45.9 million). The EU also agreed to reduce its tariff on U.S. wine from 49 cents to 39 cents per liter, a decline of about 20.4 percent. To reduce nontariff barriers, signatory countries agreed to give trademarks greater protection and to clarify technical requirements on labeling and other factors related to wine imports. Since 1983 the Wine Accord has governed U.S.-EU wine trade. The Accord expired in 1993, but has been extended annually, although the United States would like to make it permanent (Love, 1997, p. 14). Under the Accord, the United States obtained temporary derogations (waivers) from EU restrictions on some winemaking practices and certification procedures.

A common practice in the wine trade is the use of semigeneric labels such as Chablis, Burgundy, and Champagne, which can lead to trade disputes. U.S. wines generally are marketed by grape variety such as Cabernet Sauvignon or Chardonnay, but some wines are marketed under semigeneric labels. In contrast, European wines have a long tradition of labeling based on the origin of the producing region. The *appellation de origine contróllée* (AOC), a registered trade name, is the principal designation for quality wines. Each producing region's distinct quality is strictly regulated with respect to grape variety and production practices. For example, Burgundy is a well-known region of France producing red wines from Pinot Noir grapes, but the grape variety seldom appears with "AOC" on Burgundy labels. When U.S. wine exporters use semigeneric labels, consumers may assume that the label specifies the wine's origin. Thus, the EU is actively seeking to end the practice of semigeneric labeling, while the United States argues that this would be a nontariff trade barrier (Love, 1997, p. 14).

Other nontariff trade barriers also pose problems. Norway, Finland, Sweden, and Taiwan, for example, maintain state or provincial monopolies that restrict imports and wine sales. Many countries subject wine sales to import licensing, lengthy documentation processes, shipping restrictions, and additive restrictions. In Latin America, regional trading agreements give preference to member countries. The expansion of the Free Trade Area of the Americas agreement should provide an excellent opportunity to reduce tariff and nontariff barriers to U.S. wine exports.

Export Promotion

In the early 1980s, the Wine Institute urged passage of the Wine Equity and Expansion Act (WEEA), which ultimately became law in 1985. The WEA addressed barriers faced by U.S. wine producers in foreign markets (Horiuchi, 1997). Thus, the Wine Institute participates in the Market Access Program (MAP) authorized by Section 244 of the Federal Agriculture Improvement and Reform Act of 1996 (FAIR Act), and funded by the Commodity Credit Corporation (CCC). Under MAP, the wine industry and the FAS conduct foreign market development projects on a cost-sharing basis. In 1997 the Wine Institute received $3 million in export promotion funding, ranking the Institute seventh among the 64 trade organizations receiving MAP funding. About 125 California wineries participate in the Wine Institute's export development program, and two other wine trading associations also receive modest funding (USDA, FAS, 1998b).

Generally, promotion of value-added products, such as wine, has a higher return than bulk or intermediate products. Between 1996 and 1997, the values of exports of bottled table wine, 72 percent of total wine exports, jumped 31 percent, champagne and sparkling wine grew 17 percent, dessert and vermouth rose 11 percent, and grape must and other fermented beverages increased by 48 percent (Horiuchi, 1998).

Conclusions

Although wine exports have grown significantly in the 1990s, the United States holds only a 3 percent world market share and U.S. wines continue to face stiff competition from Italy, France, and Spain. Tariffs and nontariff barriers also limited market access for U.S. wine. While the United States remains a wine deficit country, supplies may increase in the near future as new vineyards commence production. These will place additional supplies on the domestic market where per capita consumption remains low. Thus, greater access to foreign markets, export promotion, and trade liberalization will be needed to promote U.S. wine exports.

SUGAR

The United States is the fourth largest individual sugar producer in the world after India, Brazil, and China, although the EU is the world's largest producing region (USDA, NASS, 1997, p. II-19). The United States produced 7.1 million metric tons in fiscal year 1997, up 9 percent from 1996 (USDA, ERS, 1997b, p. 3). In 1996 sugar beets and sugarcane ranked 19th and 26th in terms of the farm value of U.S. crop production, generating $1,211 and $833 million, respectively (USDA, NASS, 1998). Over 1,441,000 acres of sugar beets are grown on an estimated 13,731 farms in 14 states. Sugarcane production is concentrated in four states, Florida, Hawaii, Louisiana, and Texas, where cane is grown on an estimated 1,705 farms covering 829,200 acres.

Sugar production and trade in the United States have been governed by supply control policies since the 1930s, when the Jones-Costigan Sugar Act of 1934 provided for acreage restrictions, a country-by-country import quota system, and support payments to sugar producers. The FAIR Act modified the sugar program but continues supports and import quotas (Young and Westcott, 1996). Each year, the Secretary of Agriculture monitors import requirements necessary to meet U.S. domestic demand and may adjust the amount of the tariff rate quota (TRQ) at any time to ensure that there are sufficient sugar supplies in the U.S. market (Suarez, 1999a, p. 15). A prohibitive tariff of 17.62 ¢/lb (38.84 ¢/kg) effectively excludes sugar imports above the minimum TRQ unless permitted by the Secretary of Agriculture—the in-quota tariff rate is only 0.6625 ¢/lb (USDA, FAS, 1995). The Dominican Republic, Philippines, and Brazil have the larger quota allocations, representing 45 percent of the U.S. global sugar import quota (see Table 12.1). In 1996, imported sugar accounted for approximately 28.3 percent of total U.S. domestic consumption (USDA, ERS, 1997c, p. 8).

Competitive Position of U.S. Sugar

LMC International (Landell Mills Commodities Studies), based in Oxford, England, ranked U.S. sugar beet producers as having the lowest cost among the world's 38 producers for the 1992-1995

TABLE 12.1. FY1998 TRQs for Raw Cane Sugar (Metric Tons)

Country	TRQ	Country	TRQ
Argentina	48,101	Madagascar	7,258
Australia	92,846	Malawi	11,186
Barbados	7,830	Mauritius	13,424
Belize	12,305	Mexico	25,000
Bolivia	8,949	Mozambique	14,542
Brazil	162,201	Nicaragua	23,491
Colombia	26,847	Panama	32,440
Congo	7,258	Papua New Guinea	7,258
Cote d'Ivoire	7,258	Paraguay	7,258
Costa Rica	16,779	Peru	45,864
Dominican Republic	196,878	Philippines	151,015
Ecuador	12,305	South Africa	25,728
El Salvador	29,084	St. Kitts-Nevis	7,258
Fiji	10,068	Swaziland	17,898
Gabon	7,258	Taiwan	13,424
Guatemala	53,694	Thailand	15,661
Guyana	13,424	Trinidad-Tobago	7,830
Haiti	7,258	Uruguay	7,258
Honduras	11,186	Zimbabwe	13,424
India	8,949		
Jamaica	12,305	Total	1,200,000

Source: Office of the U.S. Trade Representative, 1997.

Note: United States Representative Charlene Barshefshy also announced that 25,000 metric tons (27,558 short tons) of the 50,000 metric tons (55,116 short tons) for refined sugar will be allocated to Mexico in order to fulfill obligations pursuant to NAFTA. As a result of an agreement reached with Canada, 10,300 metric tons (11,354 short tons) of refined sugar and 59,250 metric tons (65,312 short tons) of the tariff rate quota for certain sugar-containing products maintained under Chapter 17 of the Harmonized Tariff Schedule of the United States will be allocated to Canada.

period, up from second place in the previous three-year period. The U.S. sugarcane industry was ranked 27th lowest-cost among the 62 cane-producing countries (American Sugar Alliance, 1997). Overall, the United States was ranked 19th among 96 sugar producers (Todd, 1997). The United States faces stiff competition in the world market, as a comparison of international domestic world sugar prices indicates (see Figure 12.3).

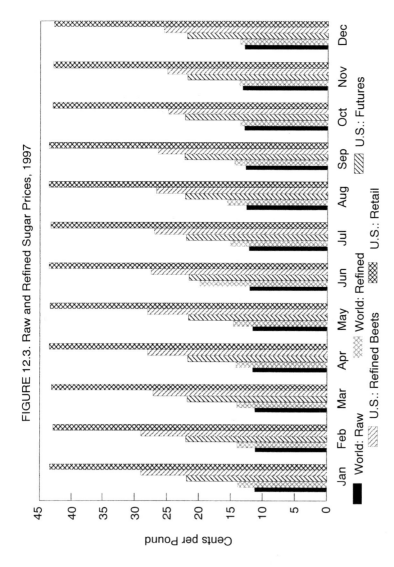

FIGURE 12.3. Raw and Refined Sugar Prices, 1997

Source: USDA, ERS, 1997b.

Sugarcane production cash costs average about $200 more per acre than beets. When opportunity costs are deducted from the returns, the residual to management and risk were negative in three out of five years (see Table 12.2). Environmental protection, such as the Everglades Forever Act, raises costs for sugar producers in the United States, according to industry sources.

Trade Policy and the Sugar Sector

In the Uruguay Round of GATT, the United States agreed to maintain a minimum annual TRQ of 1.139 million metric tons (1.256 million short tons) of raw and refined cane sugar, a level similar to the imports specified in the 1990 Farm Bill (Lord, 1995,

TABLE 12.2. U.S. Sugar Production Costs and Returns (Dollars per Acre)

Sugar Beets	1992	1993	1994	1995	1996
Gross Value of Prod.: Beets	853.26	723.47	847.39	748.52	746.61
Beet Tops	0.89	0.90	1.12	0.75	0.71
Total, Gross Value of Production	854.15	724.37	848.51	749.27	747.32
Total, Variable Cash Expenses	416.68	410.82	419.30	425.60	435.18
Total, Fixed Cash Expenses	121.52	134.71	127.04	131.92	98.3
Total, Cash Expenses	538.20	545.53	546.34	557.52	533.57
Gross Value Less Cash Expenses	315.95	178.84	302.17	191.75	213.75
Total, Economic Costs	767.43	757.75	784.67	797.86	792.28
Residual Returns to Mgt. & Risk	86.72	-33.38	63.84	-48.59	-44.96
Season-Avg. Price (Dollar/Ton)	41.32	38.98	38.80	38.19	38.19
Yield (Net Tons/Planted Acre)	20.65	18.56	21.84	19.60	19.55
Sugarcane					
Gross Value of Production	932.32	946.20	975.28	983.24	979.40
Total, Variable Cash Expenses	659.39	662.53	703.87	695.45	690.16
Total, Fixed Cash Expenses	114.47	134.69	141.16	150.42	146.58
Total, Cash Expenses	773.86	797.22	845.02	845.87	836.74
Gross Value Less Cash Expenses	159.06	148.98	130.26	137.36	142.66
Total, Economic Costs	1,038.06	1,051.63	1,093.36	1,130.72	1,125.17
Residual Returns to Mgt. & Risk	-105.14	-105.14	-118.08	-147.48	-145.77
Season-Avg. Price (Dollar/Ton)	28.10	28.50	29.20	29.50	29.50
Yield (Net Tons/Planted Acre)	33.20	33.20	33.40	33.33	33.20

Source: USDA, ERS, 1997d.

p. 65). The EU maintains higher sugar price supports than the United States but under the Uruguay Round has agreed to reduce their quantity and budgetary outlays for sugar export subsidies by 21 percent and 36 percent, respectively, relative to 1986-1990. By 2000, the EU's maximum annual allowable subsidized sugar exports were 1,277,000 metric tons, down 21 percent from the base period. While the EU spent approximately $1.740 billion on export subsidies in 1996 (USDA, ERS, 1998), it subsidized only about 24 percent of its sugar exports (USDA, ERS, 1997a).

The North American Free Trade Agreement and Sugar Policy

NAFTA establishes a "net surplus production" formula for determining market access and trade. Since the United States and Canada are not surplus producers, the market access provisions essentially apply to Mexican exports to the United States. Initially, Mexico will have duty-free access to the U.S. market of 25,000 metric tons, raw value; these will increase to 250,000 metric tons during the NAFTA transition period (Suarez, 1997b, p. 4). At the end of the 15-year agreement, there will be free trade in sugar between the two countries. Devadoss, Kropf, and Wahl (1995) concluded that access to new technology, investments, and lower-priced inputs will stimulate Mexican sugar production under NAFTA. However, the effect during the early years will be relatively small because of side agreement provisions that limit Mexican sugar exports to the United States. These side agreements have become a source of disagreement between NAFTA partners because they constrain Mexican access to the U.S. market (American Sugar Alliance, 1997). In 1996, U.S. imports of sugar from Mexico reached 27,861 tons, an increase of 35 percent over the previous year (USDA, FAS, 1998c).

The relative prices of sugar and other sweeteners in Mexico have changed since NAFTA began, leading to greater high-fructose corn syrup (HFCS) consumption (Suarez, 1997b, p. 4). U.S. exports of HFCS to Mexico have increased dramatically, reaching over 184,000 metric tons in 1997, representing more than 44 percent of HFCS utilization (Salsgiver, 1997, p. 12). Thus, in 1997, Mexico's sugar producers' association accused the United States of dumping HFCS, and Mexico imposed temporary tariffs in June 1997. In September

1997, the U.S. Corn Refiners Association protested Mexico's policies and requested World Trade Organization (WTO) consultations.

Disputes have also marked the U.S.-Canada sugar trade. During the 1980s, the United States exported over 50,000 tons of sugar annually to Canada, mostly under the U.S. refined sugar reexport program. Exports surged to over 100,000 tons in the early 1990s. In November 1995, Canada responded by imposing 69 to 85 percent antidumping duties on U.S. sugar companies. Subsequently, U.S. sugar exports to Canada fell to 20,000 tons in 1996 (Suarez, 1997b, p. 5). In September 1997, the United States and Canada reached an agreement providing Canadian access to a sugar TRQ of 10,300 metric tons, raw value, for beet sugar produced in Canada. The United States also agreed to allocate Canada 91.5 percent of the U.S. sugar-containing product TRQ of 64,709 metric tons based on historical trade patterns. Canada then dropped its challenge to the U.S. sugar-containing product reexport program.

Evaluation of the U.S. Sugar Program

A comprehensive General Accounting Office (GAO, 1993) summarization of several sugar studies indicated that a minority of producers and processors reap most of the benefits of the program. Large farmers and sugar processors receive the largest share of the benefit based on their volume of production. Hafi, Connell, and Roberts (1994) analyzed U.S. sugar policies and concluded that they have depressed world prices, increased price volatility, encouraged competition from sugar substitutes, and contributed to a loss of consumer welfare. Importing countries such as Canada and Japan have benefitted from the price-depressing effects of reduced U.S. sugar imports, while countries with sugar quotas such as the Dominican Republic, El Salvador, Panama, and the Philippines have received substantial quota rents, that is, larger returns due to selling their quotas at prices above world market levels.

Benirschka and Koo (1997) used a dynamic partial equilibrium model of the world sugar economy to simulate the potential impacts of U.S. trade liberalization in sugar. Liberalizing the U.S. sugar regime would result in declining sugar prices and a larger import market share; aggregate production would decline by only 5.9 percent with beet sugar acreage declining by about 9.0 percent, while cane sugar acreage would decline by 1.2 percent. Trade liberalization would also have repercussions for import quota holders. Quota rents of exporting countries will decrease and Caribbean sugar prices would increase between 1 to 3 percent under trade liberalization, but this modest price increase would not compensate exporters for the loss of their quotas.

Conclusions

The 1996 FAIR Act brought major changes in U.S. sugar policy, which underlies the competitive position of the industry. These include the elimination of marketing allotments, resulting in more flexible alternatives for sugar producers. The American Sugar Alliance (1998) criticizes the EU's continued use of export subsidies for sugar, alleging that much of the "free" world sugar market is a "dumped sugar market in which the United States faces unfair competition." State trading enterprises that utilize marketing boards or dual pricing systems for domestic and export sugar are another example of uneven trading practices. Finally, labor standards and environmental protection vary widely among producing countries, leading to differences in comparative costs in the world sugar market.

TOBACCO

The United States is second in world tobacco production with 995 million pounds, after China's 6.86 billion pounds. Tobacco is the sixth leading crop in the United States, with a farm value of $2.6 billion. It is grown on about 124,000 farms, in about 20 states, mainly in the southeastern United States. North Carolina and Kentucky account for about 65 percent of the acreage, while Georgia, South Carolina, Virginia, and Tennessee account for an additional 20 percent.

The Agricultural Adjustment Act of 1938 (PL 75-430) established a supply control program for tobacco that, as amended, still governs the program in the 1990s. It provides for annual tobacco marketing quotas, which are set by the Secretary of Agriculture to meet expected domestic and export demands. In 1998, the U.S. Department of Agriculture announced a national marketing quota of 637.8 million pounds, down 9.5 percent from 1997. The Act requires each major cigarette manufacturer to purchase an amount of tobacco leaf equal to at least 90 percent of its stated purchase intentions to avoid a penalty. The No-Net-Cost Tobacco Program Act of 1982 (PL 97-218) imposes an assessment on every pound of tobacco marketed. Growers and buyers each pay a portion of the no-net-cost assessment. Tax revenue from the no-net-cost assessment is used to reimburse the government for any financial losses resulting from the tobacco loan operations. Since 1994, assessments also have been levied on importers (Grise, 1995, p. 11). Tobacco also became subject to the deficit reduction requirements enacted by the Omnibus Budget Reconciliation Acts of 1990 (PL 101-508) and 1993 (PL 103-66). A marketing assessment of 1 percent of the support price is collected on every pound of tobacco marketed, equally divided between growers and buyers (Grise, 1995, p. 10; Womach, 1996, p. 4). Tobacco importers also began paying a marketing assessment fee in 1994. In 1996, this deficit reduction assessment generated $27 million which is applied toward general deficit reduction, not the tobacco program. In 1986, legislation reduced the price support from 169.9 to 143.8 ¢/lb. This reduction halted the decline in U.S. tobacco exports temporarily (imported tobacco had captured 40 percent of the domestic cigarette manufacturing market).

In 1994, a 25 percent domestic content requirement took effect under the Omnibus Budget Reconciliation Act of 1993 (PL 103-66). Several tobacco-exporting countries challenged the domestic content requirement with the GATT. A GATT panel ruled that the domestic content requirement was inconsistent with U.S. GATT obligations (Grise, 1995, p. 6). Subsequently, President Clinton proclaimed a TRQ, which replaced the domestic content requirement. Beginning September 13, 1995, 332.6 million pounds of cigarette leaf can be imported annually at the low TRQ rate, with country quotas based on historical market shares. Brazil received the largest share of the quota, 176.8 million pounds, followed by Argentina, Malawi, and Zimbabwe with 26.5 million pounds each. Imports from Mexico, Canada, and Israel were not included in the TRQ because of existing free trade agreements. In-quota tariffs on tobacco range from 11.52 to 36.0 ¢/lb., while over-quota imports are subject to a 350 percent ad valorem duty (USDA, FAS, 1995). This can be refunded if the same tobacco is used to manufacture cigarettes that are subsequently exported.

Tobacco Trade

U.S. tobacco imports nearly doubled between 1995 and 1997, reflecting the switch from a restrictive 25 percent domestic content import policy to the TRQ. In 1997 imports were $1.09 billion, with Turkey and Brazil being the leading sources for tobacco, followed by Malawi, Greece, and the Dominican Republic (see Table 12.3). However, the United States had a positive net balance of trade in unmanufactured tobacco in 1997. Tobacco exports were $1.55 billion, up 11.7 percent from 1996. The leading export markets were the EU-15, Japan, the ASEAN region, the Middle East, and the Caribbean Islands.

Competitive Position of U.S. Tobacco

U.S. competitiveness in tobacco production is closely related to average costs of production. Variable costs of producing and selling flue-cured tobacco averaged $2,052 per acre in 1997, up about 1.3 percent from 1996 (Glaze, 1997, p. 1). Ownership and general farm overhead added another $828 per acre for a total cost of production of $2,879, excluding land and quota charges. Cash receipts were $3,763 per acre, based on an average yield of 2,138 lb per acre and an average grower price of $1.76 per lb. Average costs of flue-cured tobacco production were $1.30 per pound, leaving $0.46 to cover land and quota rents (Table 12.4).

Under the current program, U.S. tobacco producers must compete with unmanufactured flue-cured imports, which averaged $1.97/lb, up 8 percent from 1996. The import price of stemmed burley tobacco averaged $1.81/lb, up 6.3 percent from 1996. U.S. leaf imports surged from 413 million pounds in 1990 to over one billion pounds in 1993. Manufacturers imported larger amounts of lower cost foreign tobacco to meet the demand for low- and mid-price cigarette brands (Capehart, 1997, p. 1). The market share of imports for flue-cured tobacco rose to 28.1 percent in 1997,

TABLE 12.3. Top Ten U.S. Export and Import Markets for Unmanufactured Tobacco ($1,000)

U.S. Exports To:	R	1993	1994	1995	1996	1997
Germany	1	162,452	178,269	217,383	192,191	227,906
Japan	2	334,102	355,831	313,665	230,864	219,580
Turkey	3	111,292	75,491	59,837	77,040	140,885
Dom. Republic	4	48,747	48,199	53,683	66,820	114,337
Belgium-Lux.	5	36,283	30,546	55,793	91,375	110,865
Netherlands	6	115,976	96,298	122,704	110,864	96,199
Thailand	7	56,430	65,749	66,004	52,139	75,260
Malaysia	8	22,337	23,489	34,951	45,496	65,302
South Korea	9	15,165	21,415	43,650	54,472	63,119
United Kingdom	10	53,879	43,299	44,063	90,842	51,622
All Countries		1,306,067	1,302,745	1,399,863	1,390,311	1,553,314
U.S. Imports From:						
Turkey	1	241,021	158,707	112,856	248,917	307,961
Brazil	2	377,333	145,737	75,835	237,306	226,867
Argentina	3	62,471	37,847	16,277	58,948	80,196
Malawi	4	66,032	34,649	30,784	47,181	79,250
Greece	5	138,046	39,777	93,021	55,343	67,573
Dom. Republic	6	11,002	11,986	16,328	27,932	56,288
Thailand	7	77,799	35,103	2,411	49,607	35,413
Indonesia	8	16,134	8,927	12,654	16,043	34,557
Mexico	9	19,523	2,059	3,553	23,917	33,027
Canada	10	27,838	9,422	10,159	12,331	23,965
All Countries		1,370,031	613,182	550,136	922,697	1,089,264

Sources: USDA, FAS, 1998d; 1998f.

TABLE 12.4. U.S. Costs of Producing and Selling Flue-Cured Tobacco (Dollars)

Item	1995	1996	1997
Cash Receipts	3,460.07	3,951.39	3,762.88
Yield (lb/acre)	1,933	2,151	2,138
Variable Costs	1,898.53	2,026.17	2,051.24
Machinery and Barn Ownership Costs	564.23	559.38	583.93
General Farm Overhead	224.04	255.83	243.62
Total Costs, Excluding Quota Rent	2,686.78	2,481.38	2,879.79
Land and Quota Charge Per Acre*	798.47	897.88	877.22
Net Returns Excluding Land & Quota	773.29	1,170.01	883.09
Net Returns	-25.18	+572.13	+5.89

Source: Glaze, 1997.

*Weighted average of cash and share rents.

up from 23.4 percent in 1995, but the share of imported burley fell by 4.8 percent to 22.8 percent. Overall, import market share increased once the TRQ came into effect because imports are cheaper, even with a 11.52 to 36.0 ¢/lb tariff.

In the export market, the United States faces stiff competition. The average U.S. export price of 316.3 ¢/lb was 85.4 percent higher than Malawian flue-cured exports and double the price of Zimbabwean tobacco (Tobacco Associates, 1998, p. 12). Exports from Brazil were priced 31.2 percent lower than U.S. leaf, while flue-cured exports from Thailand were more than 50 percent below U.S. prices. Brazil and Zimbabwe accounted for 22.5 and 21.7 percent, respectively, of the world's 2.07 billion pound export market in 1997, while the U.S. world market share was about 12.1 percent (Tobacco Associates, 1998, p. 12).

Tobacco Associates, a farmer-financed trade association established in 1947, promotes exports of U.S. flue-cured tobacco worldwide. The association provides training and technical assistance to foreign cigarette manufacturers in the use of U.S.-produced flue-cured tobacco. It also conducts seminars on U.S. leaf quality standards, promotes U.S. leaf tobacco at trade fairs abroad, hosts foreign buying missions for U.S. tobacco, and advertises in leading international tobacco publications (Tobacco Associates, 1998).

The Uruguay Round Agreement and Tobacco

The Uruguay Round of GATT resulted in a number of commitments that will benefit U.S. tobacco exports in the long run. The agreement established disciplines in the areas of market access, export subsidies, internal support, and sanitary and phytosanitary measures. Commitments include better market access, reduced tariffs, and export subsidy reductions. The EU, for example, agreed to reduce tariffs by 20 percent for unmanufactured tobacco, 36 percent for cigarettes and other manufactured tobacco, and 50 percent for cigars (USDA, FAS, 1994). It also agreed to reduce budgetary outlays for export subsidies and to reduce the quantity of subsidized tobacco exports to 2.48 million pounds, 45 percent less than was subsidized in the 1991-1992 base period.

Other countries either agreed to lower their tariffs on tobacco products or to bind tariffs at low levels. Japan, the largest single market for the United States, agreed to apply a zero-level tariff on cigarettes, consistent with the current agreement with the United States. The United States exported 67.6 billion cigarettes, worth about $1.4 billion, to Japan in 1997. Unmanufactured tobacco exports to Japan were valued at $219.5 million in 1997 (USDA, FAS, 1998e). Japan also agreed to reduce its duty on cigars from 20 percent to 16 percent, while Hong Kong agreed to bind its tariffs at a zero rate for tobacco and products.

NAFTA and Tobacco Trade

NAFTA is expected to increase U.S. tobacco exports gradually since Mexico eliminated licensing requirements for imports and substituted a 50 percent ad valorem tariff that will be phased out over a ten-year period (Grise, 1995, p. 13). U.S. licensees have traditionally had a significant share of the Mexican cigarette market, but manufacturers have predominantly used locally grown tobacco or imported leaf from non-U.S. sources, mainly Brazil and Argentina.

Evaluating the U.S. Tobacco Program

Brown and Martin (1996) analyzed the effects of a negative 10 percent shift in both domestic and export demand under a policy of quota reduction versus a policy of maintaining quota but allowing flexible prices. Welfare losses were minimized under a policy of price maintenance, the current policy, while aggregate losses to tobacco growers were greater under a policy of quota maintenance that allows prices to adjust to demand shifts. Under the current program, the marketing quota automatically declines with a shift in demand. A change in price supports that would increase competitiveness on the world market requires legislation which tobacco state legislators oppose, fearing attempts to eliminate the tobacco program (Brown and Martin, 1996, p. 451). Thus, the authors concluded that the outlook for U.S. flue-cured tobacco appeared to be one of status quo—maintaining the quota price at the expense of marketing opportunities (p. 451). In the near term, the U.S. tobacco program appears set; marketing quotas and price supports are in place and linked by a policy formula through 2002. Unmanufactured tobacco exports may expand slowly, but U.S. cigarette exports remain strong on world markets.

IMPLICATIONS FOR U.S. TRADE IN WINE, SUGAR, AND TOBACCO

Agricultural policy has a significant impact on the competitive position of U.S. wine, sugar, and tobacco products in the world market. Sugar and tobacco production and marketing are governed by a long history of price supports, supply constraints, and border protection, and these policies are not expected to change in the shortrun. In the world market, the United States faces strong competition from tobacco producers in Brazil, Argentina, Malawi, Zimbabwe, Thailand, and Turkey. World sugar supplies remain abundant and prices are relatively stable and low. NAFTA and the Uruguay Round agreements have had relatively minor impacts on trade in these commodities. Market access in sugar and tobacco is slowly changing but ad valorem tariffs and side agreements still limit trade during the phase-in period. The world market for wine is highly competitive and U.S. exports are growing, aided by MAP expenditures promoting U.S. vintages in overseas markets. Overall, the competitive position of these commodities in the world market is not expected to change significantly in the near future.

REFERENCES

American Sugar Alliance. (1997). "U.S. Trade Ambassador Confirms Validity of NAFTA Side Letter on Sugar," URL <http://www.sugaralliance.org/press.html #080597a>, August 5.

American Sugar Alliance. (1998). "Economic Impact." URL <http://www.sugaralliance.org/economic_impact.html>.

Benirschka, M. and W.W. Koo. (1997). "Liberalizing World Sugar Trade: The Impact of U.S. Tariff Rate Quota Changes." Selected paper presented at the Western Agricultural Economics Association Annual Meeting, Reno, NV, July 13-16.

Brown, A.B. and L.L. Martin. (1996). "Price versus Quota Reductions: U.S. Flue-Cured Tobacco Policy." *Journal of Agricultural and Applied Economics* 28(December):445-452.

Capehart, T. (1997). "The Tobacco Program: A Summary and Update." *Tobacco Situation and Outlook.* S&O/TBS-238. Washington, DC: Market and Trade Economics Division, Economic Research Service, U.S. Department of Agriculture, April.

Devadoss, S., J. Kropf, and T. Wahl. (1995). "Trade Creation and Diversion Effects of the North American Free Trade Agreement of U.S. Sugar Imports from Mexico." *Journal of Agricultural and Resource Economics* 20(December):215-230.

GAO. (1993). *Sugar Programs: Changing Domestic and International Conditions Require Program Changes.* Washington DC: U.S. General Accounting Office, GAO/RCED-93-84, April.

Glaze, D. (1997). "Costs of Producing and Selling Flue-Cured Tobacco: 1995, 1996, and Preliminary 1997." U.S. Department of Agriculture, Economic Research Service, URL <http://www.econ.ag.gov/Briefing/tobacco/Costs.htm>, September.

Grise, V.N. (1995). *Tobacco: Background for 1995 Farm Legislation.* Agricultural Economic Report No. 709. Washington, DC: Commercial Agriculture Division, Economic Research Service, U.S. Dept. of Agriculture, April.

Hafi, A., P. Connell, and I. Roberts. (1994). "U.S. Sugar Policies: Market and Welfare Effects." *Australian Commodities* 1(December): 484-500.

Horiuchi, G. (1997). "Backgrounder: The Wine Institute of California's International Program." URL <http://www.wineinstitute.org/communications/exports/ wi_international_background.htm>, February.

Horiuchi, G. (1998). "U.S. Wine Exports Climb to $425 Million in 1997—Up 400 Percent Over Last Decade." URL <http://www.wineinstitute. org/communications/statistics/exports/wine_1997.htm>, April 16.

Lord, R. (1995). *Sugar: Background for 1995 Farm Legislation.* Washington, DC: Commercial Agriculture Division, Economic Research Service, U.S. Dept. of Agriculture, AER-711, April.

Love, J. (1997). "The U.S. Wine Market Uncorked." *Agricultural Outlook.* Washington, DC: U.S. Dept. of Agriculture, Economic Research Service, A0-243, August, pp. 12-14.

Office of the U.S. Trade Representative. (1997). "USTR Announces Allocation of the Raw Cane Sugar, Refined Sugar and Sugar Containing Products Tariff-Rate Quota for 1997-98." Washington, DC, Press Release 98-85, September 17.

Salsgiver, J. (1997). "HFCS Trade Dispute with Mexico." *Sugar and Sweetener Situation and Outlook.* Washington, DC: Commercial Agriculture Division, Economic Research Service, U.S. Dept. of Agriculture, SSS-221, September, p. 12.

Suarez, N.R. (1997a). "Origin of the U.S. Sugar Import Tariff-Rate Quota Shares." *Sugar and Sweetener Situation and Outlook.* Washington, DC: Commercial Agriculture Division, Economic Research Service, U.S. Dept. of Agriculture, SSS-221, September, pp. 14-15.

Suarez, N.R. (1997b). "Sweetener Flows in the North American Region." *Sugar and Sweetener Situation and Outlook Yearbook.* Washington, DC: Commercial Agriculture Division, Economic Research Service, U.S. Dept. of Agriculture, SSS-222, December, pp. 4-6.

Tobacco Associates. (1998). *Fifty-First Annual Report.* Report of Kirk Wayne, President, Tobacco Associates, Inc., Raleigh, NC, March 6.

Todd, M. (1997). "The International Cost Competitiveness of the U.S. Sugar and HFCS Industries." Paper presented at the 14th Annual International Sweetener Symposium, Big Sky, Montana, August 5.

USDA, ERS. (1997a). "EU Export Subsidy Commitments Not Yet Binding, But Future Uncertain." *Europe Situation and Outlook Series.* Washington, DC: U.S. Department of Agriculture, Economic Research Service. WRS-97-5, December.

USDA, ERS. (1997b). *Sugar and Sweetener Situation and Outlook,* Washington, DC: U.S. Department of Agriculture, Economic Research Service. SSS-221, September.

USDA, ERS. (1997c). *Sugar and Sweetener Situation and Outlook Yearbook.* Washington, DC: U.S. Department of Agriculture, Economic Research Service. SSS-222, December.

USDA, ERS. (1997d). "Sugarcane Costs and Returns Data." U.S. Department of Agriculture, Economic Research Service. URL <http://www.econ.ag.gov/briefing/fbe/car/cane3.htm>, September 30.

USDA, ERS. (1998). "European Agricultural Statistics." U.S. Department of Agriculture, Economic Research Service. URL <http://mann77.manlib.cornell.edu/data-sets/international/98001/html>.

USDA, ERS. (1994). "GATT/WTO and Tobacco." U.S. Department of Agriculture, Foreign Agricultural Service. URL <http://www.fas.usda.gov/itp/policy/gatt/tobacco. htm>, June.

USDA, ERS. (1995). "WTO Tariff Schedules." U.S. Department of Agriculture, Foreign Agricultural Service. URL <http://www.fas.usda.gov/scriptsw/wtopdf/stopdf_frm.ide>.

USDA, FAS. (1998a). "European Union Extends Wine Derogations to Allow U.S. Wine Trade to Continue." U.S. Dept. of Agriculture, Foreign Agricultural Service. URL <http://fas. usda.gov/htp/hortculture/html>, February 10.

USDA, FAS. (1998b). "Market Access Program." U.S. Dept. of Agriculture, Foreign Agricultural Service. URL <http://www.fas.usda.gov/info/factsheets/mapfact.html>, January 2.

USDA, FAS. (1998c). "NAFTA Agriculture FACT Sheet: Sugar." U.S. Department of Agriculture, Foreign Agricultural Service. URL <http://www.fas.usda.gov/itp/policy/nafta/sugar.html>, April 9.

USDA, FAS. (1998d). "U.S. Imports and Exports of Tobacco." U.S. Department of Agriculture, Foreign Agricultural Service. URL <http://www.fas.usda.gov/scriptsw/bico/bico.idc?doc=624>.

USDA, FAS. (1998e). "U.S. Tobacco Leaf and Products Trade: Calendar Years 1990 through 1997." U.S. Department of Agriculture, Foreign Agricultural Service. URL <http://www.fas.usda.gov/tobacco/circular/1998/9803/specials.htm>.

USDA, FAS. (1998f). "U.S. Tobacco Trade Data." U.S. Department of Agriculture, Foreign Agricultural Service. URL <http://www.fas.usda.gov/tobacco/circular/1998/9803/table02.html>.

USDA, NASS. (1997). *Agricultural Statistics 1995-96.* Washington, DC: U.S. Department of Agriculture. National Agricultural Statistical Service. U.S. Government Printing Office.

USDA, NASS. (1998). "United States Crop Rankings—1996 Production Year." U.S. Department of Agriculture. National Agricultural Statistical Service. URL <http://usda.mannlib.cornell.edu/data_sets/crops/9X180/97180/crp-us96.txt>.

Womach, Jasper. (1996). "The Tobacco Briefing Room." Congressional Research Service. URL <http://www.econ.ag.gov/briefing/tobacco/program.htm>, December 21.

Young, C.E. and P.C. Westcott. (1996). *The 1996 U.S. Farm Act Increases Market Orientation.* Washington, DC: Commercial Agriculture Division, Economic Research Service, U.S. Dept. of Agriculture, AIB-726, August.

Chapter 13

Beef

Rudy M. Nayga Jr.
Flynn Adcock
Parr Rosson

INTRODUCTION

Beef is an important protein source in the United States and in many other parts of the world. This chapter examines trends, opportunities, and challenges facing the U.S. beef sector in today's competitive global marketplace. A discussion of changing dynamics in world beef demand is followed by an overview of world meat and fish consumption and trade. Recent trends in world beef consumption, production, and trade are discussed, emphasizing the competitive position of U.S. beef.

THE CHANGING DYNAMICS OF WORLD BEEF DEMAND

The demand for beef is shaped by a myriad of forces. As with all products, beef demand is the quantity of beef consumers are *willing* and *able* to purchase at specific periods in the marketplace. A combination of demographic and economic factors, acting interdependently or independently, can alter consumer demand. These factors include demographic shifts, changes in consumer tastes and preferences, fluctuations in incomes and relative prices, and the development of new marketing techniques and technology (Purcell, 1989; Charlet and Henneberry, 1990; Kinsey, 1992).

Demographic trends in industrialized countries are among the principal factors influencing the development of the global meat trade. Today, the number of single-person households is rapidly increasing due to declining birth rates, a higher average age of marriage, rising divorce rates, and increasing longevity (International Labor Organization, 1991).

An increasing proportion of women, especially in the industrialized world and in urban centers of the developing world, are now engaged in paid employment away from home. As a result, they have less time for cooking. Many households in a variety of countries have both spouses working outside the home. For instance, over one-half of all American women over the age of fifteen are in the workforce, leaving them less time to prepare food (Putnam and Allshouse, 1991). Because more women are in the labor market, an increasing number of families now enjoy double incomes and greater purchasing power. This allows them to buy more value-added food items of quality.

Consumers around the globe are becoming increasingly conscious of the health and nutritional aspects of their food. Understandably, they would like to avoid sacrificing taste, texture, and color as they improve their diets. Also, they would like greater convenience in their meals—whether it be shorter preparation time, or longer storage without losing freshness. Consequently, the value added in the meat processing industry and the distribution sector has continuously increased in the last two decades (OECD, 1992). The way beef and beef products are marketed today bears little resemblance to the marketing activities of twenty years ago (Lister, 1988; McGuirk, Preston, and McCormick, 1990).

Great dynamism has been noticed recently in the promotion of meat and meat products worldwide. The United States has established promotional efforts for beef products and France has instituted a promotional program with resources supplied by private business. The Netherlands, Great Britain, and other nations have followed the same path, hoping to maintain, and if possible to attract greater, consumer preference for beef.

Meats have become raw materials for sophisticated industrial elaboration and transformation. Today, beef is fighting for its place in the global marketplace with the assistance of sophisticated modern promotional media. Strategies for marketing beef are beginning to take advantage of the health consciousness of consumers by providing nutritional information at the point of sale and by promoting leaner cuts of meat and low-fat recipes.

Also, food consumption globally is becoming more similar. As incomes rise in developing nations and in Eastern Europe, the demand for animal proteins and fruits and vegetables increases while the demand for starchy foods declines. At the same time, in the affluent Western countries food saturation has largely occurred, and concerns have turned increasingly to a reduction in the consumption of fat and calories—with more grains, fruits, and vegetables being consumed (Kinsey, 1992).

The current globalization of eating habits does not mean a narrowing of market opportunities. Rather, it means that a wider variety of foods will be sold and consumed in a wider variety of places. Advances in biotechnology and its application in food processing have made it possible to produce foodstuffs with no damage to flavor, color, and nutritional value even after multiple stages of processing. Moreover, improved packaging techniques have added considerable value to the products by increasing the shelf life of the products and by reducing the risk of damage during transport of the products.

Along with competition in scientific research for new products, innovations in packaging are also affecting the beef industry's growth. Processed meat manufacturers are increasing efforts to meet consumer demands for quickly prepared, convenient products that taste good and are nutritionally balanced and "healthy." These ongoing pursuits by industry are likely to result in more microwavable products, with less fat, calories, sodium, and cholesterol, containing more essential vitamins and minerals. Granted, nutrition and safety issues are likely to change and shift over time. Nonetheless, several current beef-related issues deserve close attention, including: (1) concern over chemical residues in beef and beef products; (2) negative views of meat processing techniques and practices; and (3) avoidance of fat, calories, and cholesterol in beef and beef products.

OVERVIEW OF WORLD MEAT AND FISH CONSUMPTION, 1985-1997

World meat and fish consumption increased by 35 percent from 1985 to 1995, the last year for which fish consumption data are available, reaching 275 million metric tons (mt) (see Table 13.1). This represents an average increase of just over 3 percent per year. Of this 1995 total, 109 mt, or 40 percent, consisted of fish, 75 mt (27 percent) was pork, 47 mt (17 percent) was beef, and 38.1 mt (14 percent) was poultry. On a global basis, beef is the third most important meat in terms of consumption. Of the four meats analyzed, only beef's share of total meat consumption has decreased since 1985. For that year, total meat consumption was 204 mt. Fish accounted for 38 percent of world total meat consumption, while 27 percent was pork, 22 percent beef, and 11 percent was poultry. While pork held steady, the 5 percent decline in beef's share over the 11 years was split about equally between fish and poultry.

TABLE 13.1. World Meat and Fish Consumption, 1985-1997 (Millions of Metric Tons)

Year	Fish	Pork	Beef	Poultry	Total
1985	77.5	54.9	44.0	22.7	204.3
1986	88.2	57.1	45.8	23.7	215.1
1987	85.2	58.9	46.5	27.0	223.2
1988	90.5	61.5	45.6	26.2	228.9
1989	92.3	62.9	46.5	28.5	235.8
1990	91.5	64.3	46.3	30.5	238.1
1991	93.2	66.3	47.6	32.1	245.1
1992	98.3	68.1	46.8	33.6	252.3
1993	101.9	69.7	45.7	34.2	257.3
1994	106.1	71.2	46.4	35.9	265.3
1995	108.5	75.2	46.9	38.1	274.8
1996	***	78.1	47.1	40.1	***
1997	***	80.0	47.4	42.5	***

Sources: USDA/FAS and FAO Stat Database, 1998.

Global consumption of meat, excluding fish, has continued to increase through 1997. Of the three major meats, beef, pork, and poultry, beef ranked second to pork in total consumption with 47.4 mt, compared to 80.0 mt for pork. Poultry has continued to make significant consumption gains when compared to beef. Poultry consumption, over 21 mt less than beef use in 1985, was only 5 mt less than beef consumption in 1997.

A growing share of total world meat consumption can be attributed to trade. While meat consumption has increased 35 percent from 1985 to 1995, world meat exports increased nearly 50 percent, averaging just about 5 percent growth each year. Further, meat exports as a percentage of production grew slightly, from 5 percent to 5.6 percent over the same period. When fish is not considered, meat exports grew 52 percent from 1985 to 1997, peaking in 1995. Also, meat exports, excluding fish, increased from 5.8 percent of production in 1985 to 7.8 percent in 1997. Trade in meat becomes a more important factor influencing consumption in other countries as trade policies and demographics have changed.

TRENDS IN WORLD BEEF PRODUCTION, 1985-1997

World beef production reached its second highest level in 1997 at 48.6 mt (see Table 13.2). The United States with 11.7 mt (24 percent) ranked first, followed by the European Union at 7.8 mt (16 percent), Brazil at 6.0 mt (12 percent), China at 5.4 mt (11 percent), and the Newly Independent States (NIS) of the former Soviet Union at 3.6 mt (7 percent). Other major beef-producing countries include Argentina at 2.5 mt (5 percent), Australia at 1.9 mt (4 percent), and Mexico at 1.8 mt (4 percent).

Since 1985, however, beef production has fluctuated only slightly, from a low of 45.4 mt to a high of 49.1 mt. Total production growth from 1985-1997 was just 3.2 mt (7 percent), or an average of less than 1 percent per year. The primary reason for lagging output is that the EU, Eastern Europe, and the NIS experienced large declines in production. For instance, the EU had a decrease of 1.2 mt in annual beef production (12 percent) from 1985 to 1997, while the NIS had a decrease of 3.7 mt (50 percent). Even Eastern

TABLE 13.2. Summary of World Beef Production, Consumption, and Trade, 1985-1997 (Millions of Metric Tons)

Year	Production	Imports	Exports	Consumption
1985	45.4	2.6	3.6	44.0
1986	46.6	3.1	4.1	45.8
1987	47.9	2.9	3.9	46.5
1988	47.1	4.2	5.4	45.6
1989	47.6	4.5	5.7	46.5
1990	48.7	4.6	5.7	46.3
1991	49.1	4.5	5.6	47.6
1992	48.5	4.1	5.5	46.7
1993	46.9	3.8	5.4	45.7
1994	47.3	4.1	5.4	46.4
1995	47.7	4.0	5.2	46.9
1996	48.6	3.9	5.1	47.1
1997	48.6	4.3	5.3	47.4

Source: USDA/ERS, 1998.

Note: All values in carcass weight equivalents.

Europe experienced a decline of 1.4 mt (56 percent) in beef production. Together, these three regions accounted for a 6.3 mt reduction in beef output during this period.

Offsetting the decline in production by the EU, NIS, and Eastern Europe was increased beef output by Brazil and China. Brazil increased beef output 2.3 mt (63 percent), while China increased production by 4.9 mt (more than 1,000 percent) from 1985 to 1997. This combined increase in beef production by Brazil and China of 7.2 mt more than offset the decreases in production by the EU, NIS, and Eastern Europe. When added to the slight increases in production by the United States (.8 mt), Mexico (.4 mt), and Australia (.6 mt), as well as smaller increases in other countries, this led to an increase in world beef output of 2.7 mt.

TRENDS IN WORLD BEEF CONSUMPTION, 1985-1997

Global beef consumption closely tracks world beef production. Since beef is a perishable product that is not kept in storage for long periods of time, stocks typically remain at a low level. In 1997, total world beef consumption was 47.4 mt, compared to 48.6 mt produced. While consumption and production volumes are similar, reasons for any difference may include spoilage, stock accumulations, and discrepancies in data collection. Consumption of beef in 1997 was led by the United States at 11.8 mt (25 percent of world total), followed by the EU at 7.0 mt (15 percent), Brazil at 6.0 mt (12 percent), China at 5.3 mt (11 percent), and the NIS at 4.3 mt (9 percent). Other major beef consumers include Argentina at 2.1 mt (4.4 percent), Mexico at 1.9 mt (4.0 percent), and Canada at 1.0 mt (2.1 percent).

Analysis of these data reveals that some countries had major changes in beef consumption while others did not. The largest volume and percentage increase in consumption came in China, where the 5.3 mt consumed in 1997 represented an increase of over 1,000 percent above the 0.45 mt consumed there in 1985. This was an average increase of about 90 percent per year. Other big gainers in consumption included Japan, whose total increase for the period was 87 percent (7.25 percent per year); Brazil, which had a total period increase of 84 percent (7.0 percent per year); and Mexico with a 45 percent increase (3.75 percent per year).

In regions where beef consumption decreased, the change was not as dramatic. All of the largest declines occurred in Europe, as consumption decreased by about 36 percent from 1985 to 1997. The biggest percentage drop in beef consumption was in Eastern Europe, where it decreased by 50 percent, from 1.9 mt to 0.96 mt. The largest volume decrease occurred in the NIS, which had a decrease of 3.4 mt (44.5 percent), to 4.3 mt in 1997 as compared to their 1985 level of 7.7 mt. The EU decreased annual consumption of beef by 1.0 mt, or 13.4 percent for the same period.

Several other countries had either fairly constant annual beef consumption, or slight increases or decreases. For instance, the United States maintained a virtually constant annual consumption of beef over the period, dropping by only 50,000 metric tons (.4 percent) over the

12-year period, as consumption was 11.82 mt in 1985 and 11.77 mt in 1987. Beef consumption ranged from a high of 12.0 mt in 1986 to a low of 11.0 mt in 1993. The entire range of changes in consumption was only 9 percent, indicating only slight fluctuations in U.S. beef consumption patterns.

Per Capita Consumption of Beef

Per capita beef consumption is an important indicator of consumer demand. While the United States is the largest consumer of beef in total, it is not the largest per capita consumer. Uruguay, consuming a total of just 200,000 metric tons (t) on a carcass weight equivalent, or 140,000 t on a retail weight equivalent,* was actually the leading per capita consumer of beef in 1997, with 43.5 kilograms per person (95.8 lb/person). Argentina was second overall at 41.1 kg per person (90.5 lb/person) (see Table 13.3). Following Argentina was the United States, which consumed 30.8 kg/person (67.9 lb/person), and Australia at 28.5 kg/person (62.9 lb/person), with Brazil and Canada close behind at 25.1 kg/person (55.4 lb/person) and 23.1 kg/person (50.9 lb/person), respectively. Of the major

TABLE 13.3. Per Capita Beef Consumption, 1985-1997, Retail Weight (Pounds per Person)

Country	1985	1990	1997
United States	76.1	67.9	67.9
Mexico	26.4	32.6	30.0
Canada	72.5	55.5	50.9
European Union	34.8	32.6	28.7
Newly Independent States	42.6	59.6	30.7
Argentina	126.0	104.1	90.5
Brazil	36.4	51.4	55.4
Japan	10.0	13.7	18.4
China	0.7	1.5	6.7
Australia	62.5	59.0	62.9

Sources: USDA/ERS, 1998; CIA, 1998; FAO Statistical Database, 1998.

Note: To obtain carcass weight equivalents, divide by 0.7.

---

*Per capita beef consumption will be discussed on a retail weight equivalent basis for a more realistic comparison.

world consumers of beef, China had an average per capita consumption of just 3.0 kg/person (6.7 lb/person) in 1997.

If we examine how per capita consumption changed over the period, results are similar to total beef consumption. The largest volume increase in annual per capita beef consumption occurred in Brazil, with an 8.6 kg/person (19.0 lb/person) rise from 1985 to 1997. This represents an increase of about 52 percent during the period.

The largest percentage gain in per capita beef consumption occurred in China, whose increase of 2.7 kg/person (6.7 lb/person) translates to a 903 percent increase. The largest volume decrease occurred in Argentina at 16.0 kg/person (35.5 lb/person), a 28 percent decline. This is the same per capita percentage decrease that occurred in the NIS. The largest percentage decrease was slightly higher, at 30 percent in Canada, as a result of a volume decrease from 32.7 kg/person (72.5 lb/person) to 23.1 kg/person (50.9 lb/person).

In the United States, per capita consumption of beef fell from 34.5 kg/person (76.1 lbs/person) in 1985 to 30.8 kg/person (67.9 lb/person) in 1997, or 11 percent. However, the vast majority of this decrease occurred from 1985 to 1990, at which time per capita consumption stabilized.

TRENDS IN WORLD BEEF TRADE, 1985-1997

Trade in beef and beef products has become increasingly important over the last 15 years. In 1985, for example, about 8 percent of world beef production, or 3.6 mt out of 45.4 mt, was exported on a global basis. In 1997, almost 11 percent of production was exported, or 5.3 mt out of 48.6 mt produced. Highlighting the change toward more trade in beef is the fact that world exports increased almost 46 percent from 1985 to 1997. While beef exports remain high compared to 1985 levels, they actually peaked in 1989 at 5.7 mt.

Australia, the United States, and the EU dominate world beef trade. The single biggest exporter in 1997 was Australia, with 21 percent of the world market on a volume basis (see Figure 13.1). Australian beef exports of 1.1 mt account for about 60 percent of their production. The United States, as the world's largest beef producer, was second in exports with slightly less than 1 mt, capturing 18 percent of the world

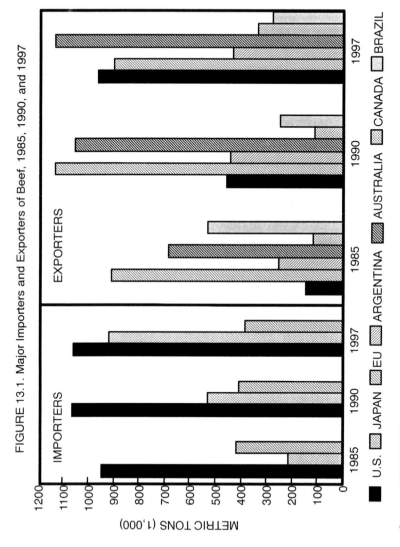

FIGURE 13.1. Major Importers and Exporters of Beef, 1985, 1990, and 1997

Source: USDA/ERS, 1998.

market and representing about 8 percent of U.S. beef production. The EU was the third leading exporter, exporting 946,000 t of beef, accounting for 17 percent of total world exports. The EU export volumes do not consider trade between EU nations. Other major exporters of beef include Argentina, Canada, and Brazil, who in 1997 exported 430,000 t; 340,000 t; and 275,000 t, respectively.

World imports of beef were 4.3 mt in 1997, with 56 percent of all imports accounted for by the Japan, the United States, and the EU. World beef imports were up 64 percent over 1985 levels, and peaked in 1990. Further, in 1985, approximately 5.9 percent of beef consumed was imported, while in 1997 that proportion rose to 9.0 percent. The rising importance of trade to beef production and consumption is emphasized by growing reliance on imported beef as a source of high-quality product.

The world's leading beef importer is the United States, which imported 1.1 mt, representing 25 percent of world beef imports and about 9 percent of domestic consumption. Following closely behind was Japan, which is highly dependent on foreign supplies of beef. In 1997, for example, Japan imported 924,000 t of beef, accounting for 64 percent of domestic consumption. Japan's share of the world market for imported beef was about 22 percent. In third place was the EU, importing about 350,000 t in 1997 and accounting for 9 percent of total world beef imports in that year.

The United States and EU, both major exporters and importers of beef, trade different types of beef based on consumer preferences, product quality, and availability. For instance, the United States imports ground beef from Australia and New Zealand, and in turn exports high-quality beef cuts to Japan and Korea. Edible offals and lower value beef cuts are exported to Mexico. Due to changes associated with the Uruguay Round of GATT, the United States imports an increasing volume of fresh/frozen higher quality beef cuts from Argentina and Uruguay.

U.S. COMPETITIVENESS IN WORLD BEEF MARKETS

The United States is a major beef exporter, with competition coming from the EU, Australia, Argentina, Brazil, Canada, and New

Zealand. Market overviews of Japan, Mexico, Korea, and Canada provide an indication of the competitive position of U.S. beef relative to these other suppliers. U.S. beef has remained competitive in the major markets of the world due to the ability of U.S. suppliers to deliver a safe, reasonably priced, quality product.

The Japanese Market for Beef

Japan is by far the largest market for U.S. beef exports, purchasing $1.4 billion worth in 1996. This represents $500 million more than the combined value of the remaining foreign markets for U.S. beef. The primary competitor for U.S. beef in Japan is Australia. The United States exported 308,000 t of beef to Japan in 1996 compared to 284,000 t from Australia; New Zealand was third at 27,000 t. The United States captured almost 50 percent of the Japanese beef import market in 1996, Australia accounted for 45 percent, and New Zealand held 4 percent (Agricultural Minister Counselor, American Embassy, Tokyo, 1997).

Upon examining retail prices paid in Japan for similar cuts of beef, U.S. beef is consistently higher priced than Australian beef, but usually well below retail prices for Japanese beef. For instance, U.S. beef sirloin sold for about $13.73 per pound in Japan during March 1998 compared to Australian sirloin at $12.18 per pound (see Figure 13.2). Japanese beef sirloin, however, was $23.71 per pound. While sirloin prices have been falling in Japan, the trend is different when the dollar-yen exchange rate is considered. The price in yen (¥) has been fairly stable over the past several years, and has actually rebounded from the lowest price in 1994. For comparison, the price for Japanese sirloin in March 1998 was ¥2,973 per pound, ¥1,772 per pound for U.S. sirloin, and ¥1,572 per pound for Australian sirloin.

The same pattern of U.S. beef being in the middle of Japanese and Australian beef generally holds true for other cuts. The March 1998 price in Japan for Japanese brisket was $10.29 per pound (¥1,328 per pound), $9.57 per pound (¥1,235 per pound) for U.S. brisket, and $6.11 per pound (¥780 per pound) for Australian brisket. Again, this pattern held for chuck and round beef cuts as well.

Even though U.S. beef is typically higher priced than Australian beef in Japan, more U.S. beef is purchased. The reason for this may be

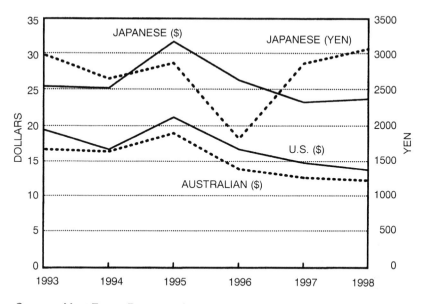

FIGURE 13.2. March Beef Prices in Japan by Source

Sources: Meat Export Research Center, 1995-1999; Federal Reserve Bank of St. Louis, 1992.

found in the type of beef produced by each country. The United States produces a more consistent and marbled grain-fed beef, which commands a price premium, while Australia relies predominantly on grass feeding, producing less marbling and less consistency. Therefore, U.S. beef commands a higher price than Australian beef in Japan due to a preference by many Japanese for a more marbled product.

Mexican, Korean, and Canadian Beef Markets

The three next largest markets for U.S. beef are Mexico, Canada, and Korea, in volume order, with Canada slightly surpassing Mexico in value. In 1996, Canada imported 90,000 t (198 million lb) of beef from the United States, 47,000 t (104 million lb) from New Zealand, and 27,000 t (60 million lb) from Australia. Imports from the United States accounted for 55 percent of the beef import market in Canada, while 29 percent came from New Zealand and 16 percent from

Australia. Although retail price data for imported beef in Canada were not available, the average price U.S. exporters received was $1.50 per pound (free alongside ship f.a.s. value). This does not reflect the price paid for retail cuts in Canada.

The United States enjoys at least two advantages over Australia and New Zealand in the Canadian market. The first is geography—the United States is obviously much closer to Canada, and therefore should be able to ship beef at a lower transportation cost and arrive somewhat fresher. Second, the North American Free Trade Agreement (NAFTA) has removed quota and tariff barriers to United States (and Mexican) beef entering Canada, while Australia and New Zealand must contend with both a quota and a tariff for quantities over a specified quota amount (tariff rate quota—TRQ) (Agricultural Minister Counselor, American Counselor, Ottawa, 1997).

The United States dominates the Mexican market for imported beef as well, although Canada and Australia provide some competition. In 1996, Mexico imported 72,000 t (158 million lb) of beef from the United States valued at $175 million, or an average of $1.10 per pound. In contrast, Australia exported 560 t (1.22 million lb) of beef to Mexico valued at only $605,000, for an average of 49¢ per pound, indicating low-value product, such as edible offals. On the other hand, Mexico imported 470 t (1.03 million lb) of beef from Canada valued at $1.4 million, or $1.34 per pound (Agricultural Minister Counselor, American Embassy, Mexico City, 1997). Thus, the United States maintains a market share of 98 percent of the Mexican beef import market. Although Canada enjoys the same access to the Mexican market that the United States does under NAFTA, Australia does not. Further, as Mexico concludes agreements with the EU and South American countries, competition for the Mexican market will intensify.

In Korea, the United States faces competition from Australia and New Zealand, but the United States ships a much higher value product than these other two nations. In 1996, the United States exported 70,500 t (155 million lb) of beef to Korea valued at $306 million, or $1.97 per pound. That same year, Australia exported 46,000 t (101 million lb) valued at $122 million ($1.20 per pound), and New Zealand exported 20,000 t (45 million lb) valued at $52 million ($1.16 per pound). On a market-share basis, the United States captured about 50 percent of the Korean beef import market in 1996, compared to 33 percent for Australia, and 14 percent for New Zealand (Agricultural Minister Counselor, American Embassy, Seoul, 1997).

Substantial government involvement in Korean agriculture is designed to protect domestic producers. Thus, the government purchases imported beef and maintains a TRQ on beef. The overall beef import quota in 1998 is 187,000 t, with an over-quota tariff of 42.4 percent. As a result of the Uruguay Round of GATT, this quota will increase until it is eventually eliminated in 2001, while the tariff will decrease by 0.4 percent per year before stabilizing at 40 percent in 2004. Further, Korea, along with the rest of the Pacific Rim nations of Asia, suffered significant economic turmoil during the Asian economic crisis, which began in 1997. It is uncertain how long and how deeply the economic downturn will affect Korean demand for imported products and how much the crisis will reduce U.S. beef exports to all Asian markets.

U.S. beef products are competitive in foreign markets for several reasons. First, U.S. beef is produced and marketed in one of the most efficient systems in the world. Low-cost feed grains, along with an abundant supply of cattle, lead to competitively priced meat products at the packing plant door. An efficient transportation system, characterized by low-cost freight rates and electronic documentation procedures, has resulted in the low-cost delivery of U.S. beef to virtually any port in the world. Second, emphasis on product quality, food safety, and production flexibility has increased the ability of U.S. suppliers to provide a well-marbled carcass to markets where marbling is desired and to provide a lean, less marbled meat to other markets. Implementation of Hazard Analysis and Critical Control Points (HACCP) standards at U.S. packing plants has resulted in an increased assurance to foreign buyers that U.S. beef is safe and wholesome, despite isolated outbreaks of *E. coli* in some products. Finally, a low-value exchange rate of the U.S. dollar has maintained competitiveness for beef even though the prices of some competitors have declined. When the dollar gains strength, as it did in 1998, this competitive position may erode, costing U.S. market share, especially in price-sensitive, low-quality product market segments.

CONCLUSION

Changing global market conditions, such as demographics, lifestyles, health concerns, and economic conditions, directly affect the U.S. beef industry. Technological advances in packaging and product preparation, new product development, and promotion also are important factors influencing beef consumption and its competitiveness relative to other meats. The growing internationalization of food processing will accelerate product innovation, increase competition, and force U.S. firms to adjust to the changing market environment. A highly consumer-focused marketing orientation will improve the competitive position of U.S. beef at home and in world markets.

The beef industry is compelled to cross the threshold of contemporary food marketing and move away from the *selling of beef as a commodity* in favor of the *marketing of meat products* targeted to the wants and needs of specific consumer segments (National Research Council, 1988; Graf and Saguy, 1991). The trend toward the globalization of food consumption patterns will not result in a narrowing of consumer tastes, but rather an increased diversity in consumption patterns within individual countries. Therefore, an enhanced knowledge of consumer tastes, preferences, and habits, which all may vary among countries and cultures, will be crucial to successful beef product marketing (Nayga and Waggoner, 1994).

REFERENCES

Agricultural Minister Counselor, American Embassy, Mexico City. (1997). "Livestock Annual Report." AGR Number MX7074, USDA/FAS.

Agricultural Minister Counselor, American Embassy, Ottawa. (1997). "Livestock Annual Report." AGR Number CA7042, USDA/FAS.

Agricultural Minister Counselor, American Embassy, Seoul. (1997). "Livestock Annual Report." AGR Number KS7045, USDA/FAS.

Agricultural Minister Counselor, American Embassy, Tokyo. (1997). "Livestock Annual Report." AGR Number JA7029, USDA/FAS.

Charlet, B. and Henneberry, S.R. (1990). *A Profile of Food Consumption Trends and Changing Market Institutions.* Bulletin No. 511. Cooperative Extension Service, Oklahoma State University, Stillwater, Oklahoma.

CIA. (1998). *CIA World Factbook.* Central Intelligence Agency. Internet, <http://www.odci.gov/cia/>.

Federal Reserve Bank of St. Louis. (1992). Exchange rate statistics. Internet <www.stls.frb.org/fred/data/exchange.html>.
FAO. (1998). *FAO Statistical Database.* Food and Agriculture Organization. Internet, <http://apps.fao.org/>.
Graf, E. and I.S. Saguy (Eds.) (1991). *Food Product Development: From Concept to the Marketplace.* New York: Van Nostrand Reinhold.
International Labor Organization. (1991). *Food and Drink Industries Committee General Report, Second Session.* Report 1, Geneva.
Kinsey, J.D. (1992). "Seven Trends Driving U.S. Food Demands." *CHOICES, The Magazine of Food, Farm, and Resource Issues* (Third Quarter), 26-28.
Lister, D. (1988). "Diets in Transition: Human Health and Animal Production. Consequences for Agriculture and Some Possible New Approaches." *Proceedings of the Nutrition Society* 47(3), 331-342.
McGuirk, A.M., W.P. Preston, and A. McCormick (1990). "Toward the Development of Marketing Strategies for Food Safety Attributes." *Agribusiness* 6(4), 297-308.
Meat Export Research Center (1995-1999). *U.S. Meat Export Analysis and Trade News.* Ames, IA: Iowa State University.
National Research Council. (1988). *Designing Foods: Animal Product Options in the Marketplace.* Washington, DC: National Academy Press.
Nayga Jr., R.M. and D.B. Waggoner. (1994). "Competing in the Global Marketplace: Issues, Trends and Challenges Facing New Zealand's Sheepmeat Industry." Agribusiness & Economics Research Unit Discussion Paper No. 137, Lincoln University, New Zealand.
OECD. (1992). *Agricultural Policies, Markets and Trade: Monitoring and Outlook 1992.* Paris: Organization for Economic Cooperation and Development.
Purcell, W.D. (1989). *Analysis of Demand for Beef, Pork, Lamb and Broilers: Implications for the Future.* Research Bull. 1-89. Research Institute on Livestock Pricing, Virginia Tech University, Blacksburg, Virginia.
Putnam, J.J. and J.E. Allshouse. (1991). *Food Consumption, Prices, and Expenditures, 1968-89.* Statistical Bulletin No. 825. Economic Research Service, U.S. Department of Agriculture, Washington, D.C.
USDA/ERS. (1998). *P, S, & D View,* database. Economic Research Service, U.S. Department of Agriculture. Internet, <www.econ.ag.gov/ProdSrvs/mostdp-sm.htm>.

Chapter 14

Poultry

Dale Colyer

Exports have become increasingly important to the U.S. poultry industry. In 1997 exports were 17 percent of total poultry production, while in 1980 and 1990 they were only 3.6 and 4.5 percent, respectively, of production. During the 1990s, exports increased absolutely and as percentages of quantities and values produced. Data for 1997 show broiler exports at 4.66 billion pounds and turkeys at 606 million pounds (Figure 14.1). Due to the economic crises in Russia and Asia, 1998 broiler exports increased only slightly (about 0.1 percent), but turkey exports declined by over 26 percent. Growth in the export market is fueled by large domestic supplies, a general easing of trade restrictions, attractive prices for U.S. products (especially for dark meat parts), development of foreign markets by individual companies and export promotion groups, and the entry of Russia and other former Soviet Union (FSU) nations as large importers of poultry products. In 1997, the major importing countries for broilers from the United States were the FSU, Hong Kong, Mexico, Japan, and Canada (USDA, 1998a). Mexico, Hong Kong, Russia, and South Korea were the leading importers of turkey meat in 1997, with 196.7, 125.6, 80.6, and 24.7 million pounds, respectively. These data indicate that substantial benefits accrue to the U.S. industry from an expanding world poultry market.

The United States is the world's leading producer of poultry meat (chicken and turkey) as well as the largest exporter (see Table 14.1). Major exporters of poultry products (primarily broilers, i.e., young chicken meat) include three highly industrialized countries (United

FIGURE 14.1. U.S. Poultry Exports

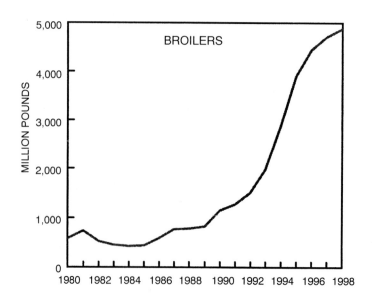

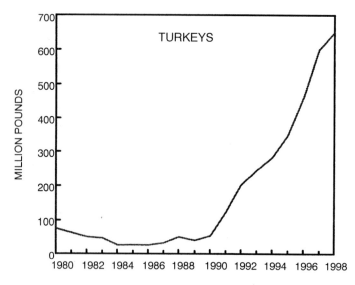

Sources: USDA, 1995; 1998a.

TABLE 14.1. World and U.S. Production and Exports of Poultry Meat, 1996

Product	Production (1,000 t)			Exports (1,000 t)		
	World	U.S.	U.S.%	World	U.S.	U.S.%
Chicken Meat	49,676.6	12,167.0	24.3	5,361.2	2,125.9	39.7
Turkey Meat	4,652.2	2,479.0	59.1	683.5	199.0	25.1
Duck Meat	2,591.2	46.0	1.8	115.6	4.5	3.9
Goose Meat	1,704.8	—	0.0	42.7	—	0.0
Other Poultry	16.1	—	0.0	—	—	—
Totals	58,640.9	14,692.0	25.4	6,203.0	2,325.4	37.2

Source: FAO, 1999.

States, France, Netherlands) and three developing nations (Brazil, Thailand, China). In recent years, the United States has accounted for about 36 percent of the volume and 22 percent of the value of world poultry meat exports; in 1980 volume and value were 22.2 percent and 17.5 percent, respectively. This difference between relative volumes and values is due to the growing dominance of cheaper dark meat parts in U.S. exports. France is the largest exporter of turkey meat, although a substantial proportion of French trade is due to intracountry trade among the members of the European Union (EU), which is estimated as two-thirds of total French exports in 1995.

POULTRY TRADE RESEARCH

Historically, relatively little research has been done on the economics of international trade in poultry products, but in recent years research activities have increased along with the growth in trade. Bishop et al. (1990), in a USDA publication, evaluated international markets for both imports and exports of poultry products as they existed during the late 1980s. Their analyses were largely descriptive and provided a good source of information for understanding the factors and conditions affecting international activities of the major importers and exporters of poultry products. Henry and Rothwell (1995) conducted a thorough analysis of international poultry production and trade during the early 1990s for the International Finance Corporation of the World Bank. Their inclusion of detailed cost data for the major poultry-producing countries is an especially important contribution to the understanding of competitiveness in the industry.

Quantitative analytical approaches including econometric models, domestic resource cost calculations, and shift-share analysis also have been used to analyze international trade in poultry products. Kapombe (1997) developed a U.S. broiler export model using the structural time series (STS) approach (see also Kapombe and Colyer, 1997). In addition to a very strong positive trend component, U.S. broiler exports were found to be negatively influenced by U.S. broiler prices and the value of the dollar but, unexpectedly, positively associated with Brazilian broiler exports. This latter finding is probably associated with rapid increases in international demand in recent years that have enabled both countries to increase their exports without affecting those of the other country; the two countries also tend to export to different markets, but compete in Japan and Hong Kong. Imports by four of the major purchasers of U.S. broilers (Canada, Hong Kong, Japan, and Mexico) were found to be influenced by exchange rates, quotas and other import policies, domestic production, broiler prices, and in the case of Canada by beef prices. Russia was not included in the Kapombe study due to the short time that the country was active in international poultry trade. However, an econometric study by Ames (1998) found a positive trend in broiler leg quarter prices, which were positively affected by the quantity of exports to Russia, but negatively affected by the 1996 embargo and the 30 percent tariff imposed on poultry imports.

Michel, Gempesaw, and Biery (1998) studied export competitiveness by use of domestic resource cost (DRC) calculations. In this procedure, costs of production are evaluated in terms of international costs and prices for each exporter in relation to the countries to which they export. For most countries where it exported, Brazil had the lowest DRCs; an exception was for exports to Japan by Thailand. Thailand tended to have lower DRCs than the United States (except for exports to Hong Kong), while the United States had lower DRCs than France and the Netherlands.

Varvello Vicente (1997) analyzed changes in poultry exports during the period 1985-1995 using shift-share analysis, a procedure that shows changes in shares of poultry exports over time, although shift-share analysis cannot indicate why the changes occurred. Although Varvello's results indicate that France was the largest gainer in turkey exports, that country does not have a cost advantage; much of its exports were within the EU and most exports outside of the area were heavily subsidized. The United States was the primary gainer in terms of shares of exports for chicken meat, primarily due to the large increases of dark meat exports to Russia and other FSU countries.

COMPETITIVE POSITION OF THE U.S. POULTRY INDUSTRY

The relative costs of production for U.S. poultry and its major competitors in the international arena affect the ability of the United States to export the product, although competitiveness is determined by a more complex set of issues than just costs. Tariffs and nontariff barriers including health and sanitary regulations, as well as consumer preferences, market shares, time of entry into markets, brands, diverse products and prices, trademarks and reputations, and related factors have important implications for the competitive position of a country's agricultural products, as well as for those of individual exporters (Bishop et al., 1990; Henry and Rothwell 1995). Improved technology is a key to competitiveness, and Vocke (1991) has shown that poultry technologies are relatively easily transferred, even to most of the less developed countries—this can be done as a whole integrated system. Although a modern poultry industry requires substantial investments to implement the technology, development of the necessary capabilities can be done with training programs and through the use of skilled field workers.

Cost Comparisons

Comparative data on costs of production and processing for broilers in the six major exporting countries are shown in Figure 14.2.

FIGURE 14.2. Broiler Production and Processing Costs

[Bar chart showing Cents/Kilogram for Brazil (~85, production ~70), Thailand (~93, production ~82), U.S. (~105, production ~77), China (~106, production ~95), France (~158, production ~123), Netherlands (~162, production ~130). Legend: Processing (white), Production (black).]

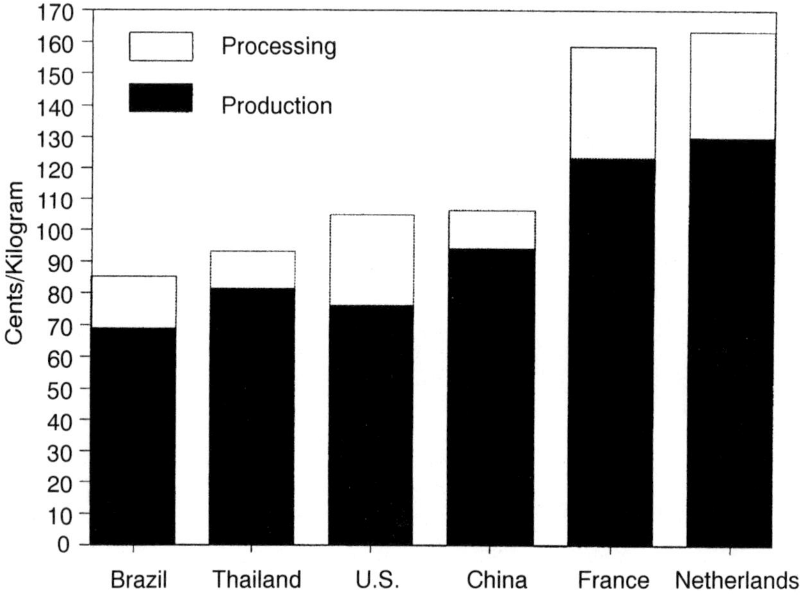

Source: Based on data in Henry and Rothwell, 1995.

Brazil is the lowest cost producer while the Netherlands has the highest costs. Thailand has slightly higher costs than Brazil and China has costs about equal to those of the United States, while those in France are nearly as high as in the Netherlands. The United States has lower production costs than either Thailand or China, but substantially higher processing costs. Brazil has both lower production and processing costs than the United States, although its processing costs are higher than those of Thailand and China. Both production and processing costs are substantially higher in France and the Netherlands than in the other four major exporting nations; these two countries are able to export substantial amounts of poultry to Germany and other countries within the European Union where imports are restricted for nonmembers; they also subsidize exports to nonmember countries.

Table 14.2 displays a breakdown for 1993-1994 broiler production costs for the six countries, grouped by the three industrialized and three developing economies. Feed costs make up the largest share of production costs, varying from 56.1 percent in the Netherlands to 72.2 percent in Thailand, while they are 64.5 percent of production costs in the United States and 68.8 percent in Brazil. Actual feed costs in terms of cents per kilogram of broilers on a ready-to-cook (RTC) basis are nearly the same in Brazil and the United States, 47.7 and 47.4 cents, respectively. Feed costs are much higher in France and the Netherlands (75.4 and 72.9 cents per kilogram, respectively), although they are smaller as a percentage of total production costs since processing and other costs are also higher in those countries. A major contributor to the higher costs in the European countries is the much higher payments to producers, 25.4 and 34.1 cents per kilogram for France and the Netherlands, respectively, compared to only 7.0 to 8.3 cents per kilogram in the three developing countries and 11.4 cents per kilogram in the United States.

Processing costs per kilogram of broiler meat are much lower in the three developing than in the three industrialized nations, ranging

TABLE 14.2. Comparative Ready-to-Cook 1993-1994 Costs for Broiler Production (Cents per Kilogram)

Cost Item	Industrialized Countries			Developing Countries		
	United States	France	Netherlands	Brazil	Thailand	China
Chicks	10.8	18.9	21.1	12.4	12.9	13.2
Feed	47.7	75.4	72.9	47.4	68.0	56.6
Grower Payment	11.4	25.4	34.1	7.0	8.3	7.7
Other Costs*	4.1	4.0	1.8	2.1	5.0	3.9
Total Production	74.0	123.6	129.9	68.9	94.2	81.4
Processing Costs	31.1	35.0	33.5	16.5	12.4	11.8
Total Costs	105.1	158.6	163.4	85.4	106.6	93.2
Feed as % of Prod.	64.5	61.0	56.1	68.8	72.2	70.2

Source: Developed from data in Henry and Rothwell, 1995.

* Veterinary, medication, transportation, administrative fees, etc.

from 11.8 to 16.5 cents per kilogram in the former compared with a range of 28.9 to 35.0 cents per kilogram in the latter. A major contributor to these large differences is lower labor costs in the developing countries, although other factors such as product mix also affect processing costs. Labor as a percentage of total processing costs for the countries included in the Henry and Rothwell (1995, pp. 37-38) study and for which data were available, varied from a low of 38 percent in Brazil to more than 66 percent in Thailand, compared with somewhat more than 50 percent for the United States. The high percentage for Thailand is due to additional processing in the form of hand cutting, deboning, and dicing. China also exports substantial amounts of further processed products that involve hand labor.

Domestic Production and Price Impacts

The United States is a relatively low-cost producer of poultry, although both Brazil and China have lower total costs. The United States, however, has an advantage based on its very large domestic industry, where white meat cuts are preferred to dark meat due primarily to health-related issues. As a consequence, breast meat is relatively expensive, while leg quarters tend to be inexpensive. In December 1997, for example, whole broiler wholesale prices in the Northeast were reported at $55 per hundredweight (cwt), boneless breast meat was $144/cwt, and breast with ribs in was $69/cwt, while leg quarters were only $28/cwt (USDA, 1998a, p. 22). Thus, in the United States there is, relatively, an excess supply of dark meat that can only be sold at low prices. This makes the prices of leg quarters very competitive vis-à-vis other countries where domestic markets are not so large and differentiated, and where preferences for light cuts are not so strong. Ames (1998), for example, found a very close association between Arkansas prices of leg quarters and poultry exports to Russia. While subsidies have not been an important factor for U.S. exports in recent years, easy credit terms were used to help finance exports to Russia.

Trade Barriers

The primary barriers to trade in poultry products are tariffs and sanitary/health regulations, although quotas, licensing, and other nontariff barriers also exist. Under the Uruguay Round of GATT, which created the World Trade Organization (WTO) in 1995, nontariff barriers are being eliminated through a process of tariffication, although health and sanitation requirements can be used to restrict imports where they are based on science—tariffs imposed in this process are to be reduced by an average of 36 percent over the six years of the agreement. This agreement also has provisions for market access that require signatories to allow imports equivalent to 3 percent (rising to 5 percent over six years) of their domestic consumption. This has resulted in the development of tariff rate quotas (TRQ), under which there are no or low tariffs for imports made within the quota but higher tariffs for those in excess of the quota—the conversion of nontariff barriers into tariffs often results in very high rates. In the EU, for example, the in-quota most favored nation (MFN) average tariff on poultry meats is 11 percent while the out-of-quota MFN rate is 43.6 percent, although under Europe Agreements with most Eastern European countries, the in-quota rate is 9 percent (USDA, 1997a, p. 7). As shown in Table 14.3, tariffs vary considerably between countries and by type of poultry and parts being imported; Hong Kong is not in the table since it is a relatively free market (USDA, 1997c). Turkeys, which are not produced in significant quantities in many countries, tend to have lower tariffs than broilers. Tariffs also vary with the type of product, whole birds compared to parts, light meat compared with dark, livers, feet, etc.

Nontariff barriers in the form of prohibitions, quotas, import licenses, health and sanitary regulations (legitimate or not), environmental requirements, and other conditions are still used despite the WTO requirement for tariffication of nontariff barriers (see Table 14.4). Health and sanitary requirements are permitted where there is a scientific basis for their utilization. Existence of poultry diseases (such as avian influenza or Newcastle disease) in flocks of the exporting country is often used as a reason to prohibit imports from that country. Similarly, salmonella or other organisms that can be spread by poultry meat are also frequently used to restrict trade; a number of countries have zero tolerances for salmonella, which effectively prohibits imports. In addition, some countries prohibit or restrict imports of mechanically deboned meat because of the existence of bone chips in that

TABLE 14.3. Tariffs on Imports of Poultry Products, 1997

Country	Product	Tariff* (Percent)
Argentina	Poultry Meat	10
Bulgaria	Whole Chickens	45
	Leg Quarters	20
Canada	Chicken	177.5
	Turkey	2.6-273
China	Poultry Meat	20
Egypt	Chicken	80
European Union	Chicken	177.5
	Turkey	3.7-273
Guatemala	Poultry Meat	15-45
Honduras	Poultry Meat	70
Hungary	Chicken/Turkey	35
India	Poultry Meat	10-40
Indonesia	Poultry Meat	15-20
Japan	Bone-in-Leg	9.5
	Whole Chicken	14
South Korea	Chicken	30.5
Mexico	All Poultry	10-260
Philippines	Poultry	30-80
Poland	Poultry Meat	30-60
Romania	Poultry Meat	30-60
Russia	Chicken	30
	Turkey	15
South Africa	Chicken	2.2 Rand/kg.
	Turkey	27
Taiwan	Chicken	40
	Turkey	15
Venezuela	Poultry Meat	20-40
Ukraine	Parts	300 ECU/t

Source: USDA, 1997c.

*Where ranges are reported, the lower rate is generally within TRQs, while the higher rate is outside the TRQs; some rates may also vary with different products.

product. Since it is difficult to separate genuine concerns about health and safety from restrictionist policies, these types of barriers are difficult to resolve and often become issues for the trade dispute settlement process.

TABLE 14.4. Nontariff Barriers to U.S. Poultry Exports

Country	Barrier
Australia	Ban due to avian influenza; sanitary requirements for processed products
Brazil	Ban due to reciprocal U.S. policy
Chile	Zero tolerance for salmonella
China	Ban on most imports for retail sale
Colombia	Import licenses (generally denied)
Czech Republic	Zero tolerance for salmonella
Dominican Republic	Unscientific sanitary measures
Ecuador	Bans due to campylobacter and avian influenza
Egypt	Ban of parts (studying halal (Islamic) certification)
El Salvador	Zero tolerance for salmonella, etc.
European Union	Ban on U.S. poultry; veterinary certification
Guatemala	Ban due to salmonella
Honduras	Restrictive certification requirements
India	Import licenses used to restrict imports
Indonesia	Requires companies to be registered
S. Korea	Uses quota auctions
Poland	Import quotas, veterinary certification, bilingual certificates
Taiwan	Ban due to avian influenza (exceptions)
Venezuela	Ban due to avian influenza
Ukraine	Bans mechanical deboned, offal, franks

Source: USDA, 1997c.

Trade Agreements

The principle trade agreements affecting U.S. poultry exports are the North American Free Trade Agreement (NAFTA) and the preceding Canadian-U.S. agreement (CUSTA). For poultry, NAFTA primarily affects trade between the United States and Mexico; the United States eliminated its tariffs on imports, but since Mexico is considered a Newcastle-endemic region, only cooked and sealed poultry products can be exported to the United States (USDA, 1997b). Mexico eliminated its requirement for import licenses and its overall tariff of 10 percent, and established TRQs for poultry products, with out-of-quota tariffs initially ranging from 133 to 260 percent, although tariff-free imports of quantities larger than the TRQs are frequently permitted. When CUSTA was incorporated into NAFTA, Canada increased its allowable broiler imports to 7.5 percent and turkey to 3.5 percent of production, but Canada also grants supplemental quotas to meet unanticipated needs. However, as a result of the Uruguay Round of GATT, the Canadians converted their MFN quotas to TRQs with high over-quota tariffs. (The United States protested this conversion, but the NAFTA dispute settlement process found the action to be consistent with NAFTA.) Overall, the impacts of NAFTA on U.S. poultry exports have been mixed, with increased exports to Mexico and initially slightly lower exports to Canada, although in 1997 exports to Canada rose sharply to new highs in response to its strong economy.

Although NAFTA increases the competitiveness of U.S. poultry by removing barriers to trade between the United States, Canada, and Mexico, trade agreements between other groups of countries (trade blocks) can have negative impacts on the ability of the U.S. poultry industry to compete. The Common Market for the Southern Cone (MERCOSUR) countries of South America, for example, enhances the ability of Brazil to export to Argentina and other member countries, but makes it more difficult for the United States to export to those countries—just as NAFTA makes it more difficult for Brazil to export to Mexico. The European Union, which has facilitated trade between member countries by eliminating internal barriers while retaining external barriers, makes it virtually impossible for the United States to export significant quantities of poultry products to EU countries. Free trade within the union enables France, the Netherlands, the U.K., etc., which tend to be high-cost producers, to export poultry products to Germany and other poultry-deficit member countries. In addition, the EU has entered into a series of agreements, known as the Europe Agreements, with several central and eastern European countries, which facilitate trade (USDA, 1997a). These agreements result in most of the EU's TRQ poultry requirements being met by agreement countries, thus largely excluding the United States from exporting to the EU.

Other Factors

A number of other factors directly or indirectly affect the ability of the U.S. poultry industry to compete in the international trade arena. These include climate, market development activities, feed supplies, and a variety of governmental policies including agricultural, environmental, and macroeconomic policies (Henry and Rothwell, 1995, pp. 41-43). Some enhance and others detract from the competitive position of U.S. poultry exporters.

Climate has implications for competitiveness through its effects on the efficiency of poultry production or production facilities and, hence, on costs. A temperature range of 60° to 70°F (16° to 21°C) tends to be most comfortable and efficient in converting feed for most commercial species of poultry (Ensminger, 1992; Say, 1987). Broilers, for example, are better suited to temperate locations within a range of 30° to 35° latitude than to either colder or warmer areas (Lasley et al., 1988; Henry and Rothwell, 1995). Some two-thirds of U.S. production is located in states in that range (Alabama, Arkansas, Georgia, Mississippi, North Carolina, and Texas). Similarly, much of Brazil's commercial poultry production is in that country's southern states, at about the same latitude as much of the U.S. production, except that it is in the Southern Hemisphere. Lasley et al. (1988, p. 18) report that the 1980s feed conversion rate in the southern region of the United States was 1.99 pounds of feed per pound of production compared to a rate of 2.04 in the northeast and 2.09 in the west. This, according to Henry and Rothwell (1995, p. 41), gives a cost advantage for the southern states of about 2.5 percent compared to the northeast. Similarly, the cost of housing is lower than in latitudes of less than 30°, which tend to have colder climates and therefore to require better insulated buildings and more heating. Construction costs in the 1980s were estimated at $4.65 per square foot in the northeastern region compared with $3.31 in the south (Lasley et al., 1988). Heating costs also tend to be higher further north, although this is offset to some degree by the need for more cooling during the summer months in the warmer regions.

The availability of large, reliable low-cost feed supplies, both feed grains and protein meal, is essential for efficient poultry production. Countries that can produce those products more cheaply and in larger quantities have an advantage compared to those that have high costs or that lack adequate land resources. It should be noted that many of the better areas for grain production are located relatively close to those with climatic factors that favor poultry production and that such a favorable location can contribute to lower feed costs due to the relatively low cost of transporting it to the poultry-producing areas. Corn and soybeans are the more common feed materials used for poultry, although other coarse grains such as sorghum (or even food grains to some degree) and other sources of protein can be substituted either wholly or partially for soybean meal. This contributes to the competitive advantages of countries such as the United States and Brazil, large producers of both corn and soybeans, compared to France and the Netherlands, which tend to be high-cost feed producers or, say, Japan, which has relatively little land to devote to feed grain production. A country such as Japan can economically import the grain to support a domestic poultry industry (Henry and Rothwell, 1995, p. 45), but other costs tend to be high, making meat imports more desirable.

Market position and marketing practices also contribute to the ability of a country's poultry industry to compete internationally. Trade can be affected by many factors including historical circumstances, marketing practices, market promotion, brand name recognition, etc. (see Stewart, 1984, for a good general discussion of new trade theories). For example, when Wampler Industries acquired Rockingham they retained the Rockingham label for export purposes, but not for domestic market products, since Rockingham had been a leader in developing poultry products for export. Export promotion activities by poultry processors as well as by the federal government also contribute to trade in many agricultural products. In 1992, for example, the federal government spent nearly $250 million promoting agricultural exports, with some 4.3 percent of that being spent on poultry products (Ackerman, 1993). Comeau, Mittelhammer, and Wahl (1996) have shown that U.S. meat promotion activities in Japan increased the demand for U.S. beef, but not for pork and poultry. They indicate, however, that such activities might be essential for maintaining market share since other countries promote their products.

CONCLUSIONS

The United States is a relatively low-cost producer of chicken and turkey meat products with large, reliable supplies of relatively inexpensive feed, a large domestic poultry industry, a highly integrated and efficient production/processing industry, a notably productive set of research institutions, and demand patterns resulting in an excess supply of dark meat parts. These form a strong long-term basis for remaining competitive in the international market. France and the Netherlands are large exporters of poultry products, but their industries have been characterized by high costs and subsidized exports to countries outside of the EU (over half of their exports are intra-Union). Reforms under GATT and WTO are changing the dynamics of agricultural production in the EU, but these are not apt to affect poultry trade patterns significantly for several years.

Brazil has a growing poultry industry and is a major competitor to the United States due, especially, to low feed and labor costs, with plentiful supplies of essential inputs although its corn supply is somewhat variable. However, the Brazilian internal market is relatively small (but growing), it has had high interest rates and taxes, and its macroeconomic policies have caused inflation and instability. Thailand is a low-cost producer, but is a net importer of feed grains and has limited supplies of protein feeds. Its large supply of low-cost labor and its proximity to large Asian markets provides some distinct advantages. China has become an important exporter of poultry, but its imports are larger, and it has a very large population to feed; it also imports soybean meal.

Large growth in international markets for poultry products has enabled most major producers to expand production in recent years, due largely to the emergence of Russia and other FSU countries as major importers of poultry meats. Several of the FSU countries have the potential to produce more of their own poultry if they can reform their agricultural economies, a situation that does not appear to be occurring to any significant degree. Thus, the United States should be able to maintain a strong export position for both chicken and turkey meats for the foreseeable future.

REFERENCES

Ackerman, K.Z. (1993). "Export Promotion Programs Help U.S. Products Compete in World Markets." *Food Review* (May-August): 31-35.

Ames, G.C.W. (1998). "Non-Tariff Barriers and Political Solutions to Trade Disputes: A Case Study of U.S. Poultry Exports to Russia." Paper presented at the annual meeting, Southern Agricultural Economics Association, Little Rock, AR, February 1-4.

Bishop, R.V., L.A. Christensen, S. Mercier, and L. Witucki. (1990). *The World Poultry Market—Government Intervention and Multilateral Policy Reform.* Staff Report No. AGES 9019. Washington, DC: Economic Research Service, USDA.

Comeau, A., R.C. Mittelhammer, and T.I. Wahl. (1996). "The Effectiveness of Meat Advertising and Promotion in the Japanese Market." R.B. 96-24. Ithaca, NY: Department of Agricultural, Resource and Management Economics, Cornell University.

Ensminger, M.E. (1992). *Poultry Science.* Danville, IL: Interstate Printers and Publishers.

FAO. (1999). *FAOSTAT Database.* Food and Agriculture Organization, United Nations. <http://apps.fao.org/lim500/nph-wrap.pl?Trade.CropsLivestockProducts&Domain=SUA>.

Henry, R. and G. Rothwell. (1995). *The World Poultry Industry.* Washington, DC: International Finance Corporation, World Bank.

Kapombe, C.R. (1997). *Broiler Exports: A Structural Time Series Approach.* PhD dissertation, West Virginia University, Morgantown, WV.

Kapombe, C.R. and D. Colyer. (1997). "An Econometric Analysis of U.S. Broiler Exports." Paper presented at the XXII International Conference of Agricultural Economists, Sacramento, CA, August 10-16.

Lasley, F.A., H.B. Jones Jr., E.H. Easterling, and L.A. Christensen. (1988). *The U.S. Broiler Industry,* AER-591. Washington, DC: USDA, Economic Research Service.

Michel, K., C.M. Gempesaw II, and C.S. Biery. (1998). *An Analysis of International Poultry Trade and Competitiveness.* FREC 98-1 Food and Resource Economics, University of Delaware, Newark, DE.

Say, R.R. (1987). *Manual of Poultry Production in the Tropics.* Wallingfram, UK: C.A.B. Int'l.

Stewart, F. (1984). "Recent Theories of International Trade: Some Implications for the South." In H. Kierzkowski (Ed.), *Monopolistic Competition and International Trade.* Oxford: Clarendon Press, pp. 84-107.

USDA. (1995). *Poultry Yearbook 1994.* Statistical bulletin 916. U.S. Department of Agriculture, Economic Research Service, Washington, DC.

USDA (1997a). *Europe: International Agriculture and Trade Reports.* WRS-97-5, Economic Research Service, U.S. Department of Agriculture. Washington, DC, December.

USDA (1997b). *NAFTA: International Agriculture and Trade Reports.* WRS-97-2, Economic Research Service, U.S. Department of Agriculture. Washington, DC, September.

USDA (1997c). *Trade Barriers for Poultry and Eggs,* mimeo. Foreign Agricultural Service, U.S. Department of Agriculture. Washington, DC, December 8.

USDA (1998a). *Livestock, Dairy and Poultry Monthly,* Electronic Version. Economic Research Service, U.S. Department of Agriculture. Washington, DC, January.

USDA (1998b). "Poultry; Semiannual Report (Brazil)." AGR Number BR8601, Foreign Agricultural Service Online, U.S. Department of Agriculture. Washington, DC, February 1.

Varvello Vicente, M. (1997). "A Trade Change Analysis of Major Poultry Product Categories in the World Market." Master of Agribusiness Management Research Report, Mississippi State University, Mississippi State, MS.

Vocke, G. (1991). "Investments to Transfer Poultry Production to Developing Countries." *American Journal of Agricultural Economics* 73(3): 951-954.

Chapter 15

Pork

William A. Amponsah
Xiang Dong Qin

INTRODUCTION

The U.S. pork industry is mainly located in the Corn Belt region of the upper Midwest where feed grains are abundant. Indeed, for many years the states of Iowa, Illinois, Minnesota, Nebraska, Indiana, and Missouri were leading producers of pork, until the fast technological development in the pork industry in the 1980s and 1990s saw the emergence of North Carolina as the second largest pork-producing state. However, Iowa still remains the number one pork-producing state. Today, the United States is one of the world's leading pork-producing countries, and the second largest pork exporter, behind only Denmark.

International trade in pork has risen significantly in recent years. The major pork-exporting countries increased their exports at an annual rate of 4 percent from 1989 to 1997 as a result of bilateral and multilateral trade agreements, income growth, and technological innovations, especially in transport and shelf-life extension. Four countries, the United States, Canada, Denmark, and Poland, accounted for 60 percent of pork exported by major pork-exporting countries in 1998. In 1999, the United States accounted for 21 percent of global pork exports (USDA, 1999).

Excluding variety meats, the leading markets for U.S. hogs in 1997 were Japan, Canada, Mexico, Russia, Hong Kong, Korea, Italy, China, the Philippines, and the United Kingdom, in that order. Together, these top ten markets imported 93 percent of U.S. pork, a total of 302,347 metric tons (worth about $995 million) out of the total exports of 324,507 metric tons (worth nearly $1.05 billion) to all markets in 1997.

Lately, the global hog market has suffered from a decline in prices, which averaged between $32 and $33 per cwt in 1998, the lowest since 1972 when prices dipped to a low of $27 per cwt. Large supplies of poultry kept hog prices hovering from $30 to $35 per cwt in 1999 (USDA, 1999). Nevertheless, lower pork prices continue to boost exports. Export volume was up 30 percent from January to August 1998 compared to the same period in 1997. But most of the increase is due to attractive prices of lower quality cuts, such as picnics and trimmings, in low-income countries such as Russia and Mexico. Russia accounts for about 10 percent and Mexico for about 20 percent of U.S. pork exports. However, just as for Japan and Korea, exports to Russia fell following the financial crisis in 1999, although sales to Russia helped boost exports in 1999. U.S. exports to Mexico are also expected to be affected, if tariff rate quotas are exceeded (USDA, 1998). U.S. pork exports rose in 1997, due in part to the early-1997 outbreak of foot-and-mouth disease in Taiwan, which spread chaos in global pork markets. However, Taiwanese pork differs from the U.S. product, offering several characteristics (such as darker meat color, tougher texture, and sweeter flavor) that are preferred by Japanese consumers. Additionally, Japan rescinded its safeguard tariff (a WTO legal mechanism to protect domestic producers from excessive imports). Nevertheless, Japanese processors did not maintain large stocks of pork (Southard, Shagam, and Haley, 1998).

U.S. pork imports were projected at 575 million pounds in 1998, down by roughly 9 percent from 1997. Increased U.S. pork production accounted for the drop in imports. In addition, the major U.S. suppliers, Canada and Denmark, focused on other markets. While Denmark helped fill the shortfall in the European Union because of the outbreak of swine fever, Canada used its competitive exchange rate advantage over the United States to gain market share in Japan.

In the following sections of this chapter, we examine the challenges and opportunities facing the U.S. pork industry and how they affect its competitiveness in the volatile global market. First, structural changes in the U.S. pork industry are discussed, followed by a description of the changing dynamics of pork consumption. U.S. competitiveness in global pork markets is discussed in the context of relatively cheaper costs of production and emerging global trade policies. This is followed by a summary and conclusion.

STRUCTURAL CHANGE IN THE U.S. PORK INDUSTRY

The United States is a leading hog producer, second only to China. The U.S. pork industry generates about 10 percent of total world supplies. Pork production is a series of complex processes of converting feed grains, high-protein feed ingredients, vitamins, minerals, and water into live hogs, and eventually pork and pork products. Feed is the major production input in the pork industry, accounting for more than 65 percent of total production costs. Since the United States is one of the largest feed grain producers in the world, its comparative advantage in pork production is due, to a large extent, to the abundance of low-cost feed grains.

The pork industry is also an important source of employment and income to many American farmers. In North Carolina, for example, hogs were the leading source of gross farm income in 1996, generating over 22.3 percent of all farm receipts that year. In 1998 hog production provided 8,139 full-time jobs to the residents of North Carolina, an increase of 88 percent over 1993 hog production employment (North Carolina Pork Council, 1998).

North Carolina's hog industry was severely impacted by the flooding following Hurricane Floyd. The *Wall Street Journal* (Brooks, 1999) reported that damages were expected to reach at least $6 billion, including losses of $700 million or more in North Carolina's agriculture industry. State officials estimated that more than 100,000 hogs drowned in the deluge. Additionally, drinking water was contaminated by hog waste from lagoons that were swamped by rising waters and then spilled into nearby streams and rivers. To date, official estimates on losses sustained by the hog industry in 1999 have not been released. However, losses from Floyd alone are estimated in the millions of dollars.

Technological Change

The 1980s and 1990s have witnessed major technological breakthroughs in the pork industry. Advances in both genetic research and production systems have contributed to the high efficiency of the U.S. pork industry. The once-devastating "pig fever" has been practically eliminated from the U.S. pork industry, and the average total per unit cost of production has decreased 5.7 percent since 1984. The average whole-herd feed conversion ratio (pounds of feed required per pound of live weight produced) of the U.S. pork industry is between 3.6 and 3.8 (compared to 7 to 10 in the U.S. beef cattle industry), and is improving continuously.

Additionally, despite high feed costs, producers have maintained lower costs in pork production by using genetic improvements to attain higher reproductive efficiency and enhanced lean muscle growth. For example, compared to the breeds of the 1950s, today's model pig has 50 percent less fat. Many hog farmers conduct their operations under controlled environments to contain the spread of diseases, capture economies of scale, and, therefore, improve production efficiency. For instance, by employing genetic improvements in the United States, the average pork production per breeding animal attained for the most common variety of hogs has grown from 1,397 pounds in 1967 to 2,587 pounds in 1997; an increase of 85.2 percent in 30 years.

Production Management

Rapid structural changes in the U.S. pork industry have also affected the manner in which hog producers conduct business. One of these changes is in the growing number of contracts between hog producers and vertical coordinators, which include feed companies and packers, as a way to reduce income risk (Kliebenstein and Lawrence, 1995; Lawrence and Kaylen, 1990; Johnson and Foster, 1994). However, Gillespie and Eidman (1998) have found that while risk is an important factor in most producers' contracting decisions, autonomy is a very significant attribute and, thus, is an important consideration when predicting the success of independent hog business arrangements. The dilemma that faces most producers these days is that hog price at the farm gate has decreased dramatically, but consumers have continued to pay high prices in retail stores, leaving the packers and the processors as the primary gainers and making hog farmers the major losers. Consumers have hardly benefited from lower prices and increased production.

Environmental Concerns

Land is the key resource in pork production because of its multiple functions in housing the animals, providing feed, and in manure utilization. The United States and Canada enjoy greater comparative advantages over their principal global competitors in pork production and exports, mainly because of their huge domestic land bases. Denmark and Taiwan, for example, are countries with small land endowments and high densities in hog inventory and human population, and public concern has been more vociferously expressed over manure utilization. Yet, even in the United States and Canada, despite large bases of sparsely populated land, public demands for stricter governmental regulation of hog industry expansion and manure disposal have risen to a level that may constrain hog production (Haley et al., 1998).

Therefore, environmental externalities have long been a lingering concern for the U.S. pork industry. Whether hogs are raised in open pastures or in totally enclosed barns, they are blamed time and again for their odors and manure spillovers. In fact, the U.S. pork industry produces only 12 to 15 percent of all livestock manure in the country, and it is abiding by the most stringent environmental regulation in the world. Concentrated hog feeding operations are held to a zero discharge standard in their management of manure. When hog manure is later applied to cropland as a fertilizer, controls are imposed to eliminate pollution of surface and groundwater.

Haley et al. (1998) further document that recent concerns have been aimed primarily at large, intensive hog operations and the threats they pose to the environment and the public's "quality of life." Although relatively small, the risk of water pollution via manure lagoon leakage or spills, along with the odors that accompany large, intensive livestock operations, have galvanized citizens at local, county, state, and federal levels to advocate for more strict regulation of hog operations. In certain states, as well, environmental concerns and efforts to restrict structural changes in the industry, such as increasing the size and concentration of operations, have become politically linked, bringing further pressure to bear on hog industry expansion.

Citizens close to new or expanded intensive hog production facilities have articulated a broad range of proposals for regulation, from heightened scrutiny by local zoning boards to statewide moratoria on new facilities. For example, in August 1997, North Carolina instituted a statewide moratorium on new or expanding hog operations. Effective retroactively from March 1, 1997 through March 1, 1999, the moratorium applies to operations of 250 head or more, although operations that rely on manure management systems other than lagoons are exempt from the moratorium. Because these and similar measures have implications for the ability of the U.S. hog industry to expand, the level of environmental regulation may become a key determinant of the future scale of the U.S. pork export industry.

Federal legislation to regulate hog operations, such as the Animal Agriculture Reform Act, introduced in Congress in October 1997, would require, among other things, livestock operations raising more than 1,330 hogs to submit a manure-handling plan to the USDA for approval. The Administration's Clean Water Action Plan will also focus attention on livestock operations and land application of manure, together with resources and actions to help protect water quality and the environment.

Genuine efforts by the U.S. pork industry aimed at curtailing environmental concerns have won some recognition from the Environmental Protection Agency (EPA). The National Pork Production Council (NPPC) and the EPA signed an agreement on November 25, 1998, which allows pork producers who volunteer to have their farms inspected under NPPC's On-Farm Odor/Environmental Assessment Program to pay lower penalties for any Clean Water Act violations discovered, reported, and corrected. The On-Farm Odor/Environmental Assistance Program, initiated in 1997, is designed to identify specific on-farm environmental and odor problems and suggest appropriate engineering, biological, and management solutions. The agreement is the first of its kind between the EPA and a major sector of American agriculture.

Hog Profitability

Profitability of the pork industry can be measured by calculating hog/corn ratio (the ratio of the per 100 pound hog price and per bushel corn price). The average hog/corn ratio in the United States in the 1990s has been about 19.14. Driven by hog price and profitability, the pork industry is characterized by 12- to 24-month business cycles. A high hog/corn ratio suggests higher profitability, which may lead to expanded pork production within 12 to 24 months. On the other hand, a low hog/corn ratio indicates contraction in pork production within the same time frame.

Previous studies have shown that the dynamic process generating the pork cycle is nonlinear (Chavas and Holt, 1991), but prediction of prices based on the historical characteristics of the pork business cycle can only be qualitatively measured (Dorfman, 1998). Figure 15.1 summarizes the time series data for the U.S. hog/corn ratios; hog prices received by farmers, and the quantities of pork produced.

Hog/corn ratios and hog prices have oscillated over the years, but have not exhibited a clear upward trend. However, pork production, although fluctuating over time, increased by almost 50 percent in the 22-year period. Relatively flat hog prices and rapidly increasing production (supply) indicate an increasing demand for pork and pork products, both in the U.S. domestic market and in international markets.

THE CHANGING DYNAMICS OF PORK CONSUMPTION

A combination of demographic and economic factors has influenced the changing demand for U.S. pork and pork products. These factors include demographic shifts, changes in consumer preferences, fluctuations in incomes and relative prices, and the development of new technology and marketing techniques (Senauer, Asp, and Kinsey, 1991; Putnam and Duewer, 1995).

Demographic changes and their attendant economic consequences are among the principal factors influencing the demand and consumption patterns of pork and pork products. One reason for the

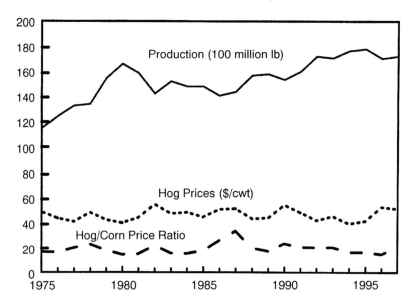

FIGURE 15.1. U.S. Pork Industry Trends

Source: National Pork Producers Council, 2000.

increasing demand is population growth. On the other hand, Americans have become wealthier and better educated and, as a result, are more health and nutrition conscious than their predecessors—they substitute more white meat (i.e., poultry, pork, and fish) for red meat (i.e., beef). Figure 15.2 illustrates the changing meat consumption patterns in the United States, including per capita consumption data for beef, pork, chicken, and fish in retail weight.

It is evident that per capita consumption of beef has decreased considerably over the past decade. Beef consumption dropped from 79 pounds in 1988 to 68 pounds in 1998. Pork consumption changed very little during the same period. However, per capita consumption of chicken increased by more than 37 percent, while fish consumption was nearly constant.

To increase pork consumption, the U.S. pork industry invested resources in producing better-flavored, juicier, and more tender pork and launched a strong marketing campaign to encourage the sub-

FIGURE 15.2. U.S. Per Capita Consumption of Meats

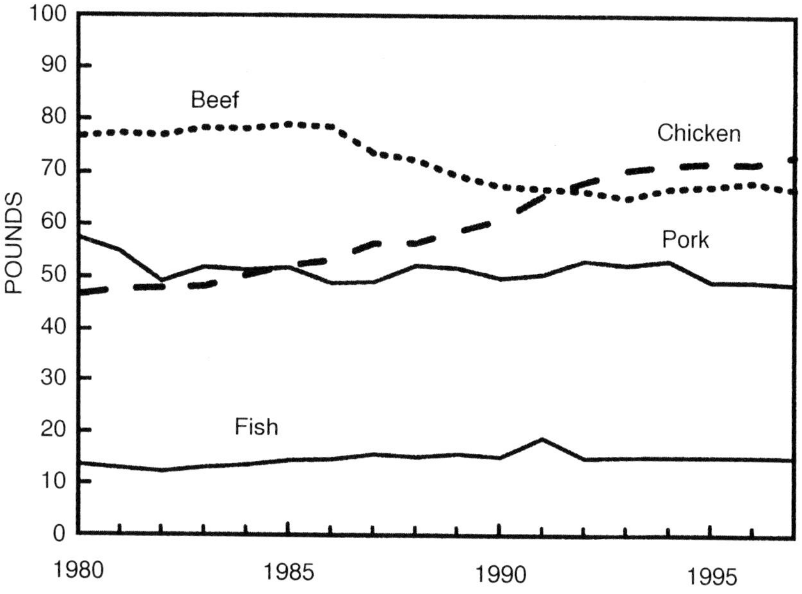

Source: National Pork Producers Council, 2000.

stitution of pork for chicken (as the other white meat). The industry also developed marketing techniques to promote pork and pork products. Major U.S. food retailers, which operate approximately 23,000 food stores, are contacted by pork industry promoters with information and ideas on how to market pork more effectively. The industry is providing the nation's top food distribution companies with promotional menu ideas in an effort to provide consumers with new ways to enjoy pork. These marketing campaigns also have placed information on major television networks. The objective of the industry is to make pork the meat of choice.

U.S. COMPETITIVENESS IN GLOBAL PORK MARKETS

A very important goal of the U.S. pork industry is gaining greater access to global markets. In 1997, over 6 percent of all U.S. pork production, 474,000 metric tons in carcass weight, were destined for export markets. In the past decade, the prosperity of the U.S. pork industry was due in part to the expansion of its exports, without which the producers' welfare would have been much reduced.

Figure 15.3 depicts an increasing growth trend for U.S. pork exports. From 1985 to 1997, the industry increased its exports by more than 4.6 times in product weight. It is natural for U.S. pork producers to hope that such an upward trend would be sustained in the foreseeable future so that the pork industry will continue to prosper. The four other top pork-exporting countries in 1997 were Denmark (470,000 metric tons in carcass weights*), Canada (410,000 metric tons), Poland (200,000 metric tons), and China (150,000 metric tons).

The competitiveness of the U.S. pork industry in global markets is strengthened by its inexpensive and high-quality products. The basic advantage in the United States is low-cost feed grains. As explained

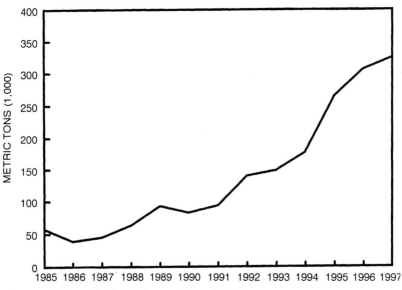

FIGURE 15.3. U.S. Pork Exports

Source: National Pork Producers Council, 2000.

---

*Excluding trade within the European Union.

earlier, the cost of production is key in explaining the flow of pork from a low production-cost region to a high production-cost region (Hayes, 1993; Clemens, Hayes, and Johnson, 1996).

A basic difficulty, however, exists in genuinely comparing pork production costs among countries. The Dutch have coordinated a research program that provides standardized cost comparisons among five European Union (EU) countries by using a standard financial model called Europor C 2.0. The result of that research is that the range of cost differences in pork production is fairly narrow among all countries studied. Costs per pound (U.S. dollars, carcass-weight basis) for the five countries are (1) France: $1.04; (2) Italy: $1.09; (3) Denmark: $1.10; (4) U.K.: $1.11; (5) Netherlands: $1.16. By comparing Dutch and U.S. production costs, the program also found that large-scale U.S. companies produced pork at 68 percent of the Dutch cost. Even average-performing, smaller-scale U.S. pork producers operated at 77 percent of the Dutch cost. The broad conclusion reached by the study is that it will be extremely difficult for the rest of the world to compete with the large pork integrators in the United States because the United States offers cheap feed ingredients and economies of scale.

Iowa State University has also released a between-country cost comparison of pork production for 1998. Their analysis shows the following costs of pork production in different countries (all costs are in dollars per pound of live weight sold): (1) Iowa mega-producer: $0.35; (2) Western Canada: $0.36; (3) Brazil: $0.38; (4) Iowa traditional: $0.39; (5) Argentina: $0.44; (6) Australia: $0.47; (7) Denmark: $0.54; (8) Hungary: $0.55; (9) Poland: $0.55; (10) Mexico: $0.58; (11) China: $0.60; (12) Philippines: $0.70; and (13) Taiwan: $0.70. Although the methodologies used by the two studies may differ, the basic conclusion is clear in that the United States possesses an absolute cost advantage over most of its global competitors in pork production. The United States also produces the highest quality, disease-free pork in the world because it has been extremely successful in eradicating pig fever disease from its territory.

Factors that may continue to pose challenges for U.S. pork export performance include the value of exchange rates, changing disposable incomes in foreign countries, and reduction of trade barriers in global pork markets. Historically, as the value of the U.S. dollar declines against the rest of the world (as it did in the late 1980s), U.S. pork and pork products become more affordable in global markets. Additionally, economic development in many countries in the 1980s and 1990s added to the growing demand for pork and pork products. Consequently, the U.S. pork industry found new markets in many developing countries.

Trade Agreements

The Uruguay Round agreement of the General Agreement on Tariffs and Trade (GATT) is expected to increase access to highly protected pork markets in Asia and Western Europe, reduce subsidized exports from the EU, and boost global pork demand by raising world incomes. According to a USDA (1994) study, higher international and domestic demand would offset rising feed costs, leading to an expansion in U.S. production at higher prices. Japan is expected to cut its minimum import price by 15 percent from current levels, raising imports for 2005 by $290-$340 million above baseline projections. South Korea is also expected to remove all nontariff barriers to imports of frozen pork and to reduce its tariff from 37 to 25 percent by 2004.

The EU is expected to convert its variable import levies into tariff equivalents and to establish a quota for 75,000 metric tons of pork imports, with 39,000 metric tons reserved specifically for high-value products such as tenderloins, boneless loins, and boneless hams. However, Eastern European producers that meet EU health and sanitary restrictions, such as Poland, are in a position to fill much of the quota, due to locational advantages and preferential access granted under EU agreements. The EU is also expected to reduce subsidized exports. Major EU suppliers, such as Denmark, are expected to decrease exports to North America and focus on maintaining market share in the Japanese market.

Following in the wake of falling tariff and nontariff trade barriers associated with the GATT and WTO negotiations, the U.S. pork industry has gained in trade. For example, U.S. pork exports to Japan increased from 12,076 metric tons in 1986 to 157,190 metric tons in 1997. Much of this increase was the result of an end to meat quotas and reductions in tariffs and levies in Japan. U.S. pork exports to South Korea have also grown, from only 3 metric tons in

1986 to 9,200 metric tons in 1997, as the country underwent trade liberalization.

Since the inception of NAFTA in 1994, U.S. pork trade with Canada and Mexico has enjoyed an increasing surplus. In 1997, Canada and Mexico became the second and third largest markets for U.S. pork exports. Their combined imports from the United States were 17 percent of total U.S. pork exports or $150 million in 1997. U.S. pork exports to the rest of the Western Hemisphere amounted to $21 million in 1997. Venezuela and the Caribbean islands are the largest non-NAFTA destinations for U.S. pork in the Western Hemisphere, where U.S. exports primarily serve the hotel and restaurant sectors. Low incomes and demand for lower-cost pork cuts, such as trimmings, pose the main constraint in increasing sales to that region. Therefore, it is not clear how much benefit could be derived for the pork industry should there be a Free Trade Area of the Americas (FTAA).

Although trade barriers such as outright bans, quotas, and tariffs designed to deter imports have been reduced in the 1990s, important sanitary barriers still remain. They are employed to deter the transmission of animal and human diseases across borders (Lynham, 1987). Because the United States maintains the highest sanitary standards in the world, it has been relatively easier for the United States to gain greater access to other markets for its pork products than for other countries to be able to export to the United States. Therefore, the United States has experienced slower growth in pork imports over the last decade. Most major categories of pork enter the United States duty-free, but the United States restricts the import of pork from regions where foot-and-mouth disease or swine fever is present.

Other Factors

Changes in the economic and political environments in the former Soviet Republics and Eastern Europe, ongoing economic reforms in China, and the new Sino-U.S. agreement loom large for U.S. pork exports. Russia is now the fourth largest destination for U.S. pork. China, along with Hong Kong, is the fifth largest export market for U.S. pork and pork products.

One challenge that the U.S. pork industry faces in international markets is the ability of other countries to eradicate pork diseases. As sanitary standards in the United States and other major markets are met, greater competition will be introduced by other low production-cost regions in South America, such as Brazil. In 1997, Brazil ranked tenth in global pork exports by supplying 57,000 metric tons in total carcass weights.

Another challenge that the U.S. pork industry faces is how to expand exports to the emerging markets in East and South Asia and Africa. Potentially, these regions could grow in wealth to provide new marketing opportunities for pork exporters, although the growth of Islam in Africa and the Middle East poses major challenges in increasing pork consumption in those regions.

CONCLUSIONS

The U.S. pork industry has experienced relative growth in production and maintained competitiveness in exports. These benefits have been enhanced in large part because of the availability of low-cost feed grains and advanced genetic technology in pork production. By producing the highest quality pork, the U.S. industry has managed to serve an ever-increasing number of consumers with different tastes and preferences both at home and abroad.

In light of current changing meat consumption patterns, the U.S. pork industry must strive to maintain its market share by promoting new pork varieties and other value-added pork products. To abide by stringent environmental regulations stemming from concern expressed by a more involved citizenry, the U.S. pork industry is faced with a major challenge of developing and implementing new environmentally friendly production processes.

Thanks to the superior quality of its products and high sanitary standards, the U.S. pork industry has been able to increase its exports around the world, and to exploit new opportunities whenever its trading partners have been willing to liberalize their economies. But more and more pork-exporting countries will become globally competitive, especially when they are able to eradicate diseases and reduce their costs of production.

Greater environmental regulation in the future could increase the costs of producing hogs in the United States, leading to lower hog production. Should U.S. consumer demand and other major exporters' production costs remain constant, imposing higher costs on land utilization for the U.S. hog industry could increase domestic pork prices and reduce U.S. competitiveness in international pork markets. Consequently, the extent to which the U.S. hog industry can retain its international competitiveness will depend, at least in part, on how governments in other pork-producing countries choose to respond to their own citizens' environmental concerns.

REFERENCES

Brooks, R. (1999). "North Carolina Hurricane Damages Rise." *Wall Street Journal,* September 21, p. A2.

Chavas, J.-P. and M.T. Holt. (1991). "On Nonlinear Dynamics: The Case of the Pork Cycle." *American Journal of Agricultural Economics* 73 (3): 819-828.

Clemens, R., D. Hayes, and S. Johnson. (1996). "How Competitive Are the U.S. Livestock and Poultry Industries?" *U.S. Meat Export Analysis and Trade News* 4 (6): 3-5.

Dorfman, J.H. (1998). "Bayesian Composite Qualitative Forecasting: Hog Prices Again." *American Journal of Agricultural Economics* 80 (3): 513-551.

Gillespie, J.M. and V.R. Eidman. (1998). "The Effect of Risk and Autonomy on Independent Hog Producers' Contracting Decisions." *Journal of Agricultural and Applied Economics* 30(1):175-188.

Haley, M., E. Jones, E.A. Jones, and L. Southard. (1998). "World Hog Production: Constrained by Environmental Concerns?" *Agricultural Outlook* (March): 15-19.

Hayes, D. (1993). "Are U.S. Agricultural Exports Undergoing a Structural Transformation?" *U.S. Meat Export Analysis and Trade News* 1 (2): 8-11.

Johnson, C.S. and K.A. Foster. (1994). "Risk Preferences and Contracting in the U.S. Hog Industry." *Journal of Agriculture and Applied Economics* 26: 2 (December): 393-405.

Kliebenstein, J.B. and J.D. Lawrence. (1995). "Contracting and Vertical Coordination in the United States Pork Industry." *American Journal of Agricultural Economics* 77 (4): 1212-1218.

Lawrence, J.D. and M.S. Kaylen. (1990). "Risk Management for Livestock Producers: Hedging and Contract Production." Staff paper no. P90-49. St. Paul: Dept. of Agriculture and Applied Economics, University of Minnesota, July.

Lynham, M. (1987). "Nontarrif Trade Barriers in the Beef Industry." *Experiment Station Technical Bulletin 261*. Tucson: University of Arizona.

National Pork Producers Council. (2000). *1999/2000 Pork Facts*. Des Moines, IA: National Pork Producers Council.

North Carolina Pork Council. (1998). *Facts About NC Pork Industry.* Raleigh, NC: North Carolina Pork Council.

Putnam, J.J. and L.A. Duewer. (1995). "U.S. Per Capita Food Consumption: Record-High Meat and Sugars in 1994." *Food Review* 18 (2): 2-11.

Senauer, B., E. Asp, and J. Kinsey. (1991). *Food Trends and the Changing Consumer.* St. Paul, MN: Eagan Press.

Southard, L., S. Shagam, and M. Haley. (1998). "Asia Crisis to Trim Prospects for U.S. Meat Exports." *Agricultural Outlook* (March): 7-9.

USDA. (1994). *Livestock, Dairy and Poultry Situation and Outlook* (November). Washington, DC: Economic Research Service. U.S. Department of Agriculture.

USDA. (1999). *Livestock, Dairy and Poultry Situation and Outlook* (December). Washington, DC: Economic Research Service. U.S. Department of Agriculture.

Chapter 16

The Food Processing Industry

Michael Reed

INTRODUCTION

The food processing or manufacturing industry (standard industrial classification [SIC] 20) is the industry that transforms agricultural products into foods and beverages for human consumption and related products, such as fats, oils, and animal feeds. It is the largest manufacturing industry in the United States, accounting for 14 percent of U.S. manufacturing output (U.S. Department of Commerce, 1999). The industry is broken into nine three-digit SIC categories: meat products (SIC-201), dairy products (SIC-202), preserved fruits and vegetables (SIC-203), grain mill products (SIC-204), bakery products (SIC-205), sugar and confections (SIC-206), fats and oils (SIC-207), beverages (SIC-208), and miscellaneous (SIC-209).* All of these items are processed, but they are not all final consumption items—intermediate products are included in SIC-20.

U.S. food processing companies employed over 1.5 million people and shipped $421.3 billion of products in 1995 (U.S. Department of Commerce, 1999). The food manufacturing sector has shown steady, but relatively slow, growth over the years. From 1987 to 1995, shipments from food manufacturers increased by 37.8 percent. Yet this slow growth masks the dynamic changes taking place in exports of food.

More recent data, from 1997, is available for gross output by three-digit SIC. This allows the breakdown of output among the

---

*The sum of all output from SIC-201 through SIC-209 equals SIC-20.

nine subindustries. The meat products subindustry output was $102.9 billion in 1997; beverages, $70.8 billion; dairy products, $59.5 billion; grain mill products, $59.4; and so on (U.S. Department of Commerce, 1999). These are huge subindustries and their output is tremendously important to cities and towns throughout the United States.

There has been a clear, upward trend in the percentage of food production exported from SIC-20 firms since the mid-1980s. In 1989, food exports accounted for only 5 percent of shipments, but the percentage exported had increased to 6.9 percent by 1995 (U.S. Department of Commerce, 1999). The value of processed food exports increased from $17.2 billion in 1989 to $29.7 billion in 1996, a 73 percent increase in seven years. Processed food imports have increased during the same period, but at a slower rate (from $18.7 billion to $24.0 billion). The United States had a trade deficit in processed food until 1991, but now the surplus has reached $5.7 billion.

U.S. EXPORTS OF PROCESSED FOODS

From 1991 to 1997, total commodity exports from the United States increased from $421.9 billion to $687.6 billion, or 63 percent (U.S. Department of Commerce, 1999). Exports of unprocessed agricultural products increased from $21.7 billion to $28.3 billion, or by 30.3 percent. Total SIC-20 exports increased by 64.2 percent, about the same percentage that all commodity exports increased, but more than double the rate that other agricultural products increased. This is evidence that the food processing industry is becoming more competitive relative to raw agricultural products.

Table 16.1 presents U.S. processed food exports by three-digit SIC code from 1991 to 1997. More meat products are exported than any other category, totaling $8.8 billion in 1997, or 8.6 percent of gross meat output. Meat product exports increased by 67.2 percent from 1991 to 1997, about the same as all SIC-20 exports. Much of this increase has been fueled by lower trade barriers and higher incomes throughout the world. Beef, pork, and poultry exports, which make up the great majority of meat product exports, are discussed in other chapters.

TABLE 16.1. U.S. Exports of Processed Foods by Three-Digit SIC, 1991-1997 (Millions of Dollars)

Code	Industry or Subindustry	1991	1992	1993	1994	1995	1996	1997
20	Food and kindred products	17,726	20,059	20,847	23,459	26,401	27,568	29,109
201	Meat products	5,262	5,842	5,880	6,936	8,435	8,878	8,811
202	Dairy products	571	851	965	883	894	850	1,066
203	Preserved fruits and vegetables	2,019	2,272	2,339	2,603	2,885	3,015	3,202
204	Grain mill products	3,233	3,476	3,710	3,951	4,093	4,289	4,351
205	Bakery products	224	295	340	368	362	378	411
206	Sugar and confectionery products	1,452	1,532	1,700	1,868	1,919	2,160	2,015
207	Fats and oils	2,455	2,877	2,681	2,951	3,799	3,681	4,693
208	Beverages and flavorings	1,261	1,460	1,558	1,890	2,104	2,161	2,338
209	Other food products	1,249	1,454	1,674	2,008	1,910	2,155	2,222

Source: U.S. Department of Commerce, 1999.

Exports of fats and oil are the second leading category of processed food exports, totaling $4.7 billion in 1997, or 19.2 percent of gross fats and oil output (the highest percentage among three-digit SIC categories). This is also the fastest-growing category, increasing by 91.2 percent over the six years. Fat and oil exports rose sharply in 1995 and 1997, particularly to China and Mexico.

Grain mill products are the third leading export category, totaling $4.4 billion in 1997, or 7.3 percent of gross grain mill output, but this is also the slowest growing category. It increased only 34.6 percent over the six years, though it did so consistently each year. These products include breakfast cereals, animal feed, and pet food.

Other fast-growing SIC-20 exports are dairy products (increasing 86.7 percent over the six years), beverages and flavoring extracts (85.4 percent), bakery products (83.5 percent), and other food preparations (77.9 percent). All of these products began from relatively small export levels, but their current totals are impressive, as shown by their 1997 levels. Further, these four SIC categories have the lowest exports as a percentage of gross output (none of these subindustries had exports greater than 5.5 percent of gross output).

Table 16.2 shows U.S. three-digit SIC exports by area of the world. The NAFTA countries, Canada and Mexico, are the leading

TABLE 16.2. U.S. Exports of Processed Foods by Three-Digit SIC and Destination, 1997 (Millions of Dollars)

SIC Code	Industry or Subindustry	NAFTA	Japan	EU-15	Asian NICs	Other Americas	Rest of World	World
20	Food and kindred products	7,203	5,188	4,232	4,753	2,738	4,997	29,111
201	Meat products	1,807	2,743	299	2,108	321	1,533	8,811
202	Dairy products	316	117	27	252	131	225	1,066
203	Preserved fruits and vegetables	1,024	677	556	432	218	295	3,202
204	Grain mill products	1,061	558	1,178	341	522	690	4,351
205	Bakery products	260	19	32	21	45	34	411
206	Sugar and confectionery products	549	243	635	209	139	240	2,015
207	Fats and oils	921	233	656	738	804	1,342	4,693
208	Beverages and flavorings	443	339	582	256	334	384	2,338
209	Other food products	822	259	267	396	224	254	2,222

Source: U.S. Department of Commerce, 1999.

destinations for U.S. processed food exports, with $7.2 billion in 1997. Japan, the only single country in the table, accounted for $5.2 billion. The rest of the world, Asian newly industrialized countries (NICs), and the EU-15, were close behind with imports from the United States of more than $4 billion.

The NAFTA countries are the largest destinations for dairy products, canned and preserved fruits and vegetables, bakery products, and other food preparations (SIC-209). The dairy and bakery products are highly perishable and bulky, so it is natural that a large percentage of U.S. exports go to proximate countries. The EU-15 is the leading destination for grain mill products, sugar and confectionery products, and beverages and flavoring extracts. Japan is the leading destination for U.S. meat product exports, while the rest of the world is the leading destination for food preparations (SIC-208).

U.S. IMPORTS OF PROCESSED FOODS

From 1991 to 1997, total commodity imports by the United States increased from $488.9 billion to $870.2 billion, i.e., by 78 percent. Imports of agricultural products (unprocessed) increased from $7.3 billion to $12.3 billion, or 67.7 percent. Total SIC-20 imports increased by 46.2 percent, much lower than the rates for all commodities and agricultural products—further evidence that the U.S. food processing industry is becoming more competitive.

Table 16.3 presents U.S. processed food imports by three-digit SIC code from 1991 to 1997. Beverage imports were larger than any other category, totaling $6 billion in 1997. Their growth rate was close to the total increase for SIC-20. They increased particularly during 1996 and 1997. Three other SIC categories had imports slightly above $3 billion in 1997—fruits and vegetables, sugar and confectionery products, and meat products. Imports for all of these categories grew more slowly than all SIC-20 imports; meat product imports fell during the six-year period. The other products with imports growing faster than all SIC-20 imports were food prepa-

TABLE 16.3. U.S. Imports of Processed Foods by Three-Digit SIC, 1991-1997 (Millions of Dollars)

Code	Industry or Subindustry	1991	1992	1993	1994	1995	1996	1997
20	Food and kindred products	15,741	16,944	16,599	17,799	18,713	21,295	23,006
201	Meat products	3,182	3,020	3,092	2,971	2,692	2,684	3,036
202	Dairy products	701	807	785	860	1,058	1,218	1,158
203	Preserved fruits and vegetables	2,286	2,616	2,339	2,515	2,610	3,052	3,180
204	Grain mill products	567	653	706	863	893	1,101	1,196
205	Bakery products	376	419	468	540	599	658	706
206	Sugar and confectionery products	2,144	2,228	2,098	2,132	2,377	2,896	3,129
207	Fats and oils	942	1,182	1,193	1,388	1,524	1,874	1,883
208	Beverages and flavorings	3,639	4,096	4,025	4,387	4,625	5,338	6,015
209	Other food products	1,905	1,922	1,893	2,142	2,334	2,474	2,701

Source: U.S. Department of Commerce, 1999.

rations (which grew by 99.9 percent over the six years); bread, cake, and related products (which grew by 87.8 percent); and dairy products (which grew by 65.2 percent).

Table 16.4 shows U.S. three-digit SIC imports by area of the world. The EU-15 countries are the leading origin of U.S. processed food imports, totaling $7.4 billion in 1997. The NAFTA countries were close behind at $6.8 billion. Other areas were at least $3 billion behind the two leading origins. Notice that Japan supplies few food products to the United States. The United States has a positive processed food trade balance with all areas of the world except the EU-15, where it has a trade deficit of $3.1 billion.

The NAFTA countries are the most important origin for all U.S. food imports except dairy, sugar and confectionery, and beverages. The EU-15 is the leading origin for dairy (mostly cheeses) and beverages (wine and beer), while South and Central America are the leading origins of sugar and confectionery products. Over 50 percent of EU-15 food exports to the United States are beverages, accounting for 17 percent of total U.S. processed food imports.

TABLE 16.4. U.S. Imports of Processed Foods by Three-Digit SIC and Origin, 1997 (Millions of Dollars)

SIC Code	Industry or Subindustry	NAFTA	Japan	EU-15	Asian NICs	Other Americas	Rest of World	World
20	Food and kindred products	6,815	242	7,365	2,215	2,650	3,725	23,012
201	Meat products	1,337	15	332	69	357	927	3,036
202	Dairy products	73	0	632	3	31	420	1,158
203	Preserved fruits and vegetables	923	63	537	603	623	432	3,180
204	Grain mill products	670	14	174	208	14	116	1,196
205	Bakery products	385	18	222	43	13	26	706
206	Sugar and confectionery products	631	8	474	322	1,043	652	3,129
207	Fats and oils	668	20	431	269	62	433	1,883
208	Beverages and flavorings	1,445	32	4,010	50	224	255	6,015
209	Other food products	683	72	553	648	283	464	2,701

Source: U.S. Department of Commerce, 1999.

THE COMPETITIVE ENVIRONMENT FOR PROCESSED FOODS

It is very difficult to characterize the competitive environment for the diverse set of products covered by this chapter. The approach chosen is to pick certain products for which data are available, and discuss important competitors in those products. In general, European countries are the most important exporters of food products, particularly processed products. This is because trade barriers are minimal, product standards are similar or identical among countries, and transportation costs are very small relative to product value. One must question, though, whether they are truly U.S. competitors if they only export to other European countries, in the sense that most U.S. exports of processed foods are moving to NAFTA and Asian countries. Nonetheless, because it is difficult to cover all countries' exports by final destination, aggregate export figures are used to judge U.S. competitors. All of the data for this section comes from the Food and Agriculture Organization (FAO, 1999) of the United Nations (UN).

Most meat products, as mentioned earlier, are covered in other chapters. Only prepared meats are discussed here. The leading prepared meat exporters are Belgium-Luxembourg, Denmark, France, and China. No country exports a large amount (Belgium-Luxembourg exported $481 million in 1997) and the United States exported only $318 million in 1997. The leading importers of prepared meats are the United Kingdom ($611 million), Japan, and the United States.

The United States is not a leading dairy-exporting country due to production costs, transportation costs, and government policies. The leading dairy-exporting country, surprisingly, is Germany with exports of $4.6 billion in 1997. It was closely followed by two other European countries, France and the Netherlands. New Zealand was the only non-European country in the top five. U.S. dairy product exports totaled $723 million in 1997. The United States was never in the top five for specific dairy products either. New Zealand was the leading butter exporter in 1997 (with $638 million), while the United States exported $27 million; France was the leading cheese exporter with $2.0 billion, while the United States exported $137 million; and

New Zealand was the leading dry milk exporter with $1.2 billion, while the United States exported $186 million.

Germany was also the leading dairy-importing country in 1997 at $3.5 billion; the United States was sixth with $745 million. Germany was the leading importer of butter and cheese. Other important dairy product importers were Italy, Belgium-Luxembourg, the Netherlands, and France. Dairy products are obviously highly differentiated because countries import and export large quantities of similar products. Market niches are particularly important for European producers and consumers, where cheese is definitely not a homogeneous product. The *appellation d'origine contrôlée* and other official product identification standards in Europe encourage these differentiated food products.

The United States is an important supplier of some grain mill products, which is natural given the large volume of crop production in the United States. The United States is the leading exporter of feeding stuffs with $4.8 billion in 1997, followed by Brazil (with $2.8 billion), Argentina, and the Netherlands. The leading importers of feeding stuffs are Japan ($2.1 billion), France, Germany, and China. The United States is also the leading exporter of pet food ($822 million), followed by France, the United Kingdom, Australia, and Germany. The leading importers of pet food are Japan ($598 million), Germany, Italy, and Belgium-Luxembourg.

Other important grain mill products for which the United States trails other exporters are wheat flour, where U.S. exports are $141 million, behind France (at $420 million), Turkey, Italy, Belgium-Luxembourg, and Germany, and breakfast cereals, where U.S. exports are $201 million, behind the U.K.'s exports of $389 million. The leading import markets for wheat flour are Algeria ($416 million), Cambodia, Libya, Yemen, and Russia, while the leading markets for breakfast cereals are France ($200 million), the United States, and Germany. There is no question that the world wheat flour trade is greatly distorted through export subsidies of the European Union and the United States.

Beverages are another important processed food item traded internationally. France is the leading beverage exporter, with $8.2 billion in 1997, followed by the U.K., Italy, Germany, and Spain. The United States is the sixth leading beverage exporter at $1.5 billion.

Much of this trade is in wine, which is covered in Chapter 12. The leading beer exporter is the Netherlands ($978 million), followed by Germany, Mexico, and the United States (at $321 million).

The United States is a huge importer of beverages, importing $6.4 billion in 1997. The other leading importers are the United Kingdom ($3.7 billion), Germany, Japan, and France. The United States is also the largest beer importer, with $1.6 billion imported in 1997. It is followed by the United Kingdom ($465 million), Italy, France, and China.

The United States is the leading exporter of miscellaneous foods (as classified by the FAO). It exported $2.8 billion in 1997, ahead of France ($2.4 billion), Germany, and the Netherlands. Germany is also the leading importer of miscellaneous foods ($1.6 billion), followed by the United Kingdom, Japan, and France.

COMPETITIVE POSITION OF THE UNITED STATES

It is obvious from the previous section that most of the trade in manufactured food takes place among more developed countries. Much of this is because European countries are very close geographically and economically, and there are no trade barriers among EU members. Additionally, importing manufactured food is more expensive than importing bulk commodities, so less-developed countries that import food tend to import lower-priced bulk commodities and process them into food items within their countries. Less-developed countries are not as active in manufactured food exports because manufactured foods must meet food safety, packaging, labeling, and other requirements of importing countries. Brand identities also play a large role in consumer purchasing decisions.

Henderson, Handy, and Neff (1996) provide the most comprehensive appraisal of processed food trade and its integration with the activities of multinational food firms. They discuss origins and destinations of U.S. food trade, the linkages between foreign direct investment (FDI) and food exports, product standards, the move toward harmonization, and the role of technical barriers in food trade.

The market for manufactured food has become increasingly differentiated over time. This move toward product differentiation and the importance of brands explains some of the trade patterns in processed foods. McCorriston and Sheldon (1991) found that over 50 percent of the four-digit SIC subindustries in manufactured food exhibit intraindustry trade, as measured by the Grubel-Lloyd index. Thus, trade in food is dominated by countries importing and exporting very similar products. In such a situation, factor endowments from the Heckscher-Ohlin-Samuelson tradition are not important in explaining trade patterns. Economies of scale and product differentiation become the important factors (Grossman, 1992; Lancaster, 1980).

Another major consideration in understanding international trade in processed foods is the existence of multinational enterprises (MNEs), where a firm has an ownership presence in more than one country. In 1995, there were 764 affiliates of U.S. food MNEs with assets of $99.6 billion, sales of $113.2 billion, and 554,000 employees (Bureau of Economic Analysis, 1997). The international sales of these U.S. MNEs were 4.28 times U.S. processed food exports in 1995. These U.S. subsidiaries (affiliates) imported $3.4 billion from the United States (or almost 13 percent of U.S. processed food exports) and exported $2.7 billion back to the United States (or over 14 percent of U.S. processed food imports).

Henderson, Voros, and Hirschberg (1996) used a 144-firm sample of food firms throughout the world in their analysis of MNE behavior. They found that smaller firms tend to export, while larger firms tend to invest to reach foreign markets. They also found that more specialized firms tend to export, while more diversified firms tend to invest, and that firms with a large domestic market share are more likely to invest than export. In addition, they found that firms were investing abroad to exploit brands and corporate goodwill.

It is well known that most of these international investments by U.S. MNEs are horizontal in nature, and the great majority of the output from U.S. affiliates is sold in the host country (Handy and Henderson, 1994). Therefore, one must understand these MNEs if one is to understand U.S. processed food exports. Unfortunately, there has been little analytical work on the operations of MNEs, particularly how their affiliate sales are related to U.S. exports of processed foods. Ning and Reed (1995) found that affiliate sales substitute for U.S. food exports, but Overend, Connor, and Salin (1995) and Munirathinam, Reed, and Marchant (1999) find that affiliate sales and U.S. food exports are complementary in some cases.

Abbott and Solana-Rosillo (1998) found that FDI by U.S. food MNEs was growing faster than exports. They also found that the choice of entry mode for international markets depends on product characteristics. They argued that FDI was often preferred by food processors, especially for beverages because of the differentiated product nature and the need to tailor products to specific markets.

Kalaitzandonakes, Hu, and Bredahl (1998) take a different approach to explain FDI versus exportation by food firms. They argue that economies of scale are a major determinant of whether the firm will export. Their case study of the pork industry suggests that exports are high because of scale economies in packing, whereas processed product exports are low because there are no scale economies in processing.

Licensing is also important for international food markets. Information is sketchy on the nature of the arrangements between firms, but Henderson and Sheldon (1992) found that over 50 percent of world's largest publicly held food manufacturers were involved in licensing. This means that U.S. food exporters must also compete with international firms that have arrangements which allow use of special recipes, brands, and other assets of large food companies.

Despite these competitive difficulties, U.S. food manufacturing firms have been successful in increasing food exports in recent years. Henderson, Handy, and Neff (1996) and McDonald and Lee (1994) attribute this to high labor productivity in the United States and the high degree of capital intensity in the food manufacturing industry. Output per food worker in 1992 was $238,000 in the United States versus $210,000 in France, $184,000 in Germany, $166,000 in the U.K., and $159,000 in Japan. Given food industry restructuring in the United States, that productivity advantage may have widened since 1992. Nonetheless, the United States still seems to be lagging other countries in terms of processed food exports.

Reed and Marchant (1992) found that the food processing industry was also lagging other U.S. manufacturing industries. They analyzed the U.S. food industry's international competitiveness relative to thirteen other U.S. manufacturing industries. Using data from the late 1980s and early 1990s, they found that the food manufacturing industry had the lowest percentage of output that was exported (5.4 percent) and had the second highest ratio of foreign-affiliate sales to U.S. exports (2.33) among U.S. manufacturing industries. So U.S. firms were much more likely to process food in their overseas affiliates than to export it from their U.S. plants. They also found that most of the investment overseas in food manufacturing facilities was horizontal in nature (where the company is performing the same activities in the host country as they perform in the home country).

TRADE AGREEMENTS, RULES, AND REGULATIONS

The Uruguay Round of the GATT dealt with processed food items in the same way that it dealt with other agricultural products. Thus, most processed foods are considered agricultural and tariffs must be reduced by an average of 36 percent for all commodities and a minimum of 15 percent for each. Export subsidies are to be reduced by 36 percent in value and 21 percent in volume for agricultural products.

Jones and Blandford (1998) argue that there is little evidence that the tariff reductions under the Uruguay Round will influence trade in processed food more than other agricultural products. They believe that most tariff reductions for processed foods will be at the 15 percent minimum. However, the Uruguay Round dealt extensively with technical barriers to trade, and this might be the area where processed food trade is most affected.

Standards and technical requirements for products are increasing over time. These requirements include rules for quality, packaging, labeling, standards of identity, and conformity assessment. Some of these regulations help increase the information flow in the marketing process and allow consumers to be informed of product origin, safety, and quality. Other technical regulations or barriers include sanitary and phytosanitary (SPS) regulations on plants and animals that ensure that traded products are not infected with harmful pests or diseases.

Some concerns exist that because tariff and nontariff barriers have been lowered due to GATT, governments are relying on product standards and technical barriers to keep imports out and domestic industries more profitable (Orden and Roberts, 1997). This could be especially true for manufactured food products because standards and technical requirements vary markedly by country. Some of these standards and technical requirements can be disguised means of protectionism, used because there are limited opportunities for other GATT-legal tariffs and nontariff barriers.

According to GATT, countries have a right to protect their domestic industries through technical barriers that will prevent the spread of harmful species. The GATT section XX regulations on health and safety allow technical barriers to preserve health and safety. Unless GATT/WTO rules are changed, countries will always be allowed to establish their own rules on product shelf life, labeling, pesticide and other chemical residues, environmental standards, and so on. However, those rules must not discriminate between domestic products and imported products.

There is no question that product standards can affect the cost competitiveness of imported food products, so it is important that these standards be justified, or imported products will be at a disadvantage. That is why national treatment is so important. The possibility that standards and technical requirements are disguised economic barriers has led to calls for a more scientific approach to justifying and establishing technical standards.

The Technical Barriers to Trade (TBT) agreement, which was enacted during the Tokyo Round, aimed at protecting the consumer against deception and fraud in product standards. The TBT agreement specified that all technical standards and regulations must have a legitimate purpose (to provide information on the product or protection from harmful pests or diseases). The cost of implementing the standard must be proportional to the purpose of the standard, meaning that the costs of implementing the standard cannot be many times the value of the standard in terms of health and safety protection. Finally, the agreement stated that if there were multiple ways of attaining the objective, the least trade-restricting method should be chosen. These are all familiar GATT ideas and expressions that are based upon the founding principles of liberalized trade.

In the early 1990s, there was increased pressure in many countries to protect domestic agricultural industries, more scientific findings related to health and food safety, more accurate detection technology to create more barriers, and more disputes associated with those barriers (Roberts, 1998). Due to weak dispute settlement procedures under GATT and because there was no structured mechanism to reach agreement on legitimate standards and regulations, the Uruguay Round of GATT focused on standards and technical barriers.

The Uruguay Round's agreement on the Application of Sanitary and Phytosanitary Measures (the SPS agreement) defined basic GATT principles behind the use of SPS barriers to trade. The first principle is that countries have the basic right to institute SPS measures that protect plant, animal, and human health when they are based on scientific principles. These measures cannot discriminate between members where identical or similar conditions prevail.

A second principle is that members should base their SPS requirements on international standards, guidelines, or recommendations when they exist, although they can institute higher standards if there is a scientific justification. This implies a need to harmonize SPS measures among countries so that international trade will flow more easily and production processes can be geared toward one set of regulations.

A third principle is that members are obliged to recognize measures adopted by other countries which provide equivalent levels of protection. This concept of equivalency means that countries should view the end product, rather than the process that is used to meet the standards.

A fourth principle is that members should base their SPS measures on a risk assessment, taking into account methodologies developed under the auspices of three international organizations. The organization most relevant for food products is the Codex Alimentarius Commission (Codex). The Codex should help move the world toward a common risk assessment methodology and harmonized standards.

These principles, along with the enhanced dispute settlement procedures, should help GATT and its successor agency, the WTO, to ensure that SPS and other technical barriers are truly used to protect health and safety. If the rules are successfully standardized and countries follow these procedures, standards and technical regulations should have a lessened role in restricting trade in processed food products. This would, naturally, be very beneficial to U.S. food manufacturers.

Hooker and Caswell (1996) argue that national-level food quality regulations lead to increased FDI and reduced trade in food products. Firms operating within the country are more likely to stay abreast of laws and regulations, and they will also be treated better (despite the national treatment provisions of the WTO). Mutual recognition will be important in promoting trade, but harmonization of standards would be even more trade-enhancing.

CONCLUSIONS

The U.S. food processing industry is experiencing tremendous growth in certain subindustries, indicating that its competitiveness is improving with time. Exports are growing rapidly for fats and oils, dairy, beverages, and bakery products. Meat products are another subindustry that has experienced tremendous increases in export value. Increasing labor productivity in the industry has been and should continue to be the key to increased export growth.

European countries are the largest exporters of processed foods, but they may not be direct competitors with the United States for many products. Canada, Mexico, and Japan, which account for 44 percent of U.S. processed food exports, are locations where competition with European food processors is less important. European countries tend to trade among themselves due to their geographic proximity, minimal barriers to trade, and similar or identical product standards.

Product differentiation and national-level quality standards are fundamental determinants of world processed food trade. It is well established that much of the trade in processed food is of an intra-industry nature, so scale economies and brands play a fundamental important role in determining trade patterns. Yet understanding the role of multinational food firms, as traders and investors, is the key in understanding the future of processed food trade.

National-level rules and regulations will play an enhanced role in processed food trade, but the moves toward harmonization (through the Codex and other organizations) and mutual recognition should mitigate their impacts. These trends in promoting scientific standards and the next multilateral round of trade negotiations can set the stage for more fluid trade in processed foods.

REFERENCES

Abbott, Philip and Juan Solana-Rosillo. (1998). "International Firms in the Manufacture and Distribution of Processed Foods." In *Global Markets for Processed Foods,* D. Pick, D. Henderson, J. Kinsey, and I. Sheldon (Eds.). Boulder, CO: Westview Press, pp. 135-159.

Bureau of Economic Analysis. (1997). "U.S. Direct Investment Abroad." Washington, DC: U.S. Department of Commerce.

FAO. (1999). *FAOSTAT Database.* Food and Agriculture Organization, United Nations. Internet, <http://apps.fao.org>.

Grossman, Gene. (1992). *Imperfect Competition and International Trade.* Cambridge, MA: MIT Press.

Handy, Charles and Dennis Henderson. (1994). "Assessing the Role of Foreign Direct Investment in the Food Manufacturing Industry." In *Competitiveness in International Food Markets,* M. Bredahl, P. Abbott, and M. Reed (Eds.). Boulder, CO: Westview Press, pp. 203-230.

Henderson, Dennis, Charles Handy, and Steven Neff. (1996). "Globalization of the Processed Food Market." Agricultural Economic Report Number 742, Economic Research Service, U.S. Department of Agriculture, Washington, DC.

Henderson, Dennis and Ian Sheldon. (1992). "International Licensing of Branded Food Products." *Agribusiness: An International Journal* 8(5): 399-412.

Henderson, Dennis, Peter Voros, and Joseph Hirschberg. (1996). "Industrial Determinants of International Trade and Foreign Investment by Food and Beverage Manufacturing Firms." In *Industrial Organization and Trade in the Food Industry,* I. Sheldon and P. Abbott (Eds.). Boulder, CO: Westview Press, pp. 197-215.

Hooker, Neal and Julie Caswell. (1996). "Trends in Food Quality Regulation: Implications for Processed Food Trade and Foreign Direct Investment." *Agribusiness: An International Journal* 12(5): 411-419.

Jones, Wayne and David Blandford. (1998). "Trade and Industrial Policies Affecting Processed Foods." In *Global Markets for Processed Foods,* D. Pick, D. Henderson, J. Kinsey, and I. Sheldon (Eds.). Boulder, CO: Westview Press, pp. 33-54.

Kalaitzandonakes, Nicholas, Hong Hu, and Maury Bredahl. (1998). "Looking in All the Right Places: Where are the Economies of Scale?" In *Global Markets for Processed Foods,* D. Pick, D. Henderson, J. Kinsey, and I. Sheldon (Eds.). Boulder, CO: Westview Press, pp. 255-276.

Lancaster, Kelvin. (1980). "Intra-Industry Trade Under Perfect Monopolistic Competition." *Journal of International Economics* 10(1): 151-176.

McCorriston, Steve and Ian Sheldon. (1991). "Intra-Industry Trade and Specialization in Processed Agricultural Products: The Case of the U.S. and the E.C." *Review of Agricultural Economics* 13 (1): 173-184.

McDonald, Stephen and John Lee. (1994). "Assessing the International Competitiveness of the United States Food Sector." In *Competitiveness in International Food Markets,* M. Bredahl, P. Abbott, and M. Reed (Eds.). Boulder, CO: Westview Press, pp. 191-202.

Munirathinam, Ravi, Michael Reed, and Mary Marchant. (1999). "Effects of the Canada/US Free Trade Agreement on US Agricultural Exports." *International Food and Agribusiness Management Review* 1 (3): 403-415.

Ning, Yulin and Michael Reed. (1995). "Locational Determinants of U.S. Direct Foreign Investment in Food and Kindred Products." *Agribusiness: An International Journal* 11 (1): 77-85.

Orden, David and Donna Roberts. (1997). "Understanding Technical Barriers to Agricultural Trade." Proceedings of a Conference of the International Agricultural Trade Research Consortium, University of Minnesota, St. Paul, MN.

Overend, C., John Connor, and Victoria Salin. (1995). "Foreign Direct Investment and U.S. Exports of Processed Foods: Complements or Substitutes?" Paper presented at the Organization and Performance of World Food Systems (NCR-182) Spring Conference, Arlington, VA, March 9-10.

Reed, Michael and Mary Marchant. (1992). "The Global Competitiveness of the U.S. Food Processing Industry." *Northeastern Journal of Agricultural and Resource Economics* 22 (1): 61-70.

Roberts, Donna. (1998). "Implementation of the WTO Agreement on the Application of Sanitary and Phytosanitary Measures: The First Two Years." International Agricultural Trade Research Consortium Working Paper #98-4, St. Paul, MN, May.

U.S. Department of Commerce. (1999). International Trade Administration. Official Statistics of Bureau of Census, Internet, <www.ita.doc.gov/industry/otea>.

Chapter 17

North American Free Trade and U.S. Agriculture

Parr Rosson

INTRODUCTION

U.S. agriculture has undergone dramatic change in the 1990s. Agricultural markets have gone from relatively tight supply/demand conditions to weak demand and ample supplies in just three crop years. U.S. trading partners have been actively pursuing the development of trade agreements, especially in the Western Hemisphere and Europe. Economic and political uncertainty have increased in the rest of the world, particularly in Asia and Central Europe, leading to market instability, falling currency values, and weaker demand for many U.S. agricultural products.

High world prices and expanding foreign economies boosted U.S. agricultural exports to a record $60 billion in 1996 (USDA, 1995-1999). Since then, exports declined 10 percent to $51.8 billion in 1998 and were forecast to decline another 7 percent in 1999 (USDA, 2000a). The financial crisis in Asia, coupled with large domestic and world supplies of some grains and oilseeds, has resulted in lower world prices and slumping farm incomes. In 1999, U.S. corn stocks were forecast to exceed the previous year's by 48 percent, making it the largest harvest since 1992-1993. U.S. wheat output was expected to generate the largest domestic supply since 1987-1988, while soybean production was expected to lead to year-end stocks more than double the levels of 1986-1987. With a relatively stable domestic demand and weak export demand, stocks are forecast to approach record levels for most major export crops.

This dramatic shift in the fortunes of much of U.S. agriculture has occurred in less than two years and is, in part, due to increased dependence on international markets.

Due to trade liberalization agreements such as the North American Free Trade Agreement (NAFTA), some U.S. producers have experienced more market access and rising exports for their products, while others have faced more import competition and lower prices. Along with new trade opportunities, some producers have experienced more risk associated with greater dependence on international markets, such as the 1995 Mexican peso devaluation and the subsequent economic recession, which reduced the demand for U.S. exports.

CUSTA AND NAFTA IMPLEMENTATION

NAFTA, implemented on January 1, 1994, is designed to liberalize trade among the United States, Mexico, and Canada. The Canada-U.S. Trade Agreement (CUSTA) was implemented on January 1, 1989, with its remaining provisions subsumed within NAFTA in 1994. In agricultural trade, NAFTA eliminates all tariff and nontariff barriers to trade between the United States and Mexico over a five-, ten-, or 15-year period. CUSTA rules apply to U.S.-Canada agricultural trade, with most duties on liberalized sectors eliminated by 1998.

Although NAFTA will have both positive and negative impacts on U.S. agriculture, it is early in the implementation process, and many factors have influenced North American agricultural trade in recent years. NAFTA is designed to expand the flow of goods, services, and investment throughout North America. In contrast to the Uruguay Round agreements (URA), which reduce tariffs and increase import quotas, NAFTA calls for the full, phased elimination of import tariffs and the elimination or fullest possible reduction of nontariff trade barriers, such as import quotas, licensing schemes, and technical barriers to trade. Each member country receives significant preferential trade status in the markets of the other NAFTA partners. Trade restrictions applied by the three countries to imports from all other countries will remain in effect unless modified through the World Trade Organization (WTO) or other trade agreements.

OVERVIEW OF AGRICULTURAL TRADE IN NAFTA COUNTRIES

Mexico became a member of GATT in 1986, and tariffs on many U.S. products were reduced from as much 100 percent to 10 to 20 percent before NAFTA. Lower tariffs, coupled with stronger economic growth in Mexico, led to an upsurge in trade. Mexico has become the third largest market for U.S. agricultural exports, purchasing food and fiber valued at $5.2 billion in 1997, up 68 percent since 1990 (USDA, 1995-1998). Major U.S. agricultural exports to Mexico include grains, meats, livestock, oilseeds and products, and dairy (USDA, 1985-1997). Exports of beef, poultry, pork, corn, and soft fruits increased during 1994, NAFTA's first year.

The sharp devaluation of the Mexican peso in 1995 reduced short-term export opportunities. U.S. exports of most food products declined 25 to 35 percent because the prices of most U.S. goods nearly doubled. U.S. imports of Mexican products surged as the peso's value dropped relative to the U.S. dollar. For example, tomato imports increased 60 percent in one year and most other vegetable exports to the United States also increased. Mexican economic recovery was rapid. U.S. agricultural exports set a record of $5.4 billion in 1996; however, U.S. imports from Mexico have also remained strong.

With a growing population, more two-income families, and diversification of diets, Mexico is poised to remain a major market for U.S. food and fiber. About two-thirds of the Mexican population is below the age of 30 and half is below the age of 20 (CIA, 1994). Consumers in these age groups are approaching their peak consumption years. An estimated 40 percent of all Mexican households have two adults working outside the home, adding to household purchasing power. Mexico has about 93 million people and a population growth rate of 2.1 percent, which will result in 126 million inhabitants by 2010. However, an estimated 84 percent of Mexicans currently have per capita incomes below $865 per year, severely limiting their purchasing power. Market potential will be limited without substantial growth in incomes, continued economic development, and improvements in the efficiency of transportation, storage, and other handling infrastructure.

U.S. agricultural imports from Mexico were valued at $4.1 billion in 1997, up 64 percent since 1990 (USDA, 1995-1999). Major imports were vegetables and live animals—mainly feeder cattle, coffee, fruits, and nuts. Competitive imports have increased 57 percent since NAFTA implementation, leading to lower prices and returns for some U.S. producers.

Canada is the second largest market for U.S. agricultural exports behind Japan, with sales increasing 48 percent from $4.6 billion in 1990 to $6.8 billion in 1997 (USDA, 1995-1999). Although Canada's population is relatively small at 29.6 million people, it is an affluent country with a per capita GNP of nearly $20,000 (World Bank, 1997). Fresh vegetables, grains, fruits and nuts, meats, oilseeds, and sugar and tropical products account for the majority of export growth (USDA, 1995-1999). Canada is the number one agricultural supplier to the United States, with imports more than doubling from $3.5 billion to $7.5 billion during the same period. Most import growth has occurred in grains, live animals, meats, oilseeds, and fresh vegetables (USDA, 1995-1999).

NAFTA IMPACTS ON SELECTED COMMODITIES

NAFTA is expected to have a positive overall impact on U.S. agriculture. USDA (1997) estimates that U.S. agricultural exports to Mexico will increase by $2.0 billion per year after NAFTA is fully implemented, while U.S. agricultural exports to Canada have increased $2.26 billion since CUSTA. These estimates may be conservative if Mexico's economy continues to experience strong recovery, growth, and infrastructure development. Trade gains in the export of grains, meats, poultry, and cotton are likely to offset losses expected in the fruit and vegetable sectors. Most changes to any particular sector will be small because of the long length of the transition period and the relatively low level of many pre-NAFTA duties.

Assessing NAFTA impacts is complex because trade, employment, and economic growth are linked. Benefits to Mexico and Canada are expected through the lower prices to consumers achieved by tariff reductions; increased access to the U.S. market, leading to higher prices for some producers; greater efficiency, for example, from streamlined border procedures; and more jobs created by economic growth and development. Benefits to the United States will occur as exports in some sectors increase, creating additional business activity and more jobs as well as lower consumer prices. Costs will be incurred in both countries when domestic production is supplanted by imports and jobs are lost in some sectors. Those losses will be fully or partially offset, however, by economic growth and job gains in other sectors.

Increased demand in Mexico and Canada will raise U.S. prices, unless there is a corresponding increase in supply. Consumers of some products will be worse off if higher prices are passed on to them. More imports increase the supply available to consumers and prices will drop, unless domestic suppliers reduce production in the face of lower prices. The net benefits from trade will depend on the balance between gains and losses and the resulting impacts on prices, jobs, income, taxes, and costs.

ANIMALS AND ANIMAL PRODUCTS

Beef and Cattle Trade

Mexican duties of 20 percent on beef and 15 percent on live cattle were eliminated on January 1, 1994, under NAFTA. Canadian duties on fresh and chilled U.S. beef were eliminated under CUSTA on an accelerated schedule. U.S. duties of $.022/kilogram on Canadian cattle were phased out under CUSTA, beginning in 1989, and were eliminated in 1993. For Mexico, all duties on live cattle from the United States were eliminated immediately upon implementation of NAFTA.

U.S. exports of beef to Mexico reached 72,000 tons in 1994, up 84 percent from 1990-1991 (USDA, 1995-1999). The Mexican market represents about 5 percent of all U.S. beef exports. U.S. exports declined to 29,000 tons in 1995 due to the devaluation of the Mexican peso and subsequent economic recession, but recovered to 59,000 tons in 1996 and reached 107,000 tons in 1997. Mexican demand for imported U.S. beef could reach 200,000 metric tons by the year 2004. Over the long term, NAFTA will have a positive impact on U.S. beef producers as more meat, fed steers, and by-products are exported to Mexico. The United States imports only small quantities of Mexican beef.

Higher prices for beef and beef by-products will have a positive impact on cattle prices, on balance. However, these changes will be small because U.S.-Mexico livestock and meat trade is largely complementary, with the United States exporting meat, meat by-products, and fed cattle, and Mexico exporting feeder calves to the U.S. market. While meat exports may boost beef and cattle prices, imports of feeder steers reduce may them.

Feedlots, meat packers, and processors should benefit from larger export volumes. It has been estimated that at least 432,000 additional fed steers will be required to meet Mexican import demand. U.S. feedlot operators would experience small gains associated with additional volume. Feed grain producers will also benefit because these steers should consume an additional 351,000 tons of feed grains.

For U.S. stocker operations, Mexican cattle are a low-cost source of supply. U.S. imports of Mexican feeder and stocker cattle have expanded from 500,000 head in 1980 to 1.2 million head in 1994 and a record 1.6 million head for 1995 (USDA, 1995-1999). Only 456,000 head were imported in 1996, compared to 669,000 head in 1997 and 720,000 in 1998, while 618,000 were imported in the first 11 months of 1999 (USDA, 2000b). Constraints to increased cattle imports are limitations on herd size due to drought and herd liquidation, which occurred in 1994-1996 and continued through most of 1997. Mexican cattle herds fell from a peak of 32 million head in the early 1990s to a low of about 25.6 million in 1998 (USDA, 1997). Some sources place the herd size near 22 million. Undesirable breed characteristics, increased demand for meat in Mexico, and compliance with some U.S. animal health regulations also limit the potential supply of imported cattle from Mexico.

U.S. beef exports to Canada have been duty-free for several years, growing from about 30,000 tons in 1987-1988 to 100,000 tons in 1995 (USDA, 1995-1999). Since then, exports have fallen to about 90,000 tons. A stronger U.S. dollar has, in part, contributed to the export decline. U.S. beef imports from Canada have increased from 88,000 tons in 1987-1988 to 273,000 tons in 1997 (USDA,

1995-1999). Although beef imports from Canada decreased slightly in 1998, they increased in 1999 and were expected to rise in 2000, since larger numbers were placed in feedlots in the fall of 1999 (USDA, 2000b).

U.S. imports of Canadian cattle increased from 375,000 head in 1987-1988 to a record 1.5 million in 1996 (USDA, 1995-1999). The large majority of imported Canadian cattle are fed steers destined for U.S. meat packing plants. Imports declined to 1.4 million head in 1997 due to declining cattle inventories in Canada caused by herd liquidation, which began in 1996 and continued through 1998 (USDA, 1997). Consequently, the Canadian cattle herd will likely decline to 13 million head by the end of 1998, down from 13.4 million in 1996. Expansion of feedlot and meat packing plant capacity in Western Canada will likely result in some additional U.S. imports of fed cattle, but also could increase Canadian demand for feeder calves.

Pork and Hog Trade

Mexico implemented ten-year tariff-rate quotas (TRQ) on both hogs and pork under NAFTA. A TRQ is a transition measure implemented by the government of an importing country to protect domestic producers from import competition and provide domestic consumers with access to limited quantities of imports at world prices. A specified quantity is imported duty-free. When the quota is reached, it is nonbinding, allowing imports to continue, but a duty is charged on all over-quota imports. While a tariff may be placed on imports which fall below the specified quantity, that rate is often very small or zero.

Mexican duties of 20 percent on hogs and pork will be eliminated concurrently with the quotas under the TRQs. Canada eliminated duties on pork and hogs under CUSTA, allowing the United States duty-free access under an accelerated schedule. Most pork enters the U.S. market duty-free from Canada, but due to endemic hog cholera, imports from Mexico must be cooked and packed in airtight containers. The United States, however, has maintained a countervailing duty (CVD) on hogs from Canada since 1984. CVDs are used to offset export subsidies used by another country. Beginning in November 1996, the CVD was eliminated for slaughter sows, boars, and feeder pigs. Due to hog cholera, all hogs imported from Mexico are subject to a 90-day quarantine, which effectively eliminates hog imports.

U.S. pork exports to Mexico have increased from 22,000 tons in 1990-1991, peaking at 50,000 tons in the first year of NAFTA, declining, then recovering to 30,000 tons in 1997 (USDA, 1995-1999). During Mexico's peso crisis, 1995-1996, the largest declines in imported pork occurred for fresh, chilled, and frozen product. This decline reflected reduced purchasing power among consumers. Lower value prepared and preserved pork products led the recovery in exports in 1996-1997. Although U.S. slaughter hog exports to Mexico expanded in 1994, the peso crisis severely limited demand in subsequent years, leading to a drastic reduction in exports. Even so, Mexico represents the largest market for U.S. hogs, accounting for 95 percent of total U.S. hog exports in most years. Slaughter hog exports recovered slightly in 1996-1997, but recovery of breeding hog exports has been impeded by high feeding costs due to drought and lack of access to credit.

The United States is a net importer of pork and hogs from Canada. U.S. exports of pork are destined primarily for the population centers of eastern Canada, while Canadian pork and hogs are exported from western Canada to the U.S. West Coast. In recent years, more exports of hogs have originated in Ontario. U.S. pork exports to Canada have grown from nil in 1987-1988 to 40,000 tons in 1997, while U.S. pork imports from Canada declined 20 percent to 189,000 tons during the same period (USDA, 1995-1999). U.S. imports of slaughter hogs and feeder pigs have been stimulated by a relatively weak Canadian dollar, high U.S. prices, and expanding hog production in Canada. U.S. hog exports to Canada are limited, reaching only 2,800 head in 1997 (USDA, 1995-1999).

Poultry Trade

Mexico and Canada both utilize TRQs to control poultry meat imports. Under NAFTA, the Mexican quota was initially set at 95,000 tons, with over-quota duties ranging from 133 to 260 percent. Prior to the URA, Canada used import quotas to support the supply control system. The Canadian import quota for broilers was set at 6.3 percent of the previous year's production, with the turkey quota at 2 percent of the current output. Under CUSTA, quotas were increased to 7.5 percent for broilers and 3.5 percent for turkeys. Under the URA, Canada converted its quotas to a TRQ with over-quota duties ranging from 182 to 280 percent.

Even though both Mexico and Canada employ TRQs on poultry, U.S. exports have expanded since the early 1990s as duty-free quota amounts have been increased by both governments to accommodate greater demand for poultry meats. U.S. broiler exports to Canada have tripled since 1987-1988, reaching 60,000 tons in 1997. This trade expansion is due in part to the fact that Canada now allows the importation of poultry products as part of the TRQ set up under the URA. Products were precluded under the previous global quota system. U.S. turkey exports to Canada have fallen from 8,000 to 2,200 tons since 1987-1988. This decline has been partially offset, however, by an increase in exports of live birds.

U.S. poultry exports to Mexico have quadrupled from 51,000 tons in 1990 to 207,000 tons in 1997, while U.S. exports to Canada have more than doubled to 80,000 tons (USDA, 1995-1999). About 55 percent of the exports to Mexico are broilers and 34 percent are turkeys. About 80 percent of all U.S. poultry exports to Canada are fresh or frozen broilers. U.S. exports to Mexico are expected to continue increasing as trade barriers are lowered and economic development proceeds. Moderate gains in U.S. employment should occur as poultry output increases, mainly in the processing and service sectors, because growers can add capacity without much additional labor.

Dairy Products Trade

Under CUSTA, U.S. duties on dairy products were eliminated, but import quotas remained. Canada agreed to reduce import tariffs, but most quotas were retained to protect the integrity of the supply management program for dairy. U.S.-Canada dairy trade was excluded from NAFTA negotiations, and U.S. dairy quotas were converted to TRQs under provisions of the URA. The United States is a net exporter of dairy products to NAFTA countries, with a trade surplus of $145 million in 1996 and $179 million in 1997. Mexico accounts for 90 percent of this surplus dairy trade.

Under NAFTA, the United States granted Mexico TRQs for milk powder (422 tons) and cheese (5,550 tons), with over-quota duties ranging from 69.5 to 80 percent. Mexico granted the United States duty-free access to 40,000 tons of milk powder with an over-quota duty of 139 percent. All other Mexican duties on dairy products ranging from 20 to 40 percent are scheduled for ten-year phaseout. Mexico and Canada excluded dairy trade from NAFTA.

NAFTA creates the potential for more commercial exports of U.S. milk powder and evaporated milk to Mexico. During 1997, nonfat dry milk (NFDM) exports to Mexico were 2,013 tons, down from 34,000 tons in 1994 (USDA, 1995-1999). Evaporated and condensed milk exports increased sharply from 1,500 tons in 1990-1991 to 30,000 tons in 1996, but fell to 1,250 tons in 1997. The declines in U.S. exports reflect price and income sensitivity of Mexican consumers, coupled with reductions in U.S. dairy subsidies and restrictive trade practices of CONASUPO, the Mexican state buying agency. In the past, Mexico has imported large quantities of subsidized dairy products from the United States. In the short run, U.S. government export programs will continue to be an important factor affecting the level of Mexican dairy imports. Over time as the Mexican market is liberalized and per capita incomes grow, U.S. dairy product exports should increase.

U.S. cheese producers will probably benefit from NAFTA because the Mexican cheese import licensing scheme has been replaced with a 20 percent tariff to be eliminated over ten years. This will open a larger market for U.S. cheese and cheese products. U.S. cheese exports to Mexico have grown from 1,000 tons in 1990-1991 to 4,732 tons in 1996 and to 5,585 tons in 1997.

Lower Mexican tariffs on fluid milk, which will decline from 10 percent to zero over ten years, will create a larger market for U.S. products, resulting in expanded export demand and higher producer prices. Fluid milk processors will gain from greater demand for fluid milk and added access to the Mexican market. U.S. consumers may face slightly higher milk prices in some areas.

U.S. dairy exports to Canada have increased tenfold since 1987-1988, reaching almost $120 million in 1997 (USDA, 1995-1999). "Soft products" such as ice cream and yogurt account for about 55 percent of U.S. dairy exports, while whey exports account for

25 percent. Cheese and other products account for the balance of U.S. dairy exports. The increase in U.S. dairy export growth to Canada reflects the strength of demand, especially considering import duties ranging from 200 percent to more than 300 percent on most U.S. exports. U.S. dairy imports from Canada consist primarily of soft products, reaching $70 million in 1997. Cheese and milk imports from Canada have grown moderately.

GRAINS AND OILSEEDS

Feed Grains Trade

Trade in corn, grain sorghum, barley, and oats has been liberalized under CUSTA and NAFTA. U.S. and Canadian duties on bilateral corn trade were eliminated on January 1, 1998, following a ten-year phaseout under CUSTA, while U.S. duties on Mexican corn were eliminated immediately under NAFTA. Corn trade is being liberalized slowly by Mexico due to its political and social importance. Under a TRQ, a minimum of 2.8 million tons of corn may enter Mexico duty-free each year.

About 6.3 million tons of U.S. corn entered Mexico duty-free in 1996, exceeding the quota by more than 3.0 million tons due to drought in Mexico and a short crop (USDA, 1995-1999). NAFTA has resulted in Mexico becoming the third or fourth largest market for U.S. corn behind Japan, Korea, and, in some years, Taiwan. The duty-free quota level increases 3 percent annually and, along with the 180.6 percent duty on over-quota corn imports, will be eliminated after a 15-year transition period.

The United States and Canada immediately eliminated duties on grain sorghum upon implementation of CUSTA. Mexico's 15 percent seasonal tariff on grain sorghum was eliminated upon implementation of NAFTA, while the United States reciprocated.

During 1997, 2.1 million tons of U.S. grain sorghum were exported to Mexico. Mexican demand for U.S. feed grains has increased as more grains have been fed to cattle, hogs, and poultry. Grain sorghum exports have declined, however, due to increased competition from corn. Sorghum acreage in Mexico has increased due to domestic policy changes that made corn production less profitable, leading to more sorghum production. Mexican sorghum imports have declined, even after tariff elimination, due to changes in animal rations favoring corn. However, supply and demand conditions in the rest of the world are the dominant factors affecting U.S. sorghum prices.

U.S. duties on barley from Canada were phased out and eventually eliminated under CUSTA, while U.S. duties on Mexican barley were eliminated immediately under NAFTA. Mexico granted the United States a 120,000 ton TRQ for barley, with duties ranging from 128 to 175 percent. Canada, under CUSTA and later NAFTA, agreed to a ten-year tariff elimination on U.S. barley. Under the URA, this was converted to a TRQ with duties more than 100 percent, which were eliminated on January 1, 1998.

U.S. grain exports to Canada have quadrupled since 1990-1991, with total grain exports expanding from 600,000 tons to 2.5 million tons in 1997 (USDA, 1995-1999). Corn exports account for about half of the export growth, followed by mixed feed products, rice, feed grain products, and other grains.

U.S. imports of Canadian barley have been increasing since the late 1980s (USDA, 1997). During 1994, imports reached nearly 2 million tons, reflecting a downturn in U.S. production (USDA, 1995-1999). About two-thirds of all imported barley was for feed use, a historically high share due to less feed grain output in the United States.

U.S. barley exports to Mexico have increased since NAFTA, reaching 270,000 tons in 1996, up from 92,000 tons in 1994. Although Mexico was the leading importer of U.S. barley in 1996, exports declined to 126,000 tons in 1997. Nearly 90 percent of U.S. barley exports to Mexico are used for malting beer, while the remainder goes into the production of animal feed. Much of the Mexican beer produced with U.S. barley is exported back to the United States.

Although Mexico imposes a 10 percent duty on oats, to be phased out over ten years under NAFTA, and Mexican oats enter the United States market duty-free, the U.S.-Canada oat trade became duty-free when CUSTA was implemented in 1989. As a result of this trade liberalization, the U.S.-Canada oat market is one of the most highly integrated. As U.S. oat production has fallen, imports have risen, making the United States the world's largest importer. While Canada is the largest oat exporter to the U.S. market, Canadian production has increased only slightly in recent years. Due to relatively free trade, NAFTA has had only limited impacts on the U.S. oat sector.

Wheat Trade

The United States and Canada eliminated duties on bilateral wheat trade under CUSTA on January 1, 1998. U.S. duties on Mexican common wheat also ended on January 1, 1998, while duties on Mexican durum will be eliminated over ten years. Mexico's 15 percent duty on all U.S. wheat will be phased out over ten years.

U.S. wheat exports to Mexico have expanded from 321,000 tons in 1990 to 1.5 million tons in 1996, declining to 1.1 million tons in 1997 (USDA, 1995-1999). NAFTA will result in slight gains for some U.S. input suppliers and grain elevators as trade volume expands. Intense competition from Canada, however, has curtailed U.S. export growth in recent years. Price impacts solely due to NAFTA will be small because U.S. wheat prices are determined by many global factors, including weather and foreign economic conditions.

U.S. wheat imports from Canada increased from 372,000 tons in 1987-1988 to nearly 2.2 million tons in 1997 (USDA, 1995-1999). Due to an import surge in 1994, a one-year TRQ was implemented by the United States. Import growth resumed after the TRQ expired. It is estimated that U.S. wheat exports to Mexico have been about 1 percent higher with NAFTA and that U.S. imports of Canadian wheat have been 5 percent higher due to CUSTA and NAFTA (USDA, 1997).

Rice Trade

U.S. import duties on rice from $.0069/kilogram to $.033/kilogram will be eliminated under NAFTA over a ten-year period. Mexican duties of 20 percent on brown and milled rice and 10 percent on rough rice will be phased out over ten years as well. Canadian duties on brown and broken rice are zero. Duties on other rice were eliminated on January 1, 1998, under CUSTA.

U.S. rice exports to Mexico have expanded moderately as Mexico lowered its duties on rough and milled rice (USDA, 1995-1999). Since 1990-1991, U.S. rice exports have increased from 105,000 tons to 385,000 tons in 1997. The Mexican market represents about 10 percent of total U.S. rice exports. Mexico's phytosanitary ban on Asian rice has helped U.S. rice become a dominant product. While most Mexicans prefer high-quality long-grain rice produced in the United States, the ban provides an advantage to U.S. rice and precludes competition from high-quality long-grain rice produced in Thailand.

U.S. rice exports to Canada increased from 126,000 tons to 171,000 tons during the same period (USDA, 1995-1999). The United States is the dominant supplier of high-quality long-grain rice to Canada. NAFTA has given U.S. rice a slight price advantage in Canada, but most competing rices are relatively low quality. It is likely that most future competition for the high-quality long-grain market in Canada will come from Thailand, which may realize a stronger competitive advantage due to recent currency devaluations in Asia.

Oilseed Trade

Soybeans and soybean products dominate North American oilseed trade, accounting for 60 percent of U.S. oilseed trade with Canada and 80 percent of U.S. oilseed exports to Mexico. U.S.-Canada and U.S.-Mexico soybean trade was relatively duty-free before CUSTA and NAFTA. Under CUSTA, the United States eliminated relatively low duties on rapeseed, rapeseed meal, flaxseed, and canola oil. Similar concessions were granted to Mexico under NAFTA. Mexico agreed to a ten-year phaseout of its 10 percent seasonal duties on soybeans and other duties ranging from 10 to 20 percent on soybean meal and soybean oil. Before CUSTA, Canada did not have duties on soybeans, soybean meal, rapeseed, or other oilseed meals. Duties ranging from 7.5 percent on soybean oil to 10 percent on other vegetable oils were eliminated under a ten-year phaseout completed in 1998.

U.S. oilseed and oilseed product exports to Canada have increased 500,000 tons since 1987-1988, reaching 1.4 million tons in 1997

(USDA, 1995-1999). Soybean meal accounts for about 45 percent of total U.S. oilseed and oilseed product exports to Canada, followed by soybeans with 14 percent, other vegetable oils, and other oilseeds. Canada is the largest market for U.S. soybean meal, accounting for 10 percent of total U.S. soybean meal exports in 1997. Rising poultry production in Canada has stimulated demand for soybean meal and soybeans, with the majority of increased bean needs being met by domestic production and most additional meal needs being supplied by imports from the United States.

Canada is the major supplier of oilseeds and oilseed products to the United States, accounting for more than half of all U.S. oilseed and product imports in 1997. Oilcakes and meals represent the largest product category, tripling since 1987-1988 (USDA, 1995-1999). Oils and waxes, other oilseeds, and flaxseed account for most other import growth. As Canada expands domestic crushing capacity, it is expected that U.S. imports of some oilseed products, such as rapeseed, will decline.

U.S. oilseed and oilseed product exports to Mexico have doubled since 1990-1991 (USDA, 1995-1999). Soybean exports have accounted for about 80 percent of the growth in trade. Soybean meal and soybean oil are minor exports. Much of the demand growth for oilseeds in Mexico is directly related to greater production of poultry, fed cattle and hogs, and dairy products. The peso devaluation, coupled with greater competition from sunflower seeds, led to lower U.S. exports in 1995. Trade growth has since recovered, resulting in record export levels of both soybeans and soybean meal in 1997. The long-term success of U.S. oilseed trade with Mexico depends upon broad-based income growth and the transformation of the Mexican diet from plant protein staples to alternative sources of animal protein.

HORTICULTURAL PRODUCTS

Vegetable Trade

NAFTA has had mixed impacts on the U.S. vegetable industry. U.S. imports of Mexican products such as broccoli, cucumbers, and onions, which have been protected by high duties (17 to 35 percent ad valorem), have increased. U.S. fresh and frozen vegetable imports from Mexico increased 75 percent, from 1.2 million metric tons in 1990-1991 to 2.1 million tons in 1997 (USDA, 1995-1999). Mexico is the number one supplier of fresh vegetables to the United States, accounting for 63 percent of the total.

During this same period, U.S. tomato imports from Mexico nearly doubled to 661,000 tons, while cucumbers increased 75 percent to 286,000 tons, cauliflower was up to 203,000 tons, lettuce imports increased fourfold to 10,748 tons, squash was up 10 percent to 135,116 tons, pepper imports more than doubled to 256,000 tons, and onion imports expanded one-third to 218,000 tons.

Fresh vegetables and fruits imported from Mexico must comply with U.S. marketing order provisions *unless* such produce is destined for processing, in which case it is exempt. However, U.S. producers are concerned that as more Mexican produce is allowed into the market, the potential will increase for low-quality products to enter U.S. fresh marketing channels, driving down prices.

U.S. vegetable exports to Canada have doubled since 1987-1988, reaching 1.3 million tons in 1997. Canada is the number one market for U.S. fresh vegetable exports, accounting for three-fourths of the total. U.S. exports of canned and frozen vegetables have expanded sevenfold, exceeding 180,000 tons in 1997 (USDA, 1995-1999).

Potatoes, lettuce, and onions account for one-half of total U.S. fresh vegetable exports to Canada. Tomatoes, carrots, celery, and broccoli represent about one-quarter of U.S. exports, while other fresh vegetables account for the balance. With the exception of celery and broccoli, other U.S. fresh vegetable exports to Canada have increased since 1987-1988.

Canada is the second largest supplier of vegetables to the U.S. market, behind Mexico. U.S. imports of Canadian vegetables accounted for 29 percent of total U.S. vegetable imports. Potatoes were the number one import, representing 65 percent of the total from Canada (USDA, 1995-1999). Carrots, other vegetables, and vegetable products accounted for the balance of vegetable imports from Canada. All vegetable imports from Canada have increased under CUSTA.

Fruit Trade

U.S. exports of noncitrus fruit have increased as Canada and Mexico have lowered import duties. Fresh noncitrus exports to Canada expanded from 220,000 tons in 1987-1988 to 596,000 tons in 1997, making Canada the number one export market and accounting for 37 percent of total U.S. noncitrus exports (USDA, 1995-1999). Canned and dried fruits were 40,000 and 20,000 tons, respectively in 1997. U.S. fresh melon exports to Canada have increased more than 10 percent since 1995, reflecting growing demand for fresh produce in recent years.

U.S. imports of fresh fruits from Canada totaled 114,000 in 1997, representing only 5 percent of the total (USDA, 1995-1999). Apples accounted for 55 percent of total U.S. fresh fruit imports from Canada. Berries accounted for most of the balance of imports.

U.S. exports of fresh deciduous fruits to Mexico increased from 57,000 tons in 1990-1991 to 173,000 tons in 1997. Mexico is the third largest export market, behind Taiwan. Apples and pears account for 70 percent of the trade volume. Greater sales to the Mexican market should occur as incomes rise and consumers develop a preference for U.S. products. Sanitary and phytosanitary restrictions, however, may limit trade to some degree.

Mexico is the top supplier of fresh fruits to the U.S. market, accounting for 46 percent of the total in 1997 (USDA, 1995-1999). Melons, citrus, and mangoes account for 78 percent of total U.S. fruit imports from Mexico. Under NAFTA, some U.S. seasonal import duties on oranges and grapefruit were eliminated in 1994, while other duties will be phased out over ten years. Because of the low level of protection prior to NAFTA and the long phaseout period, it is expected that NAFTA will have no impacts on U.S. citrus producers. U.S. imports of fruits and preparations from Mexico have increased to more than 1.0 million tons since NAFTA. Melon imports were up by more than one-third to 439,000 tons in 1997, even though import duties on melons have remained relatively high. Citrus fruit and mango increased to 165,000 tons and 161,000 tons, respectively. Agribusinesses associated with the U.S. fruit and vegetable industries can expect both positive and negative impacts. Suppliers of seed, fertilizer, and other chemicals may experience minor losses as the production of some crops declines. Specialized infrastructure, such as packing sheds, may gain as imported volume expands and capacity increases; other operations may survive only by relocating nearer production areas in Mexico. Both farm and nonfarm labor can expect some lost employment opportunities as imports of Mexican produce expand.

OTHER COMMODITIES

Cotton Trade

The U.S.-Canada cotton trade was duty-free before CUSTA and NAFTA took effect. Under NAFTA, the United States implemented a cotton TRQ for Mexico of 46,000 bales, which will be phased out over ten years. Mexico's 10 percent duty on cotton imports is under ten-year NAFTA phaseout. U.S. duties on textile trade between the United States and Canada were eliminated on January 1, 1998, under CUSTA terms. U.S. import quotas were expanded and duties lowered for NAFTA partners under a five-year program. Some of Mexico's textile duties were eliminated immediately under NAFTA, while duties on about 97 percent of U.S. textile exports to Mexico were eliminated on January 1, 1998.

The United States has experienced large gains in both cotton and textile exports to Canada since NAFTA. In fact, a trade surplus exists with Canada in both raw and processed products (USDA, 1997). U.S.-Mexico textile trade is more complementary than textile trade with Canada and is complicated by recent domestic farm program changes in Mexico. U.S. exports to Mexico have traditionally consisted of raw cotton, fabric, and other intermediate products, while imports include finished products from Mexico. Mexico's cotton production has fluctuated widely due to higher support prices for corn, water constraints, and rising domestic use, which has stimulated prices. The 1994-1995 peso devaluation and the subsequent recession, along with the Asian financial crisis, all have combined to influence the Mexican cotton/textile complex and trade with the United States.

U.S. cotton producers should benefit from increased exports as Mexico's 10 percent duty is reduced over ten years. Since 1990-1991,

U.S. cotton exports to Mexico have expanded from 149,000 bales to 1.0 million bales in 1997 (USDA, 1995-1999). Despite rising Mexican production, mill use and consumption have been expanding. Recent investment in Mexican textile mills appears to favor U.S. cotton, while the Asian crisis has led to lower exports from those countries due to lack of credit and instability in their cotton industries. While U.S. cotton exports have benefitted, it is uncertain how strongly and when Asian textile exports will resume. U.S.-Mexico textile trade is probably less affected by NAFTA than currency fluctuations and other economic forces.

Asian instability and NAFTA also have led to increased U.S. exports to Canada. U.S. cotton exports to Canada expanded from 147,000 bales in 1987-1988 to a peak of 308,000 in 1996, then declined to 248,000 bales in 1997 (USDA, 1995-1999). U.S. cotton export growth averaged 10 percent annually over this period. With an affluent society and a growing market for both apparel and raw products, Canada should remain a strong market for the near term.

Mexico has gained greater access to the U.S. market as Section 22 import quotas have been replaced by a TRQ under NAFTA. U.S. cotton imports from Mexico reached 47,000 bales in 1996, up from almost nil in 1990, but fell sharply to 890 bales in 1997. Due to U.S. trade liberalization commitments under the URA, however, cotton imports from other countries reached 754,000 bales in 1996. While NAFTA has accelerated the move toward a more open market for North American textile trade, the URA phaseout of the Multi-Fiber Arrangement (MFA) will allow import protection for the most sensitive textile sectors until 2005.

Sugar Trade

U.S.-Mexico sugar trade is under a NAFTA TRQ that phases out trade barriers over a 15-year period, while U.S.-Canada sugar trade was regulated by CUSTA and became duty-free in 1998, but U.S. import quota provisions apply to Canadian sugar. Under NAFTA, Mexico has duty-free access to the U.S. market for up to 25,000 tons through 1999, provided that Mexico is a net surplus producer. Otherwise, Mexico is limited to a TRQ of 7,258 tons. Mexico's maximum TRQ increases to 250,000 tons in the years 2000-2008. Under NAFTA, U.S. sugar provisions are reciprocal and apply to

U.S. exports to Mexico, while tariffs are scheduled for elimination in 2008.

The Mexican duty on high-fructose syrup (HFS) will be reduced from 15 percent to zero over ten years. Mexican nontariff barriers to sugar-containing products have been converted to tariffs and will be eliminated over ten years. Reciprocal duty-free access exists between Mexico and the United States for refined sugar produced in either country.

In recent years, Mexico has imported up to 1.4 million tons of sugar, with the United States supplying 20 to 25 percent of the Mexican import market, mostly under the sugar reexport program. Duties of $.16 per pound on Mexican over-quota sugar imports will be reduced and eventually eliminated. In 1997 U.S. sugar exports to Mexico were 28,697 tons, down 4 percent from 1996 and nearly equal to those of 1994 (USDA, 1995-1999). U.S. sugar exports to Canada have decreased substantially over the past two years, but Canadian imports of U.S. sugar products continues to increase. Mexico has increased sugar production in recent years and is believed to be largely self-sufficient. U.S. imports of Mexican sugar were 30,551 tons in 1997, almost double 1995 levels, but down 28 percent from 1996.

U.S. and Mexican duties on HFS (6 to 15 percent) will be phased out over ten years. This should allow U.S. agribusinesses to export additional HFS as per capita incomes in Mexico increase and import demand for sweeteners expands. Although there are seven wet millers of corn in Mexico, it appears unlikely that capacity exists to keep pace with projected HFS demand without additional investment in infrastructure and increased corn production.

Peanut Trade

Canada eliminated all import duties on peanuts under CUSTA. Before NAFTA, Mexico had no quantitative restrictions on peanuts, but had a 20 percent duty on peanut butter, which will be eliminated by 2003. The United States imposed a TRQ on Mexican peanuts under NAFTA, which will be eliminated by 2008. Although there are no limits on peanut butter imports from Mexico, NAFTA rules of origin require that Mexican products be made of NAFTA-grown peanuts. U.S. imports of peanut butter from Canada are governed by a TRQ that limits imports of Canadian peanut butter and peanut paste to 14,500 tons, with the balance of the TRQ allocated to Argentina.

NAFTA was projected to have a small positive impact on U.S. peanut producers. U.S. exports to Mexico have more than doubled since NAFTA. While the peso depreciation in 1995 halted the growth in U.S. peanut exports to Mexico for that year, the speedy recovery resulted in a peak of 24,700 tons in 1996, with 1997 remaining near that level (USDA, 1995-1999). It is likely that higher consumer incomes in Mexico will lead to increased demand for U.S.-grown raw peanuts and peanut products, stimulating additional exports. However, on a percentage basis, U.S. imports from Mexico have increased even more, reaching 4,023 tons in 1996.

Canada has no domestic peanut production, making imports necessary. U.S. exports to Canada have fluctuated widely since NAFTA, but were 25 percent higher from 1994 to 1996 than in the three years preceding the agreement. Exports to Canada are somewhat limited by U.S. peanut program provisions that require peanut product exports to be manufactured from quota peanuts while allowing the export of "additionals," or peanuts grown outside of U.S. production quota limits.

CONCLUSIONS

While NAFTA created one of the world's largest free trade areas, assessing trade impacts is complex. It is also early in the transition period to more open markets, and many non-NAFTA factors may have a greater impact on U.S. agricultural trade than NAFTA. Trade, employment, and economic activity are interdependent and trade gains will not be made without some costs, as labor-intensive agricultural sectors face more competition from imports and must adjust.

Overall, U.S. agriculture stands to gain more than it will lose as trade barriers are lowered. Due to the long transition to freer trade, the relatively low level of pre-NAFTA duties, and the complementary nature of trade in many sectors, NAFTA's impacts on most sectors will continue to be slight to moderate. Livestock, meats, feed grains, dairy, cotton, soft fruits, and processed foods are examples of U.S. sectors that will benefit. Some labor-intensive fruit and vegetable producers have been adversely affected by NAFTA and are adjusting to the impacts. Even U.S. vegetable producers will receive some benefits over the longer term, as stronger economic growth in Mexico results in greater demand for many products. NAFTA also secures previous trade gains that have already benefitted many sectors of U.S. agriculture.

Benefits to the United States will occur as exports in some sectors increase, creating additional business activity and more jobs, and through lower consumer prices. Costs will be incurred when domestic production is supplanted by imports, prices fall, and jobs are lost in some sectors. Short-run adjustment problems and dislocation costs in the United States can be expected to rise in sectors that were protected by high pre-NAFTA duties, restrictive import quotas, and are labor intensive. Those losses will be offset, however, by economic growth and job gains in other sectors.

Increased demand in Mexico and Canada will raise U.S. prices, unless there is a corresponding increase in supply. Consumers of some products will be worse off if higher prices are passed on to them. More imports will also increase the supply available to consumers and prices will drop, unless domestic suppliers reduce production due to lower prices. The net benefits from trade will depend on the balance between trade gains and losses and the resulting impacts on prices, jobs, income, taxes, and costs.

Benefits to Mexico and Canada are expected through lower prices to consumers due to tariff reductions and the elimination of highly restrictive import quotas on some products, along with increased access for some producers to the U.S. market. It is important to note that while lower tariffs and less restrictive import quotas will expand trade, U.S. agriculture is now part of the global economy. Farm prices are determined by the interaction of supply and demand on the world market. Farm incomes depend not only upon NAFTA, but upon many other factors emanating from both the domestic and the international markets.

REFERENCES

CIA (1994). *World Factbook, 1994*. Washington, DC: Central Intelligence Agency of the United States.

USDA (1995-1999). *Foreign Agricultural Trade of the United States*. Economic Research Service, United States Department of Agriculture. Washington, DC.

USDA (1997). *International Agriculture and Trade Reports, NAFTA Situation and Outlook Series,* September. United States Department of Agriculture. Economic Research Service, Washington, DC.

USDA (2000a). *Livestock, Dairy and Poultry Situation and Outlook.* Economic Research Service, U.S. Department of Agriculture, Washington, DC, January.

USDA (2000b). *U.S. Agricultural Trade Update.* Economic Research Service, U.S. Department of Agriculture, Washington, DC, January.

World Bank. (1997). *World Development Report 1997.* New York: Oxford University Press.

Chapter 18
Biotechnology and Competitiveness
Nicholas Kalaitzandonakes

INTRODUCTION

Competitiveness is the ability to profitably create and deliver customer value. Firms create customer value through product offerings whose perceived benefits and price trade-offs compare favorably to those of competitors and of substitutes (Kennedy et al., 1997). Accordingly, cost leadership or product differentiation strategies may be pursued. Firms improve their competitiveness, however, only when they appropriate parts or all of the customer value they create. When the value created by firms migrates to other parts of the supply chain or is transferred to the customer, gains in their competitiveness are transitory. Hence, competitiveness studies must focus on both value creation and value distribution.

One of the most fundamental sources of competitiveness in the economy is technical and organizational innovation (Abbot and Bredahl, 1994). This chapter will present the competitiveness implications of agrobiotechnology, a new technological platform that will enable value creation in various parts of the agribusiness supply chain for years to come. The means by which value is both created and appropriated are therefore of interest in this analysis.

AGROBIOTECHNOLOGY AND VALUE CREATION

Agrobiotechnology has promised new products and processes for over two decades. This promise is finally beginning to materialize. To date, 48 transgenic crop products and almost 100 biopesticides have been approved for commercialization (James, 1997). First-generation agrobiotechnology products have been crops with herbicide tolerance and resistance to particular insect pests. Second-generation products, transgenic plants with enhanced quality traits, are fast approaching commercialization (Kalaitzandonakes and Maltsbarger, 1998).

Even the optimists among biotechnology proponents have been caught off guard by its extremely fast adoption rates at the farm level. In 1999, just four years after commercial introduction, almost 50 percent of the total U.S. corn, soybean, and cotton acreage was planted with transgenics (Kalaitzandonakes, 1999). Fast adoption rates are consistent with initial empirical evidence on value creation. Herbicide-tolerant and insect-resistant crops have created significant on-farm value, mainly through cost savings from reduced insecticide and herbicide use and improved risk management in pest control (Carlson, Marra, and Hubbell, 1997; Gibson et al., 1997; James, 1997).

It is possible, however, that on-farm economic benefits from first-generation agrobiotechnologies are transitory. If fast adoption rates result in oversupplied commodity markets, prices could fall precipitously, ultimately resulting in loss of value. This may or may not be true in the case of agrobiotechnology. Most of the first-generation agrobiotechnologies are not strongly yield-increasing but are instead input-reducing. In such a case, the "technological treadmill" effects may be limited and the value created through first-generation agrobiotechnologies may be sustainable.

Irrespectively, the possibilities for value creation are more tantalizing in the case of second-generation biotechnologies. Quality-enhanced crops, such as corn with high oil or lysine content and soybeans with high oleic acid or sucrose content, are being developed and in some cases marketed on a limited scale. Biotechnologies targeting quality enhancements have initially focused on large markets. Livestock feed uses over 75 percent of world corn and soybean production. Antinutritional factors, however, constrain the value of these feeds. Soybean meal, for example, is the most commonly used protein supplement in animal feed, but is nutritionally constrained by trypsin inhibitors. Processing overcomes some of these constraints but also downgrades its feed value. Removal of nutritional inhibiting factors through genetic engineering eliminates processing degradation.

Food markets have also been targeted. Three-fourths of the vegetable oils consumed worldwide derive from soybeans, canola, sunflower, and palm. All commodity oils are rich in polyunsaturated fatty acids that are prone to degrading oxidation during storage. Preservation by hydrogenation forms trans-fatty acids with properties similar to those of fully saturated fats associated with high cholesterol. Genes that encode enzymes in soybeans, canola, and other crops are being used to modify their fatty acid profiles.

Bioengineering may also create value through entirely new products. Genetically engineered crops are being developed to produce bioplastics, enzymes, and enhanced nutritional and pharmaceutical agents—known as nutraceuticals. Table 18.1 reports some key technologies, their developmental stage, and estimates of their potential value. Estimated values reflect both the size of the anticipated markets and the degree of technological advance. Most quality modifications have focused on corn, soybeans, and canola. Wheat, alfalfa, rice, sorghum, sunflower, and other crops are also being genetically modified for improved quality traits, but at a slower pace.

TABLE 18.1. Pipeline of Selected Quality-Enhanced Biotechnologies

Product	Technology	Developmental Stage	Value
Corn	High Oil	Commercial	High
	High Oil	Precommercial	Moderate
	High Lysine	R&D	High
	Low N-Fertilizer Need	R&D	Moderate
	Low Phytate	Precommercial	Low
	Modified Starch	R&D	Moderate
	Phytomanufacturing*		
Soybeans	High Oleic	Commercial	Moderate/High
	Improved Protein	Precommercial	High
	High Stearic	Precommercial	Low
	Phytomanufacturing*	R&D	Low
Canola	High Laurate	Precommercial	Low
	High Oleic/Low Linoleic	Precommercial	High
	High Saturates	R&D	Low
	High Erucid	Precommercial	Low/Moderate
	Phytomanufacturing*	R&D	Low

*Phytomanufacturing, also known as molecular farming, involves production of substances at the molecular level (e.g., enzymes).

Quality-enhanced crops create value through (a) product differentiation, (b) a parallel decommodification of markets, and (c) increased opportunities for value-adding services. Quality enhancements imply delivery of differentiated crops that are more valuable than commodity crops to specific end users. To a feeder, high quality is associated with high protein content. To a wet miller, high quality means large quantities of starch. Creating crops with bundles of attributes highly valued by particular end users is therefore the cornerstone of value creation through second-generation agrobiotechnologies. The unique conditions of various end users imply a wide distribution of reservation prices for quality-enhanced crops. Hence, possibilities for increasingly sophisticated market segmentation exist, leading to parallel expansion of market value (see the case of high oil corn for an example [Figure 18.1]).

The logistical particularities of value-enhanced crops are also of interest. While identity preservation is a logistical cost, it creates opportunities for new markets and/or new services (e.g., production labels, warranties, branding) and, again, parallel market value expansion. Identity preservation also leads to a more uniform, higher quality product due to improved handling and repeated testing procedures. Product uniformity is highly valued by end users as it leads to processing efficiencies.

The larger the number of quality-enhanced crops that become commercial, the faster crop markets will be decommodified. Reactionary or defensive strategies may also hasten decommodification. For instance, consumer resistance to genetically modified organisms (GMOs) in Europe has created a lucrative market for GMO-free grains produced in the U.S. Similarly, certain players in the grain processing and distribution industry are inventing new strategies that build on their existing core competence, that of logistics. Through the use of information and testing technologies they extract additional value by identity-preserving commodity grains produced at specific geographic regions known to yield certain nutritional and quality grain profiles (Ebbertt, 1998). Such strategies, in effect, increase the number of market segments to be tended and the rate of decommodification.

FIGURE 18.1. Quality-Enhanced Crops and Value Creation: The Case of High-Oil Corn (HOC)

> HOC has been the most visible quality-enhanced grain to reach market so far. HOC averages 6-8 percent oil content compared to 3 percent for conventional corn varieties. It also exhibits increased levels of crude protein and amino acids. Virtually all high-oil corn varieties are marketed under the OPTIMUM® brand developed by DuPont. In the United States, OPTIMUM was first introduced in 1992 and was planted on approximately 1.5 million acres in 1998, indicating a brisk interest by end users.
>
> The value of HOC technology stems from projected savings in supplemental fat and improved digestibility and feed efficiency. This value varies with the prices of substitutes (e.g., white grease), complementary inputs, and the final products. In 1998, 70 percent of all HOC was produced for local consumption. The remaining 30 percent was produced under contracts for export, sold typically at premium prices to countries where fat additives are in short supply. Widely differing market values between domestic and export markets encourage a two-tiered market segmentation strategy. In the domestic market, the technology is broadly distributed through standard market transactions. For the export market, DuPont has developed a tightly coordinated system. More specifically, DuPont contracts with farmers who receive a fixed premium over commodity prices. Supply-chain logistics are managed through a strategic alliance with Continental Grain. HOC may be delivered to the contracting elevator or stored on-farm for an extra premium payment. Buyer calls for delivery of grain stored on the farm allow elevators to coordinate storage and transportation and manage capacity during peak harvest operation times.
>
> Deliveries coordinated through Continental Grain utilize grain stocks from elevators and on-farm storage to fulfill export agreements developed by DuPont. Grain delivered to one of over 80 export contracting elevators is shipped in segregated loads to a port location where grain is packaged for delivery in 50,000-bushel loads. Near infrared technology is used to assess nutritional composition at each delivery point in the chain. The grain is often analyzed for oil content and other characteristics up to three times at elevator, rail, and barge port facilities. Plant identity is preserved in production and marketing supply chains. If comingled with commodity crops, quality differentials and value would be lost.

VALUE APPROPRIATION AND FIRM AND INDUSTRY COMPETITIVENESS

With market value expanding through the commercialization of agrobiotechnology products, two key questions arise next: how may such value be distributed along the agrifood chain? And, what kind of assets may facilitate value claims? These are key questions in our case since value distribution along the agrifood chain determines the competitiveness impacts of agrobiotechnology on different firms and industries. Naturally, one must begin to address this question by analyzing the potential value share of the biotechnology companies that are responsible for the innovation and value creation in the first place.

The Role of Intellectual Property and Distribution Assets in Value Claims

The ability of biotechnology firms to appropriate value from their innovations is decided principally on two conditions: the degree of imitability and the strength of intellectual property rights available to the innovation (Teece, 1987). A variety of factors make technical innovations more or less difficult to imitate. Accumulated knowledge and management experience in research, regulatory compliance, scaling up, partnership formation, procurement, and production are all significant to bringing technology from the laboratory to the market and are often difficult to imitate. If a technology can be easily codified and copied, imitators can appropriate a significant share of the value created from the innovation. Firms use intellectual property rights (e.g., patents and trade secrets) to protect their technology from imitators. Intellectual property rights that are not strongly defensible lead to weak appropriability.

Generally, if technical innovations are strongly appropriable, and if the complementary assets required for commercialization (e.g., manufacturing capability, marketing and distribution networks) are not specialized, then the innovator can ordinarily contract or make open market transactions for these assets' services while capturing the bulk of the innovation value. If innovation is only weakly appropriable and if the complementary assets are specialized to a narrow range of potential uses, however, then the owners of complementary assets may capture a large portion of the value created through the innovation. In this case, innovators must contract or vertically integrate to gain control of such assets or lose innovation value to an outside supplier. There are, of course, numerous market procurement and coordinating strategies between these two extremes that may conform to varying degrees of innovation appropriability and control of complementary assets.

Claims on Value Generated by Agrobiotechnology

First-generation agrobiotechnology has proved to be only weakly appropriable (Joly and DeLooze, 1996; Kalaitzandonakes and Bjornson, 1997). Intellectual property rights for fundamental technologies have overlapped and have been heavily contested. A number of multiparty intellectual property disputes for key technologies, such as insect and herbicide resistance, reached the courts over the last few years and led to the invalidation of key patents (Hayenga, 1999). Similarly, a relatively large number of firms brought similar commercial products to the market, suggesting substantial imitation (Kalaitzandonakes and Bjornson, 1997).

High-quality proprietary germplasm, the delivery instrument for agrobiotechnology and hence a key complementary asset, has proven to be in a stronger position than biotechnology know-how. In short supply and subject to significant development lags, germplasm commanded a significant share of the value forthcoming from weakly appropriable agrobiotechnologies. Under such conditions, vertical integration into the seed business and ownership of germplasm became a primary strategy of agrobiotechnology firms to profit from their innovations. Of course, a part of the value created from biotechnology has been transferred to seed assets in the form of lofty prices paid in recent mergers and acquisitions (see Table 18.2).

In sum, few agrobiotechnology firms, those with strong intellectual property rights positions, were able to claim significant value from first-generation agrobiotechnologies either in the form of up-front acquisition payments or through product commercialization (Kalaitzandonakes and Marks, 1999). Downstream firms (e.g., seed breeders, seed merchandisers, and farmers) claimed the remaining value.

The degree of appropriability of second-generation agrobiotechnologies is still unclear as claims on intellectual property rights continue to come forward. Significant investments are currently being made in genomics research by leading biotechnology firms (see Table 18.3). Such research races are indeed as much about establishing strong intellectual property positions as they are about developing new concepts and technologies. What is certain is that the resulting degree of

TABLE 18.2. Selected Mergers and Acquisitions in the Biotechnology, Seed, and Processing Industries, 1993-1998

Company One	Country One	Company Two	Country Two	Cost $ (Millions)
DuPont	US	Pioneer (pending)	US	9,400
Monsanto	US	DeKalb	US	1,900
Monsanto	US	Delta & Pine Land (pending)	US	1,700
DuPont	US	PTI	US	1,500
Monsanto	US	Cargill Seed Int	US	1,500
Monsanto	US	Holdens	US	1,000
AgrEvo	Germany	PGS	Belgium	770
Monsanto	US	Calgene	US	700
Monsanto	US	PBI	UK	550
Dow	US	Mycogen	US	522
ELM	Mexico	Asgrow	US	330
Monsanto	US	Asgrow Agronomics	US	240
Monsanto	US	Agroceres	Brazil	220
Monsanto	US	Agracetus	US	150
ELM	Mexico	DNA Plant Technologies	US	N/A
ELM	Mexico	Hung Seed, Choongang Seed	South Korea	100
Zeneca	UK	Mogen	Netherlands	70
Mycogen	US	United AgriSeeds	US	26
Mycogen	US	Dinamilho	Brazil	12
DuPont	US	Dalgerty	UK	N/A
DuPont	US	Hybrinova	UK	N/A
DuPont	US	Cereals Innovation Center	UK	N/A

Source: University of Missouri, 1999.

appropriability of second-generation biotechnologies will determine how value will be distributed along the agrifood chain and, hence, will have significant competitiveness implications.

Efforts to coordinate technology and complementary assets targeting second-generation biotechnology are also underway. Some variants of coordination have emerged, suggesting possible schemes for value distribution. Mycogen, a U.S. biotechnology and seed company, develops, produces, and delivers proprietary, high-oleic acid sunflower seeds exclusively to AC Humko, the world's largest marketer of edible oils. Similarly, DuPont, through its bid to acquire Pioneer Hi-Bred and its acquisition of Protein Technology International (which has over 70 percent market share of the food-quality soybean protein market), is preparing for tight coordination from seed to the

TABLE 18.3. Selected Genomics Alliances and Investments, 1996-1998

Company 1	Company 2	Research Area
Novartis		Foundation—Novartis Agricultural Discovery Institute
Novartis		Foundation—Novartis Institute for Functional Genomics
Monsanto	Millennium Pharmaceuticals	Genomics
DuPont	Lynx Therapeutics	Plant genomics
AgrEvo (Hoechst)	Gene Logic	Plant genomics
Bayer	Paradigm Genetics	Bioinformatics
Pioneer	Curagen	Agronomic traits
Monsanto	Gene TraceR	Plant genomics
Pioneer	Oxford Alycosciences	Proteomics
Ceres	Genset	Bioinformatics
Dow	Biosource Technologies	Functional genomics and bioinformatics
DuPont	Acacia Biosciences	Trait descriptions and reaction to chemical compounds
DuPont	Curagen	Bioinformatics
Monsanto	IBM	Bioinformatics
Monsanto	Incyte	Plant genomics
Novartis	Chiron	Combinatorial chemistry
Novartis	Combichem	Combinatorial chemistry
Novartis	Incyte	Plant genomics
Pioneer	Human Genome Science	Corn genomics
Rhone-Poulenc	Gene Logic Inc.	Pharmogenetics
Zeneca	Incyte	Genomics

Source: University of Missouri, 1999.

end user. Monsanto and Cargill are preparing to jointly develop and commercialize quality-enhanced bioengineered crops targeting the feed and other processing industries. Their joint venture combines Monsanto's extensive capabilities in biotechnology and seed with Cargill's global processing infrastructure and marketing and logistics capabilities.

NETWORK COMPETITIVENESS

The transformation of the U.S. and international crop markets will increasingly demand a new concept of competitiveness. Quality-enhanced crops must be produced and distributed within tightly coordinated identity-preserved chains to maintain value. Such value materializes at the end of the chain and must be allocated to all participants through up-front negotiation and conjecture. As a result, interdependencies abound.

Increased coordination along with market segmentation resulting from the introduction of quality-enhancing agrobiotechnogies will ultimately transform crop market structure. Networks will emerge to coordinate intellectual property rights and distribution assets for particular end uses (Nohria and Eccles, 1993; Penov and Kalaitzandonakes, 1999). These networks will likely transcend firm, industry, and national boundaries. And given the inherent interdependencies in the competitive position of members, factors that determine whole-network competitiveness will be increasingly relevant.

The transformation of the market will take place gradually and must be facilitated by additional innovation besides biotechnology. Information technologies and the Internet will likely play a key role allowing firms to keep up with increasing complexity and information flows. Electronic databases, precision farming, and testing technologies (e.g., near-infrared spectroscopy) will yield data and information on production and supply chain practices along with crop qualities, which will in turn define their functional characteristics and potential value. Electronic data interchange (EDI) and Internet-based tools will facilitate distributive accesses of such information, with almost no time and space constraints.

Online trading, buying, contracting, account accessing, and logistics management will be essential tools for the transformation of the market. Hence, the ability of network partners to efficiently use and leverage such resources will be essential components of their individual and collective competitiveness. Access to intellectual property rights and key (often location-specific) distribution assets will be fundamental to network competitiveness. Effective network management practices will also become important, as management of structures with distributed decision making is still an emerging field (Miles and Snow, 1992).

Economies of scale and scope will continue to be significant determinants of competitiveness within these new market structures. The sources of such economies are consequential. Economies of scale and scope in R&D are very large and favor further consolidation of agrobiotechnology firms. Such economies may not be nearly as important in the future in processing and distribution. Within a heavily segmented market, it is possible that small niche players, regional, national, and a few large global firms could all be competitive in their respective, though interconnected, submarkets. Sources of some scale and scope economies in processing and distribution may be related to large fixed investments in information technology infrastructure and advantages from geographic diversification.

IMPACTS OF AGROBIOTECHNOLOGY ON NATIONAL COMPETITIVENESS

With multinationals currently dominating the agrobiotechnology industry, is it relevant to discuss the impacts of biotechnology on national competitiveness? Surprisingly, ownership of intellectual property rights and coordination among firms in the agrobiotechnology industry has been largely contained within national or regional boundaries. It is therefore possible that innovation rents are contained within such boundaries, with implications for national and regional competitiveness.

U.S. firms have an overwhelming leadership position in agrobiotechnology patents. An estimated 93 percent of all U.S. agrobiotechnology patents issued until the end of 1998 in the United States were owned by U.S. firms or, in a few cases, by U.S. universities and public research institutes. Similarly, U.S. firms and universities own all of the top 25 U.S. agrobiotechnology patents, as defined by the number of times cited by other agrobiotechnology patents.

U.S. multinational firms with major ownership position in agrobiotechnology patents (e.g., Monsanto, DuPont, and Dow Agrosciences), have aggressively implemented expansive strategies and have made significant acquisitions of distribution assets in the European Union and South America (see Table 18.2). In the process they have secured leading positions in the global seed and biotechnology markets. The same firms are also heavy investors in genomics research (see Table 18.3). If patent ownership position in second generation-agrobiotechnology and beyond turns out to be proportional to current genomics R&D investments, U.S. firms should continue to hold their current dominant positions well into the future.

The expansive overseas acquisitions and joint ventures of U.S. multinational firms suggest that agrobiotechnology will be quickly transferred across national borders. Brazil, Argentina, Europe, India, and China are key targeted markets. Under such conditions, it is difficult to predict how innovation rents will be distributed among geographic regions. However, if strong patent and distribution asset positions continue to be claimed by U.S. firms, a large portion of innovation rents from agrobiotechnology will likely accrue within national borders, as the bulk of knowledge assets for such firms are located in the United States.

CONCLUSIONS

Agrobiotechnology has already created substantial value and, where available, it has been embraced quickly by farmers. It is expected that the value created by agrobiotechnology will continue to expand in its second commercialization phase and beyond. The competitiveness implications of agrobiotechnology, however, must be understood not only by how much value it creates but also by how it distributes value among firms and industries along the agrifood supply chain.

The relative strength of technology and distribution assets will continue to be determinant on value allocation and hence on competitiveness. Because of the increasing need to coordinate complex configurations of technology and distribution assets in order to create and appropriate value from agrobiotechnology, more or less loosely structured competitive networks of firms will emerge. Many of the recent structural changes in the agrifood supply chain are about the formation and configuration of such competitive networks and should therefore be followed with interest.

Traditional commodity crops will continue to be an important part of the market as commodities subject to coarse grades and standards will fulfill the bulk of the demand for the next several years. Over time, however, supercommodities and specialty crops will assume an increasingly important role. Demand for enhanced qualities and functional characteristics will be augmented by agrobiotechnology research. Discovery and innovation for more effective ways of making use of such enhanced qualities will also be important.

To be sure, important issues remain. Agrobiotechnology is facing increasing resistance in certain parts of the world, with uncertain outcomes (Zechendorf, 1998; Joly and DeLooze, 1996). Similarly, agrobiotechnology, like any other technical innovation at early developmental stages, is subject to uncertain outcomes that may expand or negate commercial value. This analysis has assumed no unanticipated material (positive or negative) technical outcomes or breakdown in public acceptance of agrobiotechnology.

REFERENCES

Abbot, P. and M. Bredahl (1994). "Competitiveness: Definitions, Useful Concepts and Issues." In *Competitiveness in International Food Markets*, M. Bredahl, P. Abbot, and M. Reed (Eds.), Boulder, CO: Westview Press, pp. 11-35.

Carlson, G., M. Marra, and B. Hubbell (1997). "Transgenic Technology for Crop Protection: The New 'Super Seeds'." *Choices* (Third Quarter): 31-36.

Ebbertt, J. (1998). "The Impacts of Biotechnology on the Grain Processing Industry." *AgBioForum*, 2(1): 78-80. Retrieved February 1, 1999 from the World Wide Web: <http://www.agbioforum.missouri.edu>.

Gibson, J., D. Laughlin, R. Luttrell, D. Parker, J. Reed, and A. Harris (1997). "Comparison of Cost and Returns Associated with Heliothis Resistant Bt Cotton to Non-Resistant Varieties." In *1997 Proceedings, Beltwide Cotton Conferences*, P. Dugger and D. Richter (Eds.), National Cotton Council, Memphis TN.

Hayenga, M.L. (1999). "Structural Change in the Biotech Seed and Chemical Industrial Complex." *AgBioForum* 1(2): 43-55.

James, C. (1997). *Global Status of Transgenic Crops in 1997*. ISAAA Briefs No. 5. Ithaca, NY: The International Service for the Acquisition of Agri-biotech (ISAAA).

Joly, P. and M. DeLooze (1996). "An Analysis of Innovation Strategies and Industrial Differentiation Through Patent Applications: The Case of Plant Biotechnology." *Research Policy* 25: 1014-1027.

Kalaitzandonakes, N. (1999). "A Farm Level Perspective on Agricultural Biotechnology." Paper presented at the USDA Outlook Forum, Washington, DC, February.

Kalaitzandonakes, N. and B. Bjornson (1997). "Vertical and Horizontal Coordination in the Agro-biotechnology Industry: Evidence and Implications." *Journal of Agricultural and Applied Economics* 29(1): 129-139.

Kalaitzandonakes, N. and R. Maltsbarger (1998). "Biotechnology and Identity Preserved Supply Chains: A Look at the Future of Crop Production and Marketing." *Choices* (Fourth Quarter): 15-18.

Kalaitzandonakes, N. and L. Marks (1999). "Biotech Merger Mania and Intellectual Property Rights." *Farm Chemicals*.

Kennedy, P.L., R.W. Harrison, N.G. Kalaitzandonakes, H.C. Peterson, and R.P. Rindfuss (1997). "Perspectives on Evaluating Competitiveness in Agribusiness Industries." *Agribusiness: An International Journal*, 13(4): 385-392.

Miles, R. and C. Snow (1992). "Causes of Failure in Network Organizations." *California Management Review* 34(4): 53-72.

Nohria, N. and R. Eccles (1993). *Networks and Organizations.* Boston, MA: Harvard Business Press.

Penov, I. and N.G. Kalaitzandonakes (1999). "Plant Biotechnology, Firm Strategies and Industrial Structure." Paper presented at the Southern Agricultural Economics Association Annual Meetings, Memphis, TN.

Teece, D. (1987). "Profiting from Technological Innovation: Implications for Integration, Collaboration, Licensing and Public Policy." In *The Competitive Challenge*, D. Teece (Ed.), Cambridge, MA: Ballinger, pp. 185-200.

University of Missouri (1999). *Agrobiotechnology Database.* Department of Agricultural Economics, Columbia, MO.

Zechendorf, B. (1998). "Agricultural Biotechnology: Why Do Europeans Have Difficulty Accepting It?" *AgBioForum* 1(1): 8-13.

Chapter 19

Findings and Implications

Dale Colyer
Curtis M. Jolly

INTRODUCTION

The strong export performance of U.S. agriculture in recent years is an indication that the sector is highly competitive in international markets. However, as the analyses of the various products contained in the preceding chapters show, there are large variations in competitiveness among individual commodities and groups of commodities. These results also are an illustration that the situation is in constant flux as technology and related factors affecting the productivity of individual products change in response to research, as producers and exporters evaluate and seek alternatives to expand trade, as economic and agricultural policies of the United States and other countries are modified, as existing international agreements are amended or new ones developed, as the international political situation changes, as diverse rates of economic, political, and social development occur, or as other countries expand production or adopt improved technologies to enhance their competitive positions.

IMPORTANCE OF AGRICULTURAL EXPORTS

Agricultural products were 8.9 percent of total U.S. exports in FY 1997 and 8.2 percent for 1998 (USDA, 1998b, p. 52; USDA, 2000); agricultural trade declined in 1998 due to weak demand in Asia. However, the trade balance for agricultural products is positive while that for other products is negative. In FY 1997, $53.7 billion worth of agricultural products were exported out of a total for U.S. exports of $627.7 billion. The balance of trade for the agricultural sector was a positive $1.6 billion compared to a negative $238.1 billion for total exports.

The total volume of agricultural exports increased from 131.1 million units (primarily metric tons) in 1977-1978 to 145.1 million units in 1996-1997, a 10.6 percent increase (USDA, 1998a). The total value of agricultural exports increased from $27.3 billion to 57.4 billion, a 110.2 percent increase; much of the increase in the value of exports was due to higher prices (inflation), but a substantial portion was the result of the shift toward exports of higher valued products, including processed products.

The United States produces and exports a large variety of agricultural commodities including bulk, semiprocessed, and highly processed products. The USDA publication *Foreign Agricultural Trade of the United States (FATUS)*, utilizes the harmonized codes and lists around 1,000 different agricultural products, under 12 general categories, that are exported by the United States (USDA, 1998a); several subclasses of beef, pork, and poultry and many other types of products contribute to that large total. However, a relatively small number of these account for a substantial share of total agricultural exports. Wheat and flour, feed grains (mostly corn), and oilseeds (mostly soybeans) made up over two-thirds (68 percent) of the volume and 33 percent of the value of agricultural exports in FY 1996-1997 (USDA, 1998a, p. 36). These amounts were down from over 80 percent of the volume and 55 percent of the value of exports in the 1970s, indicating that U.S. agricultural exports have become more diversified and that higher per unit value and value-added products have become more important in the export environment.

Thus, the relative composition of U.S. agricultural trade has been shifting in recent years. The physical volume of all agricultural trade increased substantially between 1987 and 1997, with the index of total trade rising from 99 in FY 1987 to 145 in FY 1997 (USDA, 1998a, p. 38). Within these totals, however, there were large shifts among the individual categories of products, including increases in processed products. Total animal products, for example, more than doubled, with the index of export volume rising from 103 in FY 1987 to 234 in FY 1997, compared with nearly constant exports of grains and feed (95 in FY 1987 and 96 in FY 1997). Within the animal products sector, poultry was the largest gainer; its index rose from 98 to 633 during that time span. Red meat exports also rose, going from an index of 102 in FY 1987 to 340 in FY 1997. However, the quantities of dairy products exported declined substantially, with their index dropping from 93 in FY 1987 to just 46 in FY 1997. Exports of oilseeds, fruits and vegetables, and other products also increased. Cotton and tobacco exports have fluctuated considerably but were somewhat higher in 1997 than in 1987, although tobacco exports were lower in 1997 than they had been in 1984.

The leading destinations for U.S. agricultural exports in FY 1997 were Japan, Canada, Mexico, South Korea, and Taiwan—the Netherlands, China, Hong Kong, Germany, and the United Kingdom round out the list of top ten importers (USDA, 1998a, p. 51). Japan alone accounted for 18.7 percent of the value of total U.S. agricultural exports. Canada and Mexico (due in part to NAFTA) together accounted for another one-fifth of the total and the top ten countries imported 63.2 percent of total U.S. agricultural exports.

Exports constitute significant proportions of the annual output for substantial numbers of many major and minor agricultural products and, thus, are very important in the determination of agricultural production, prices, and income. Table 19.1 gives information on the production and exports of major agricultural products for 1997 (USDA, 1998b). Over 40 percent of the wheat and rice crops were exported, 36 percent of the cotton, and 19 percent of the corn. For soybeans over 37 percent were exported, while about 13 and 20 percent of soybean oil and meal production, respectively, were exported, meaning that around half of the crop was exported. Among the animal products, about 17 and 11 percent of broiler and turkey production were exported while over 8 and 6 percent of beef and pork production were sold abroad; even these latter percentages are important in helping determine the prices of the products as well as determining farmers', processors', and other workers' employment and incomes.

TABLE 19.1. Production and Exports of Major U.S. Agricultural Commodities

Product	Units	Production	Exports	% Exported
Red Meats	Million lb	43,358	3,184	7.3
Beef	Million lb	25,490	2,136	8.4
Pork	Million lb	17,274	1,044	6.0
Poultry	Million lb	32,964	5,646	17.1
Broilers	Million lb	27,041	4,664	17.2
Turkeys	Million lb	5,412	596	11.0
Wheat	Million lb	2,285	1,001	43.8
Rice	Million lb	171.3	78.4	45.8
Corn	Million lb	9,293	1,795	19.3
Soybeans	Million lb	2,382	882	37.0
Soybean Oil	Million lb	15,752	2,037	12.9
Soybean Meal	1,000 tons	34,210	6,994	20.4
Cotton	Million lb	18.9	6.9	36.5

Source: USDA, 1998b.

MARKET SHARES

Although U.S. agricultural exports have been growing in volume and value, their share of world agricultural exports declined in the 1980s, after having reached historic highs in the 1970s; they have been relatively constant in recent years, at around 14 percent of the world value of agricultural exports (Figure 19.1; see also Gehlhar and Vollrath, 1997). Several important agricultural commodity exports account for a much higher percentage of the total world trade in those products. U.S. soybean exports still make up about two-thirds of total world trade, while U.S. corn and wheat exports generally make up 40 percent or more of world exports; the latter two products fluctuate from year to year depending on world and U.S. production levels, prices, and other demand- or supply-related factors. Recently U.S. poultry exports have been around 40 percent of the world total, a proportion that has increased substantially; it was only around 20 percent in the late 1980s. The fractions of U.S. exports for corn, wheat, and soybeans, however, have declined since the 1970s, when U.S. agricultural exports were more dominant than at present. Countries such as Brazil and Argentina, as well the European Union, have increased their relative shares for several products in recent years.

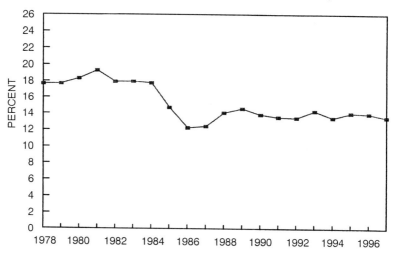

FIGURE 19.1. U.S. Proportion of World Agricultural Exports

Source: FAO, 1998.

FACTORS AFFECTING COMPETITIVENESS

The factors affecting the competitive position of U.S. agricultural products vary considerably, with the prices of some products being competitive (relatively low) due to favorable factor endowments or climate, others due to efficient production methods and technologies, or as a result of long-term dominance of the United States in total world production of the product. Other products are competitive because of their high quality, or due to factors such as closeness of the foreign market, strong positions due to a long tradition of exports, preferences of importers, trade agreements such as NAFTA, or for various other reasons.

U.S. agricultural and economic policies also contribute to the competitiveness of exports, including support programs, research investments, relatively open market polices, export promotion and development (including limited use of subsidies), and relatively large imports of both agricultural and nonagricultural products, which provide dollars to potential importers of U.S. products.

On the other hand, economic, agricultural, and trade policies of other countries also affect the competitive position of a number of U.S. products, some favorably and others unfavorably. Many trade barriers still exist, both tariff and nontariff; some countries still subsidize their exports; and others have state trading companies or other policies that enable them to affect the types, sources, and quantities of their agricultural imports. Under the World Trade Organization (WTO), established by the agreements reached in the Uruguay Round of GATT negotiations, participating countries have agreed to provide greater market access (generally through tariff rate quotas), convert nontariff barriers to tariffs (a process called tariffication), reduce domestic agricultural supports and export subsidies, and to justify sanitary and phytosanitary restrictions on trade with scientific evidence. While these barriers are still permitted to protect human health, prevent importation of harmful diseases and pests, etc., they must be justifiable by science and are subject to challenges by producing countries that believe they are not based on science but, instead, are being used to protect domestic producers. Thus, the United States and Canada, for example, have challenged the EU's exclusion of hormone-fed beef. There are numerous other challenges to specific provisions for excluding a number of products by various countries, including some provisions imposed by the United States (Roberts, Josling, and Orden, 1999; USDA, 1998c, pp. 27-33).

Livestock and Livestock Products

The export of livestock and livestock products has been growing relatively rapidly due both to an expanding world market and to the ability of U.S. producers to compete with other producers. The most rapid increases have been for poultry products, especially broiler meat, followed by pork and then beef. The United States produces relatively little lamb and mutton and is a large net importer of those products. The exports of dairy products have declined as the involvement of the U.S. government in buying and disposing of surpluses has diminished. Advantages of U.S. beef, pork, and poultry industries include large domestic production and consumption bases, a large and relatively cheap supply of feed grains (and forages in the case of beef), strong research and technological innovations, and generally high-quality products that are processed efficiently. The primary disadvantage of the United States vis-à-vis many other exporters—with the exception of most European countries—is high labor costs, although much of this is often offset by greater labor efficiency. Nonetheless, the processing of many animal products requires high labor inputs, which can make U.S. products less competitive. This disadvantage is notable in the cases of Brazil, Thailand, and, lately, China, which all have low wage rates and are able to process poultry and other products at lower costs than the United States.

Grain Crops

The United States is a major exporter of feed grains, oilseeds, wheat, and rice. Major competitors include Argentina, Australia, Brazil, Canada, and the European Union. Although except for rice the United States remains the largest exporter of most grains, its share of the market has decreased in recent years as Brazil, Argentina, and other developing countries have expanded production and exports and as European countries have subsidized production and exports of several of these products. For rice, the United States is not one of the larger producers, but it does play a significant role in international trade, despite being a relatively high-cost producer. Thailand dominates rice trade, but with a lower-quality product. The primary U.S. advantages in rice are the production of a high-quality product and the availability of a dependable source for export. For corn, soybeans, and wheat the United States is a large, efficient producer of high-quality products. Its post-World War II domination of international trade in these products resulted in advantages based on having established reputations as well as marketing channels and facilities, but other countries have been able to compete for a larger share of these markets due to lower-cost production and aggressive marketing practices; multinational corporations have contributed to the broadening of the base of exporting countries as they have sought cheaper sources to remain competitive. U.S. grain exports, however, have continued to grow as international demands for agricultural products have tended to increase at a relatively rapid rates. Soybean exports have been affected by the use of substitutes, especially for soybean oil, as canola (rapeseed) and palm oil have tended to replace some uses of soyoil. The exportation of feed grains and soybeans is also affected by the growth in exports of meat products since some producers, notably Japan, have replaced some of their domestic meat production with imports and therefore tend to import less feed grain and protein supplements.

Other Crops

The other crop and crop product exports evaluated in this book include fruits, vegetables, peanuts, wine, sugar, and tobacco. The competitiveness among these varies considerably due to their diverse nature and the widely varying situations under which they are produced and marketed throughout the world. For some, such as sugar, the United States is not a major exporter (although it is competitive in costs for beet sugar and it competes successfully in the high-fructose corn syrup market). For tobacco, while exports are important for U.S. producers, the country is not one of the low-cost exporters; Brazil, Malawi, Thailand, and Zimbabwe, for example, produce and sell tobacco at much lower costs than the United States, but the United States is more competitive in the export of manufactured tobacco products, especially cigarettes. The United States is an important exporter of fresh fruits, especially citrus (oranges and grapefruit) and apples, but are net importers of orange and apple juices. The reduction of trade barriers by several Asian countries has opened opportunities for greater trade in fresh fruits. Fresh vegetables are traded primarily between the United States, Mexico, Canada, and a few other relatively close countries due to their perishability; much trade is complementary with respect to season and other factors. While production costs in Mexico are less than in the United States, other costs tend to offset this disadvantage. For cotton, exports of the raw product are very important, accounting for around 36 percent of production. However, the United States imports large quantities of cotton textiles, implying that the value added by processing it into thread and cloth and manufacturing clothing and other textile products is captured by other, generally less developed, countries with lower labor costs. The United States has been a major exporter of peanuts although its market share has declined. U.S. production has been maintained through high-priced support programs that are being phased out, which is apt to have negative impacts on the industry.

Processed Products

The food processing industry is the largest manufacturing industry in the United States, and exports of processed agricultural products are growing and becoming more important, a situation that is desirable since additional processing adds value to exports and creates additional employment and income. The country is also a net exporter of processed foods (see Figure 19.2). From 1991 to 1997 the U.S.'s exports of manufactured foods increased at an annual rate of 7.2 percent; imports increased at a 5.4 percent annual rate. Although the contribution to exports of manufactured foods remains relatively small, the percentage of exports of processed foods surpassed the contribution of raw products to total exports. Exports of processed food products increased by 73 percent from 1989 to 1996, while exports of unprocessed foods increased by 30.3 percent from 1991 to 1996. Canada, Mexico, and Japan are the main importers of U.S. processed food products.

FIGURE 19.2. U.S. Imports and Exports of Processed Foods, 1991-1997

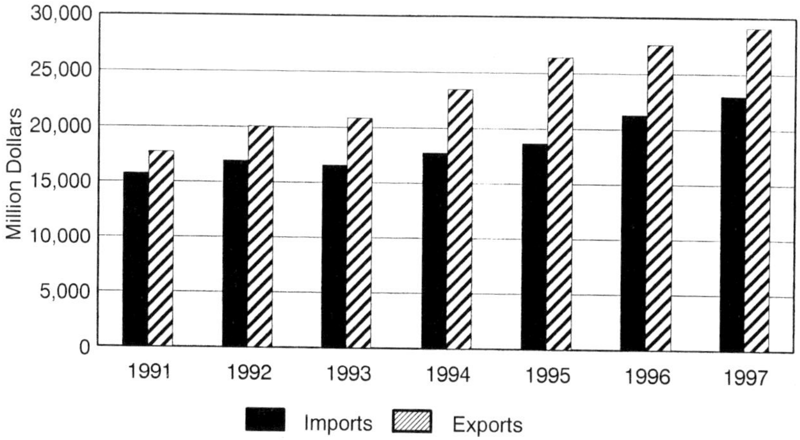

Source: USDA, 1998a.

Meat and poultry products remain the leading processed food products exported. Meat product exports increased by 67.2 percent from 1991 to 1997, about the same as all SIC-20 exports (see Chapter 16). The exports of processed meat products were influenced by lower trade barriers and higher incomes throughout the world. The United States also leads in the exports of fats and oils, and milled grain products. Other products making a significant contribution to U.S. processed food products exports are dairy, bakery products, and beverages.

The U.S. competitive position in the processed food products industry is difficult to establish because of the role played by multinational enterprises (MNEs), and the importance of foreign direct investments in the industry. A number of MNEs produce foods abroad; they import processed foods from the United States and also export food from their foreign plants to the United States (see Chapter 16). It also is difficult to tell whether the processed foods imported into the United States are consumed domestically or re-exported. The difficulty in discerning the U.S. competitive position in the processed food industry is exacerbated by trade destination. The EU countries, which produce and export large quantities of processed foods, sell their products mostly in other EU markets, whereas the United States sells more of its processed products to Canada, Mexico, and Japan.

The United States seems to be very competitive in the processed food industry with respect to labor productivity and indices of real trade for high-value-added products. Output per worker in 1992 was $238,000 in the United States versus $210,000 in France, $184,000 in Germany, $166,000 in the U.K., and $159,000 in Japan (see Chapter 16). The indices of real trade for high-value-added products were 95.2, 100.4, and 108.2 for 1995, 1996, and 1997, respectively, for the United States, compared to 98.3, 100.1, and 110.9 for U.S. competitors for the same years (USDA, 1998d). (It should be noted that the processed food products referred to previously may not be identical to high-valued products.) Changes brought about by the WTO will tend to result in greater market access for U.S. processed food products. The gains will accrue as the EU countries reduce subsidies paid to producers on exports. Other gains will result as trade barriers are eliminated, and the export market becomes more transparent. Processed food products are expected to benefit from NAFTA. Canada and Mexico remain the largest market for U.S. processed products, and exports destined to these markets will grow as trade barriers are reduced.

Globalization

The food and fiber industries, like many others, are part of the ongoing process of globalization, i.e., the international integration of markets (see Henderson, Handy, and Neff, 1996). This phenomenon is characterized by direct investment in and/or expansion of firms in countries other than the home country, by development of affiliations with firms in other countries, and by other forms of participation, such as licensing or franchising, as well as increased trade. It involves, substantially, production and trade of branded products, with increased intraindustry trade, i.e., simultaneous imports and exports of similar products by a country. Consumers tend to benefit from this process through a larger variety of products and increased competition within local markets. However, other impacts can vary depending on the direction of the flows and related factors. Thus, domestic food processors may suffer when foreign firms establish plants that compete in the national market. These plants will tend to increase employment and lower domestic prices; they often increase the demand for domestically produced farm products but may import raw materials.

TECHNOLOGY

The development and rapid adoption of genetically modified organisms (GMOs) in agriculture have become important and controversial issues in agricultural production and the international trade in such products; some 48 transgenic crops and nearly 100 bioengineered pesticides have been registered (see Chapter 18). These are becoming the improvements in technology that will enable U.S. producers to remain and/or become more competitive, replacing the mechanical and biological/chemical innovations that had contributed to increases in U.S. agricultural productivity in the decades following World War II.

Traditional approaches to technology development have been characterized by diminishing returns, while the new technological platform provided by biotechnology is restoring increases in productivity. The potential of the agricultural sector to continue to better feed the increasing world population may be enhanced by this new platform, although many innovations are directed at product or marketing improvements rather than yield enhancements. Development of these new technologies tend, however, to be expensive, and many are being produced by private firms that must be able to appropriate much of the benefits to justify the investments; however, farmers and others will not utilize the innovations unless they also benefit.

The introduction of these GMOs, however, has created problems in international trade due to consumer reluctance to consume products that they think might produce harmful effects, fear of their effects on other organisms, distrust of science, or impacts on cultural, religious, or other beliefs (USDA, 1998c, pp. 34-35). Prohibitions, restrictions, labeling requirements, and other barriers have been erected to impede the international flow of genetically modified commodities and/or those produced utilizing pesticides created using biotechnology procedures. While, under WTO rules, technical and sanitary/phytosanitary barriers must have a science-based justification, consumers in many countries do not trust the ability of science to fully and adequately evaluate the safety and effectiveness of these new technologies. A factor is that GMOs were not specifically addressed in the negotiations of the Uruguay Round, and it has not been determined if regulations applying to technical barriers to trade and sanitary/phytosanitary issues apply to the GMOs. These will, undoubtedly, be addressed more specifically in the next round of negotiations.

EFFECTS OF TRADE AGREEMENTS

Since World War II, the development of world trade has been enhanced by several sets of negotiations under the General Agreement on Tariffs and Trade, although agricultural trade was not a major issue until the recent Uruguay Round, when the WTO was created. In addition to the GATT agreements, several regional and bilateral agreements have important impacts on the competitiveness of U.S. agriculture. One of the foremost for many years has been the European Union, by which a number of European countries have united their economies to eliminate all barriers to trade within the group, while constructing a common set of barriers against countries that are not participants (although under the Lomé agreement many developing countries, former European colonies, are granted special trade access). These have included tariff and nontariff barriers to trade under their Common Agricultural Policy, domestic price supports, and export subsidies for agricultural products.

Also, a large number of regional trade organizations and agreements have important impacts, actual or potential, on the competitiveness of U.S. agricultural producers. These include MERCOSUR (Common Market of the Southern Cone Countries) and the Andean Free Trade Region in South America, and the North American Free Trade Agreement of Canada, Mexico, and the United States (which incorporated the previously negotiated and implemented Canada-United States Trade Agreement—CUSTA).

World Trade Organization

Historically, GATT has moved the world to more open trading, although the process tended to except agricultural products until the Uruguay Round. Although this agreement may foreshadow freer trade and minimum access barriers in agriculture, trade in agricultural products is still characterized by many protectionist provisions. The major impacts of the current agreement on agriculture include tariffication of nontariff barriers, reductions in tariff rates, minimum access provisions with lower or no tariffs, reductions in export subsidies, and reductions in domestic agricultural support programs. In addition, sanitary/phytosanitary regulations, while allowed for protection of human health or prevention of imports of harmful species, pathogens, pests, and diseases, must be based on science. Finally, the dispute settlement process set up under the WTO provides for a more complete and, hopefully, equitable method for settling trade disputes, for agricultural trade issues as well as more general trade disagreements.

North American Free Trade Agreement

NAFTA has both positive and negative effects on U.S. agriculture, but the overall impact is expected to be positive. The NAFTA agreement, signed between the United States, Mexico, and Canada in 1994, called for the phased elimination of tariffs and the complete elimination or reduction of nontariff barriers (see Chapter 17). Canada is the second largest market for U.S. agricultural exports, behind Japan. Sales increased from $4.6 billion to $6.8 billion between 1990 to 1997, a 48 percent change. Canada is the primary (and a net) exporter of agricultural products to the United States, with $7.5 billion in sales to the United States 1997, a 100 percent increase from 1990. Canada's major exports are grains, live animals, oilseeds, and fresh vegetables.

Since the initiation of NAFTA, U.S. agricultural exports to Mexico have escalated, making Mexico the third largest export market for U.S. agricultural products, with purchases of $5.2 billion in agricultural products in 1997, up 68 percent from 1990. Major U.S. agricultural exports to Mexico include grains, meats, livestock, oilseeds and products, poultry, and dairy. It is expected that these trends will continue as U.S. food producers and processors gear their industries toward exports to NAFTA member countries. The United States is a net exporter of agricultural products to Mexico; U.S. imports from Mexico totaled $4.1 billion in 1997, a 64 percent increase since 1990. Major imports include vegetables, live animals (mainly feeder cattle), coffee, and nuts.

The effects of NAFTA on individual crops and products vary. Trade gains in exports of grains, meats, poultry, and cotton are likely to be offset by expected losses in the fruit and vegetable industries. Increased export demand for U.S. products can result in an increase in domestic prices, unless a surplus already exists to cover the increased demand from NAFTA countries. Thus, prices of beef and beef products in the United States are expected to increase slightly. Increases in exports of hog products to Canada and Mexico are expected under NAFTA, although the United States remains a net importer of pork and pork products from Canada. Exports of poultry products to Mexico have quadrupled since 1990, and exports of broilers to Canada tripled between 1987-1988 and 1997.

Within NAFTA, the United States is a net exporter of grains, dairy products, oilseeds, and rice, while importing large quantities of vegetables and horticultural crops, trends expected to continue in the future. The United States has experienced significant growth in cotton sales to Canada while fiber trade with Mexico tends to be complementary; the United States exports raw cotton, fabric, and other intermediate products to Mexico, while importing finished products. U.S. peanut exports to Mexico almost doubled under NAFTA, while exports to Canada have fluctuated.

Other Trade Agreements

In addition to WTO and NAFTA, the United States participates in the Asia Pacific Economic Cooperation forum (APEC) and the Free Trade Area of the Americas (FTAA), and has bilateral agreements with Israel and Japan (Burfisher and Jones, 1998). APEC involves 21 Pacific Rim countries and has the objective of developing free trade by 2010 (2020 for developing country members). FTAA involves, potentially, all Western Hemisphere countries (34) and will, if completed and implemented, eliminate all tariffs and other intra-hemispheric trade barriers. The 1985 Israel-U.S. Free Trade Area agreement has the objective of eliminating tariffs and other barriers to trade between the two countries but allows each country to protect agricultural products considered sensitive and important to the country; Israel has protected a vast array of agricultural products under this escape clause, but a 1996 agreement has liberalized trade further and requires that nontariff barriers be converted to tariffs, and implements tariff rate quotas and other provisions intended to benefit the U.S. agricultural sector. A 1988 agreement with Japan only eliminated beef import quotas as well as those on oranges, orange juice, and eight other horticultural products. It had been preceded by other agreements that opened the Japanese beef market to the United States (and other countries). Japanese imports of beef have increased and U.S. producers have benefitted, although Australia and other exporters also have gained; the U.S. share has remained relatively constant.

CONCLUSIONS

The analyses of competitiveness presented in this book clearly indicate that the United States is in a strong position with respect to a large share of its agricultural products and that the country can expect to continue to benefit from a positive agricultural trade balance. Despite this it also must be realized that the world's economic and trading systems are dynamic, that many factors can affect the competitive position of any individual product, and that the existence of a competitive edge, currently or in the past, does not guarantee that one will prevail in the future; several shifts in the leading exporters of many agricultural products have occurred in recent years and more can be expected in the future. The United States has a large and highly productive agricultural resource endowment, including a human resource base, a favorable climate, and highly productive technologies that have enabled it to become and remain competitive in the production and trade of a large number of agricultural products. Despite the favorable endowments, maintaining its competitive position is determined largely by the development of improved technologies that enable its farmers to produce high-quality products efficiently and its agribusiness industries to process and market those products efficiently. Many other countries are developing or adapting such efficient agricultural production and processing technologies and/or learning to utilize quality control procedures to enhance their abilities to compete in international markets. Furthermore, the world's economic and trade policy frameworks are being transformed as globalization produces new rules and requirements for competing in international markets. The United States is in a good position to compete in this new environment, due primarily to its strong public and private agricultural research systems. To do so, however, it is essential to continue to be the world's leader in developing and implementing technological innovations.

REFERENCES

Burfisher, Mary and Elisabeth A. Jones. (Eds.). 1998. *Regional Trade Agreements and U.S. Agriculture.* Agricultural Economic Report Number 771, Economic Research Service. Washington, DC: U.S. Department of Agriculture.

FAO. 1998. Electronic Database, <http://apps.fao.org/lim500/nph-wrap.pl?Trade.CropsLivestock>, accessed March 13, 1999. Food and Agriculture Organization.

Gehlhar, Mark J. and Thomas L. Vollrath. 1997. *U.S. Export Performance in Agricultural Markets.* Technical Bulletin 1854. Washington, DC: Economic Research Service, U.S. Department of Agriculture, February.

Henderson, Dennis R., Charles R. Handy, and Steven A. Neff. 1996. *Globalization of the Processed Food Market,* Agricultural Economics Report 742. Washington, DC: Economic Research Service, U.S. Department of Agriculture.

Roberts, Donna, Timothy E. Josling, and David Orden. 1999. *A Framework for Analyzing Technical Trade Barriers in Agricultural Markets.* Technical Bulletin 1876. Washington, DC: Economic Research Service, U.S. Department of Agriculture, March.

USDA. 1998a. *Foreign Agricultural Trade of the United States: Fiscal Year 1997 Supplement.* Washington, DC: Economic Research Service, U.S. Department of Agriculture.

USDA. 1998b. *Agricultural Outlook.* Washington, DC: Economic Research Service, U.S. Department of Agriculture, December.

USDA. 1998c. *Agricultural in the WTO,* WRS-98-4 Washington, DC: Economic Research Service, U.S. Department of Agriculture, December.

USDA. 1998d. *Agricultural Outlook.* Washington, DC: Economic Research Service, U.S. Department of Agriculture, May.

USDA. (2000). "Total Value of U.S. Agricultural Trade." U.S. Department of Agriculture, Economic Research Service. <http://www.econ.ag.gov:80/briefing/agtrade/htm/Data.htm>, accessed January 22.

Index

Page numbers followed by the letter "f" indicate figures; those followed by the letter "t" indicate tables.

1978 Farm Bill, 146
1990 Farm Bill, 185
2SLS, 107

Abbott, P.C., 26, 262, 293
Abooki Bahiigwa, 105, 117
AC Humko, 300
Acacia Biosciences, 301t
Ackerman, K.Z., 230
Ad valorem tariffs, 149, 150, 194
Additionals, 146, 289
Advertising, 7, 20
Aflatoxin, 151
AgCanada, 38n
AgrEvo, 300t, 301t
Agricultural Adjustment Act of 1938, 188
Agriculture
  balance of payments, 3
  balance of trade, 308
  competitiveness, 1, 4, 11-23, 27, 125, 144, 311, 322
  commodities, 21, 37, 112, 144
  demand, 15, 269
  economies, 231
  economists, 14
  education, 5
  exports
    and competitiveness, 99
    history, 20
    importance, 307-309
    and NAFTA, 83, 310

Agriculture, exports *(continued)*
    promotion, 230
    share, 2, 25, 26, 99, 155, 307, 308, 309
    and strong dollar, 46
    value, 2, 3f, 308
    volume, 308, 310t
    world market share, 37, 38n, 311f
  factors of production, 35
  GDP, 27, 33, 34f, 35
  growth
    European, 33-35
    sources of, 31f
    U.S., 28, 29
  imports, 3, 255, 272
  industries, 265
  markets, 37, 38n
  output, 29, 32, 35, 37
  policy
    Argentina, 110, 116
    Brazil, 116
    and competitiveness, 194, 228
    domestic, 12, 129, 319
    European, 319
    international, 13, 177, 307, 312
    revisions, 46
    strategic, 23
    U.S., 5, 43, 116, 177, 307, 311
  prices, 21, 29, 32, 37, 38
  production, 13, 55, 177, 231
  productivity, 5, 35, 317
    TFP, 30, 32, 35, 36f, 37

325

Agriculture *(continued)*
   products, 1, 4, 8, 221, 230, 319
      differentiated, 13, 14, 21
      processed, 26, 251, 315-317
   programs, 12
   research, 5, 11
      R&D, 32, 37, 38
   sector
      European, 28
      intensive, 289
      primary, 32
      potential of, 318
      U.S., 1, 3, 8, 11, 14, 28
   terms of trade, 26
   trade, 1, 4
      agreements, 6, 8, 12, 13, 318-321
      composition of U.S., 308, 309
      destinations, 309
      and dispute settlement, 319
      and European Union, 6
      liberalization, 95
      and NAFTA, 112, 114, 167, 270-272, 289, 320, 321
      negotiations, 13, 84, 99
      policy, 164, 311, 312
      surplus, 11
      and tariffs, 112
      world, 4, 13
      WTO provisions, 13, 164, 167
   treadmill, 14
Agri-food chain, 297, 300
Agrobiotechnology, 293, 294, 298, 299, 303-305
Agroceres, 300t
Agrocetus, 300t
Ahearn, M., 5f, 144
Alabama, 139, 229
Alberta, 61
Algeria, 63, 64, 68, 259
Alipoe, D., 126
Allshouse, J.E., 200
American Sugar Alliance, 183, 186, 188, 194
Amponsah, 153, 235
Andayani, Sri R.M., 160, 164

Animal
   health regulation, 274
   products, 273, 308, 309, 313
   protein, 201, 283
Animal Agriculture Reform Act, 240
Antidumping, 171
APEC (Asian Pacific Economic Cooperation), 76, 77, 326
Appellation d'origine contrôlée (AOC), 180, 259
Apples
   juice, 161, 162, 163, 314
   production, 155, 157, 161, 163
   Red Delicious, 165
   trade, 6, 162, 163-166, 285, 314
Appropriability, 299, 300
Apricots, 155, 157, 161
Archer Daniels Midland, 111
Argentina
   and agrobiotechnology, 304
   beef, 203, 206, 207, 208f, 209
   cotton, 128, 132
   grain mill products, 310
   peanuts, 140, 141t, 143, 149t, 150, 151, 289
   pork, 245
   poultry, 226t, 228
   rice, 84t, 85, 86, 97
   soybeans
      costs, 109, 111
      competitiveness, 100, 101, 102, 104, 111
      consumption, 102
      market share, 101t
      model, 106
      policy, 110, 116
      production, 100
      productivity, 107, 108
      trade, 100, 112, 113
   sugar, 183t
   tobacco, 189, 191t, 193, 194
   U.S. competitor, 313
   wheat
      competitiveness, 76
      export credit, 68

Argentina, wheat *(continued)*
    exports, 59, 64f, 65, 66
    model, 71-75
    production, 59, 61, 62t, 63
    trade, 60
Arizona, 157
Arkansas, 79, 80, 81
Arkansas Global Rice Model, 86-96
Arnade, C.A., 33, 35
ASEAN (Association of Southeast
    Asian Nations), 103, 112,
    114, 190
Asgrow, 300t
Asp, E., 241
Asparagus, 170
Australia
    beef
        competitiveness, 206, 209
        and Japanese market, 210
        production, 203, 204, 211
        trade, 208f, 209, 211, 212, 213,
            321
    cotton, 121, 123, 128f, 133
    fruit, 161, 166
    grain mill products, 259
    pork, 245
    poultry, 227t
    rice, 84t, 85, 86, 88
    sugar, 183t
    TFP growth, 35, 36f
    U.S. competitor, 313
    wheat
        competitiveness, 75, 76
        exports, 60, 64f, 65, 66, 69
        model, 71, 72, 73, 74
        production, 59, 61, 62t, 63, 75
Australian Wheat Board, 68
Average cost, 18, 53, 190
Ayuk, E., 126

Bailey, K.W., 63, 69
Bakery products, 251, 254, 255, 256,
    266, 316
Balance of payments, 3

Balance of trade, 1, 190, 308
Ball, E., 33
Bangladesh, 79, 97
Banque Francaise du Commerce
    Exterieur, 68
Barbados, 183
Barkema, A., 15
Barker, R., 53, 55
Barley, 279, 280
Barshefshy, C., 183t
Basmati, 94, 96
Beef
    competitiveness, 199, 209-213
    consumption, 199, 202, 203, 204t,
        205-207, 242, 243t
    demand, 199, 230
    exports
        and NAFTA, 271, 273-275, 320
        share, 2, 309
        volume, 105, 204t, 208t, 252, 312
    feed required, 238
    and forages, 312
    and health, 242
    hormone-fed, 312
    imports, 105, 204t, 208t, 321
    industry growth, 201
    marketing, 200, 201, 214
    prices, 220
    producers, 203, 204
    production, 45, 204t
    promotion, 7, 199
    protein source, 199
    trade, 207-209
    U.S. sector, 199
Beer, 256, 260
Belgium, 300t
Belgium-Luxembourg, 191, 258, 259
Belize, 183
Bell peppers, 170, 173, 174f
Benirschka, M., 187
Beverages
    exporters, 259
    exports, 253, 254t, 266, 316
    and FDI, 262
    fermented, 181

Beverages *(continued)*
  imports, 250, 255, 256, 257t, 260
  SIC-208, 251
  subindustry output, 252, 253, 255
  wine and beer, 256
Bierlen, 83, 97, 98
Biery, 220, 232
Bilateral
  agreements, 318, 321
  competitiveness, 27
  trade, 135n, 235, 279, 281
Bioengineering, 295
Biopesticides, 293
Bioplastics, 295
Biosource Technologies, 301t
Biotechnologies
  adoption, 14
  appropriability, 299, 300
  companies, 298
    mergers, 300t
    Monsanto, 301
    multinationals, 303
  and competitiveness, 8, 12, 293
  and consumer acceptance, 115
  and corn, 52, 53
  development, 14
  and food processing, 201
  know how, 299
  and productivity, 318
  proponents, 294
  and quality, 294, 295t
  second-generation, 294, 300
  and soybeans, 115
  and trade barriers, 318
  and U.S. advantage, 299
  and value creation, 299
Bishop, R.V., 219, 221
Bjornson, B., 299
Blakely, L., 126
Blandford, D., 263
Boars, 276
Boskin, M.J., 35
Brand advertising, 20
Bray, J.D., 52

Brazil
  beef
    production, 203, 204
    consumption, 205, 206, 207
    trade, 208f, 209
  biotechnology mergers, 300t, 304
  competitive advantage, 313
  cotton, 121, 122
  feeding stuffs, 259
  fruit, 157
  orange juice, 160
  market share, 309
  pork, 245, 248
  poultry
    and climate, 229
    competitiveness, 230, 231
    cost of production, 222, 223, 224
    exports, 220
    nontariff barriers, 227t
    trade agreements, 22
  rice, 84t, 85, 86, 88, 93, 95, 97
  soybeans
    bound tariffs, 113t
    costs, 109, 110, 111
    competitiveness, 107
    exports, 105
    policy, 110, 116
    price, 106
    production, 100, 101, 102, 105,
      108, 116
    productivity, 108
    trade agreements, 112
    yield, 107t
  sugar, 182, 183t
  TFP growth, 35, 36f
  tobacco, 189, 190, 191t, 192, 193,
    194, 313, 314
Bread, 60, 63, 256
Bread wheat, 60
Breast meat, 224
Bredahl, M.E., 16, 26, 171, 262, 267,
  293
Brisket, 210
British Columbia, 68

Broccoli, 170, 283, 284
Broilers. *See also* Chicken
  competitiveness
    and climate, 229
    and costs, 221-224
    and feed supplies, 230
  costs of production, 221-224
  exports
    Brazilian, 220
    model, 220
    quantity, 218f, 220, 277, 310
    share, 1, 310
  import quotas, 276, 277
  importers, 220
  prices, 220, 224
  trade
    agreements, 227, 228
    barriers, 225-227
Brooks, R., 237
Brown, D.K., 51, 193, 197
Buckley, K.C., 171
Bulgaria, 226
Burfisher, M., 58, 32
Burgundy labels, 180
Burley, 190, 192
Burma, 86
Bureau, 39
Buse, R., 126

Cake, 256
Calgene, 300t
California
  apples, 161
  citrus, 157, 158
  cotton, 119, 121
  rice, 79, 80, 92, 94
  wine, 178, 181
Calvin, L., 166
Cambodia, 259
Campbell, G.R., 54

Canada/Canadian
  beef
    competitiveness, 209, 210
    consumption, 205, 206, 207
    hormone fed, 312
    trade, 209, 211, 212
  and biotechnology, 114
  cotton, 121, 126, 131, 321
  fruit, 160, 162, 164, 165
  grain exports, 313
  and NAFTA, 49, 112, 270-290,
    319, 320
  oilseeds, 106
  peanuts, 141, 143t, 149, 150, 151,
    321
  pork
    comparative advantage, 239
    costs, 245
    exports, 244
    imports, 236
    and NAFTA, 247, 320
  poultry, 217, 220, 226t, 227, 228
  processed foods, 253, 266, 315, 317
  rapeseed, 105
  rice, 84t, 96
  soybean tariffs, 113t
  sugar, 186, 187
  TFP growth, 35, 36f
  tobacco, 191t
  U.S. exports to, 114, 309
  vegetables, 170, 314
  wheat
    Board, 60, 62t, 64t, 65t, 68
    exports, 59, 64t, 65t, 66
    model, 71-75
    trade, 60
    trade agreements, 69
    Transportation Act, 68
    U.S. imports of, 65
  wine, 178, 179f
Canada Transportation Act, 68
Canadian Wheat Board, 60, 62t, 64t,
  65t, 72

Canola
    and AGLINK model, 106
    and biotechnology, 115, 295
    and health consciousness, 103, 114
    imports from Canada, 114
    and peanuts, 139
    and unsaturated fats, 103
    U.S. duties on, 282
    and world markets, 116
CAP (Common Agricultural Policy), 6, 33, 38n, 319
Capehart, T., 190
Cargill, 111, 300t, 301
Caribbean, 85, 188, 190, 247
Carley, D.H., 140, 141, 147
Carlson, G., 294
Carter, C.A., 51
Caswell, J., 266
Cathcart, W., 126
Cattle
    and beef industry, 238
    Canadian, 273, 275
    feeder, 272, 320
    Mexican, 274, 279, 283
    prices, 274
    supply, 213
    Trade, 273
Celery, 284
Census of Agriculture, 139
Central American Common Market, 112
Cereals Innovation Center, 300t
Ceres, 301t
Chablis, 180
Champagne, 180, 181
Chardonnay, 180
Charlet, B., 199
Chavas, J.-P., 241
Chavez, E.C., 87
Cheese, 256, 258, 259, 278, 279
Chemical
    costs, 43, 109t, 145, 146
    residues, 201
    resistance, 52, 115
    technology, 32
Chen, C., 144, 147

Cherries, 155, 157, 161, 163, 164, 165
Chicken. *See also* Broilers
    consumption, 242, 243f
    in diets, 45
    exports, 221, 231
    meat, low cost, 217, 219t
Chile peppers, 170
China
    beef
        consumption, 205, 206t, 207
        production, 203, 204
    biotechnology market, 304
    corn, U.S. exports, 51
    cotton, 120, 121, 122, 123
    imports
        beer, 260
        fats and oils, 253
        feed stuffs, 259
    and low wages, 313
    meat exports, 258
    peanuts
        costs, 144, 145, 146, 152
        exports, 142
        origin, 139
        production, 140, 141t
        WTO status, 151
    pork, 235, 237, 244, 245, 247
    poultry
        costs, 222f, 222, 223
        exports, 224, 231
        nontariff barriers, 227t
        production, 219
        tariffs, 226t
    rice
        and Arkansas rice model, 86, 88
        market access, 83
        tariffs, 84t
        U.S. competitor, 97
    soybeans, 100, 101t, 106, 113t
    sugar, 182
    tobacco, 188
    trade restrictions, 83
    U.S. agricultural exports to, 309
    vegetables, 170
    wheat, 59, 67, 71, 73

Chiquita bananas, 7
Chiron, 301t
Cholesterol, 201, 295
Cigarettes, 189, 190, 192, 193, 314
CIMMYT, 55
Citrus, 155, 156t, 157-161, 165, 285, 314
Clean Water Action Plan, 240
Clemens, R., 245
Cleveland, O.A., 122, 127
Clinton, (President) Bill, 189
Coconut oil, 102, 103
Codex, 151, 265, 267
COFACE, 68
Coffee, 272, 320
Colombia, 93, 132, 183t, 227t
Combichem, 301t
Comeau, A., 230
Commodity
  programs, 49
  trade, 13
Commodity Credit Corporation, 67, 181
Comparative productivity advantage, 28, 35
Competitive advantage
  fear of losing, 11
  and market share, 22
  in peanuts, 147
  in poultry, 230
  and production costs, 18
  and productivity, 4, 28
  in rice, 282
  and value creation, 16
  in vegetables, 173
  in wheat, 73, 74, 76
CONASUPO, 278
Confectionery products, 255, 256
Congo, 183t
Connell, P., 187
Connor, J., 262
Consumer preferences, 200, 209, 221, 241
Consumers
  and apples, 164
  benefits, 104, 166

Consumers *(continued)*
  and beef, 199, 205
  biotechnology
    acceptance, 14, 114, 115
    resistance to, 296, 318
  and cheese, 259
  competition for, 28
  and cotton, 120, 123, 125, 131, 132, 134
  demand, 15, 48, 100, 201, 206
  and differentiated products, 14, 20
  and health, 103, 114, 201
  and imports, 2
  income, 21, 115, 117, 285
  informed, 263
  low income, 50
  loyalty, 19
  and marketing, 214
  Mexican, 278
  and pork, 236, 243, 236
  preferences, 103, 200, 209, 214, 221, 241, 248
  prices, 37, 164, 239, 272, 273, 278, 290
  protection, 264
  purchasing decisions, 260
  and soyoil, 102
  supply, 273
  and tortillas, 56
  utility maximizing, 87
  welfare, 38, 172, 187
  and wine labels, 180
Continental Grain, 297f
Cook, M., 16
Cooke, S.C., 107
Corn
  animal rations, 280
  and biotechnology, 294, 295t, 296, 297f, 301
  competitiveness of
    and biotechnology, 52-55
    and CUSTA, 279
    industry, 41-42
    and market development, 43-45
    and policy, 46-49

Corn *(continued)*
  export prices, 56
  exports, 301t
  and GMOs, 151
  hog/corn ratio, 241
  hybrid varieties, 7
  high fructose syrup, 186, 187, 314
  international policies, 51
  Japanese imports of, 108
  and Mexico, 279
  and NAFTA, 279
  oil, 103
  and poultry, 229, 230
  production, 310t, 313
  stocks, 269
  subsidies, 109
  support prices, 286
  and trade agreements, 49, 50
  U.S. exports
    Canada, 280
    market share, 42t, 310
    Mexico, 271
    value, 308
    volume, 310t
  wet milling, 288
Corn Belt, 61, 110, 235
Cory, D., 125
Cost advantage
  and competition, 17, 18, 37
  of corn, 56
  of pork, 245
  of poultry, 229
  of soybeans, 110
  of vegetables, 73
Cost-reducing technologies, 23
Costa Rica, 84t, 93, 183t
Cote d'Ivoire, 183t
Cotton
  acreage variability, 133
  competitiveness, 128-132
  exports
    growth, 287
    share, 309, 310t
    volume, 119-125, 286, 287, 310t, 314, 320

Cotton *(continued)*
  fiber markets, 132
  and globalization, 134
  import quotas, 286, 287
  imports, 119-125, 287
  industries, 287
  land, 145t
  producers, 286, 287
  production, 119, 310t
  products, 134
  trade
    CUSTA, 286
    future, 132
    and GATT impacts, 133
    NAFTA, 272, 286, 289, 320
    research, 125-128
    world, 120f
  U.S. exports, 120f
Cottonseed, 102, 103, 106, 139
Couscous, 60, 63
Cramer, G.L., 81t, 82t, 83, 87
Cuba, 68, 84t, 157
Cucumbers, 170, 173, 174f, 284
Culver, D., 144
Curagen, 301t
CUSTA
  agricultural provisions, 112
  and beef, 273
  and broilers, 277
  and cotton, 286
  and dairy, 277
  and feed grains, 279, 280
  and implementation, 270
  and oilseeds, 282
  and peanuts, 288
  and pork, 275
  and regional trade, 319
  and rice, 282
  and sugar, 287
  and trade increases, 272
  and vegetables, 284
  and wheat, 69, 281
Czech Republic, 227

Dairy products
  CUSTA, 277-279
  in diets, 115
  exports, 245t, 258, 271, 309, 312, 316
  growth, 253
  imports, 256t, 257t
  and Mexico, 283
  and NAFTA, 114, 255, 256, 277-279, 289, 320, 321
  output, 252
  SIC-201, 251
  trade, 277
Dalgerty, 300t
Davis, B., 116
Debt crisis, 4
Deciduous, 155, 163, 285
Deepak, M.S., 176
Deficiency payment, 46, 108
DeKalb, 300t
DeLooze, M., 299
Delta & Pine Land, 300t
Demand
  curves, 17
  elasticities, 15, 126
Demographics, 199, 200, 203, 214, 241
Denmark
  GDP growth, 33, 34f
  land endowment, 239
  pork
    costs, 245
    exports, 235, 244, 246
    supplier, 236
  TFP growth, 36f
  wine, 179f
Devadoss, S., 186
Devaluation
  in Asia, 282
  effect of, 22
  of Mexican peso, 172, 173, 175, 270, 271, 273, 283, 286
Differentiated products, 12, 13, 21, 63, 262

Dinamilho, 300t
Diseases
  and biotechnology, 115
  of cotton, 122
  foot-and-mouth, 236, 247
  human, 247
  and pork, 238, 245, 248
  and trade barriers, 263, 264, 312, 319
Dispute settlement, 227, 229, 267, 268, 325
DNA Plant Technologies, 300t
Dobson, W.D., 54
Domestic
  content requirement, 189
  resource cost, 220
Dominican Republic, 182, 187, 190, 191t, 227t
Donald, J., 126
Dorfman, J.H., 241
Dornbush, R., 43
Douglas, C., 173, 175
Dow, 300t, 301t, 303
Drabenstott, M., 15
Dreyfus, 111
Dried fruits, 285
Dubey, A., 35
Duck meat, 219t
Duewer, L.A., 241, 250
Duffy, P., 126, 127
DuPont, 297f, 300, 301t, 303
Dutch, 245

E. coli, 213
Eastern Europe
  beef, 203, 204, 205
  income, 210
  pork, 246, 247
  quota rates, 225
  tariffs, 114
Ebbertt, J., 296
Eccles, R., 302
Economic Research Service, 25

Economies
  agricultural, 231
  developed country, 131
  developing country, 52, 223
  emerging, 175
  European Union, 38, 319
  liberalization of, 248
  production, 17
  of scale, 238, 245, 262, 266, 302, 303
  of scope, 19, 302, 303
  of size, 18, 21, 160
  southern rural, 140
  unstable, 175
  world, 1
Ecuador, 86, 93, 183t, 227t
Education, 5
EEP, 61, 74
Efficiency
  and competitiveness, 5
  feed, 54, 297
  gains
    agricultural, 12, 32
    sectoral, 28
    transfer of, 33
  and genetic research, 238
  growth through, 38n
  labor, 313
  marketing, 87
  and meat packing, 18
  and online trading, 302
  operational, 15
  and policies, 6
  processing, 296, 312
  production, 7, 147, 164, 213, 229, 230, 238, 311
  and resource use, 4
  technological, 32
  transportation, 271
  of waterways, 111
Eggplant, 173, 174f
Egypt, 63, 64t, 68, 86, 88, 226t, 227t
Eidman, V.R., 238
El Niño, 85, 86, 101
El Salvador, 93, 183t, 187, 227t

Electronic data, 302
ELM, 300t
Embargoes, 20
England, 182
Ensminger, M.E., 229
Environment
  and biotechnology, 115
  competitive effects, 258, 322
  concerns, 239, 240, 249
  controlled, 238
  economic, 247
  export, 308
  and externalities, 239
  market, 214
  policies, 5, 12, 22, 228
  political, 247
  and production
    agriculture, 133
    efficiency, 6
  regulations, 104, 248, 249
  standards, 264
  technological, 11
  and trade barriers, 225
Environmental Assistance Program, 240
Environmental Protection Agency, 240
Epperson, J.E., 166
Equivalency, 265
Erucic acid, 114
Ethridge, D., 122, 126, 127
Europe
  agriculture
    growth, 33-35
    sector, 28
    TFP growth, 37
  beef consumption, 205
  broiler costs, 223
  colonies, 319
  competitive advantage, 11
  corn, 50, 53
  cotton, 126
  dairy products, 258, 259
  differentiated products, 13
  economic growth, 49

Index   335

Europe *(continued)*
  food product exports, 258
  fruit costs, 166
  and GMOs, 296
  and grain crops, 313
  and labor costs, 313
  pork markets, 246
  processed food trade, 260, 266
  rice exports, 95, 96, 97
  soybean markets, 111
  trade
    agreements, 269, 313
    with, 56
  U.S. multinationals, 304
  and wheat, 64, 71t
  and wine, 180
European Union
  agricultural
    growth, 33
    support, 33
  and beef, 204, 205, 206t, 207, 208t, 209
  common agricultural policy, 33, 38n
  comparative advantage, 35
  cotton imports, 121f, 122t, 123, 127
  export
    credit, 68
    subsidies, 66, 67, 76, 188, 259
  and feed grains, 313
  and GATT, 50
  and hormone-fed beef, 312
  maize tariffs, 50
  market share, 38n, 102, 310
  and oilseed producers, 106
  and pork, 226, 244n, 245, 246
  and poultry, 219, 221, 222, 226t, 227t, 228
  and processed foods, 255, 256, 257t, 316
  and rice, 86, 88
  and soybeans, 100, 101t, 102, 114
  and sugar, 156
    tariffs, 84t, 113t
    textile industry, 131
    TFP growth, 35

European Union *(continued)*
  and tobacco, 190, 192
  trade
    agreements, 6, 112, 212, 319
    barriers, 260
    concessions, 50, 84
  U.S. multinationals, 303
  and wheat
    exports, 65t, 66, 74t
    flows, 64f
    model, 71, 72, 73
    production, 59, 62t, 72t
    shadow price, 75t
  and wine, 178, 180
Everglades Forever Act, 185
Exchange rates
  and beef competitiveness, 213
  and broiler exports, 220
  and Canada, 236
  and cotton costs, 129
  dollars/yen, 210
  firm influence on, 22
  and macroeconomics, 6, 21
  and monetary policy, 105
  overvalued, 117
  and pork exports, 245
  and real prices, 21
  and soybeans, 103, 105, 109t
  and tomato industry, 172
Exporting
  certificate, 67
  earnings, 1
  promotion
    of competitiveness, 11, 21, 311
    of differentiated products, 20
    policies, 66
    of poultry, 217, 230
    of wheat, 59, 71, 73, 74
    of wine, 181
  restitutions, 6
  subsidies
    Argentina, 72
    Australia, 72
    countervailing duties, 275
    elimination of, 106

Exporting, subsidies *(continued)*
    EU, 66, 67, 75, 165, 319
    and market share, 59, 60, 66
    sugar, 186, 188
    tobacco, 192
    and Uruguay round, 67, 263, 312, 319
    and U.S., 66, 67, 75, 179, 259
    and wheat model, 71
    and wine, 180
    and world price, 21
    and WTO rules, 84
    taxes, 106, 116, 127
Export Credit Guarantee, 82
Export Enhancement Program, 67, 82
Externalities, 19, 239

FAIR Act
    corn, 46, 47
    cotton, 129
    market access, 181
    peanuts, 143, 147, 149, 152
    policy, 12, 129
    rice, 81, 89
    sugar, 182, 188
    U.S.
        competitiveness, 108
        production effects, 134
FAO, 38n
Farm
    gate, 101, 238
    price, 81, 91t, 94, 110, 119, 290
    programs, 4, 152
    Sector, 2
FAS, 181
FATUS, 308
FDI, 262, 266
Federal Agricultural Improvement and Reform Act, 12, 129, 143
Feed
    and agricultural policies, 116
    animal, 251, 253
    and bioengineered crops, 301
    by-product exports, 42

Feed *(continued)*
    companies, 238
    conversion, 229, 237, 238
    corn, 52, 53
    costs
        high, 230, 238, 246
        low, 213, 223, 229, 231, 237, 244, 245, 248
    efficiency, 52, 54, 297
    exports, 259, 308, 309, 313
    foreign markets, 42, 43, 45, 56
    gluten, 42
    grain
        area, 108
        imports, 53
        and NAFTA, 289
        trade, 279-281
    high lysine, 52, 53
    hog, 239
    livestock, 41, 100, 290, 294
    materials, 230
    and midwest, 235
    peanut meal, 141
    prices, 54
    producers, 56, 274
    protein, 42, 52, 231
    quality, 44
    rapeseed meal, 114
    rations, 56
    soymeal, 102
    supplies, 228, 312
    users, 54
    wheat, 63
Feeder
    cattle, 272, 274, 320
    pigs, 276
Feedlots, 274, 275
Fertilizers
    costs
        corn, 43
        peanuts, 145
        rice, 81, 96
        soybeans, 109
    inputs, 30
    low N-fertilizer need, 295t

Fertilizers *(continued)*
  manure as, 259
  suppliers, 285
Fiji, 183t
Financial crisis, 76, 236, 269, 286
Finland, 180
Fish
  consumption, 199, 202, 242, 243f
  meal, 53, 103
  oil, 102
Fixed costs, 18, 96, 109, 110
Flaxseed, 282, 283
Fletcher, S.M., 140, 141
Fliginger, C. John, 171
Florida
  citrus, 157
  competitiveness, 173
  methyl bromide use, 172
  orange juice, 160
  peanuts, 139
  sugar, 182
  tomatoes, 171
  winter vegetable costs, 174f
Florunner (peanut), 146
Flue-cured tobacco, 190, 191t, 192, 193
Fluid milk, 278
Food and Agriculture Organization, 258
Food and Drug Administration, 114
Food crisis, 2
Food processing
  and biotechnology, 201
  companies, 251
  and exports, 37, 252-255
  growth of, 30t, 32, 33, 266
  and imports, 255-257
  and intermediate inputs, 32, 37
  internationalization of, 214
  largest industry, 315
  and other manufacturing, 262
  and prices, 37
  primary input costs, 33, 38
  SIC-20, 251

Foreign corn market, 46
Foreign direct investment, 260, 316
Foreign feed market, 42, 43, 44, 45, 56
Foreign market demand, 21
Foreign markets
  access, 26, 181
  barriers to, 181
  beef, 210, 213
  competitive edge in, 38
  corn, 46
  cotton, 133
  development of, 217
  invest in, 261
  market share, 16
  and phytosanitary standards, 85
  wine, 181
Foreign Agricultural Trade of the United States, 308
Foster, K.A., 238
France
  agricultural prices, 33
  apples, 161
  beef, 200
  beverages, 26
  dairy, 258, 259
  DRC for poultry, 220
  export credit, 68
  feed exports, 259
  fruit, 157
  GDP growth, 33, 34t
  grain mill products, 259
  output per worker, 262, 316
  peanut imports, 143t
  pet food, 259
  pork, 245
  poultry, 219, 222, 223t, 228, 230, 231
  prepared meats, 258
  productivity, 35
  technology research, 53
  TFP growth, 36f
  turkey, 219, 221
  wheat production, 62
  wine production, 177, 178, 180, 181

Free trade
  agreements, 76, 131, 189
  and APEC, 85, 321
  areas, 6, 69, 289
  and economic rents, 171
  and European Union, 228
  free trade areas, 6, 289
  and GATT, 6
  models, 71, 72t, 73, 74, 75t
  and NAFTA, 6
  scenarios, 76
  and sugar, 186
  and value added, 134
Free Trade Area of the Americas
  cotton, 130, 132
  peanuts, 144, 150, 151, 152
  pork, 247
  rice, 85
  U.S. participation in, 321
  wheat, 76
Freeman, H.A., 139, 141
Fresh fruits, 165, 166, 284, 285, 314
Friedland, J., 111
Frozen vegetables, 169, 170, 284
Fruit fly, 166
Fruit juices, 165
Fruits
  citrus, 155, 157-161
  costs, 166
  deciduous, 155
  demand for, 169
  duties, 285
  exports, 285
  imports, 285
  Mexico as top supplier, 285
  and NAFTA, 285, 289, 320
  noncitrus, 161-164
  policy impacts, 164-166
  production, 155
  trade, 155-157, 166, 167, 272, 285

Gabon, 183t
Gallagher, P., 46, 104
Garett, 176

GATT (General Agreement on Tariffs and Trade)
  and agriculture, 6, 165, 319
  and beef, 209, 213
  and competitiveness, 116, 318
  and corn, 49, 50
  dispute settlement, 265
  and European Union, 231
  and fruit, 165
  and global market place, 144
  market access, 116, 165, 312
  and Mexico, 271
  negotiations, 46
  and nontariff barriers, 225, 265
  and peanuts, 140, 143, 147, 148t, 149, 150
  and pork, 246
  and poultry, 225, 228
  and reduced tariffs, 116
  and rice, 83, 84t
  and soybeans, 99, 104, 112
  and sugar, 185
  and technical barriers, 264, 265
  and tobacco, 189, 192
  and wheat, 59
  and wine, 178
  and zero-for-zero approach, 106
GDP (gross domestic product)
  agricultural, 25, 27, 29, 32, 33, 34t, 35
  and competitiveness indices, 38n
  defined, 27
  growth, 25, 27, 30t, 33, 34f
  real, 27, 32
  U.S., 29
Gehlhar, M., 26, 38, 310, 328
Gempesaw, C.M., 220
Gene Logic Inc., 301t
Gene TraceR, 301t
General Accounting Office (GAO), 187
Genetic engineering, 52, 294, 295
Genomics, 299, 301t, 303, 304
Genset, 301t
Geography, 4, 6, 212

Georgia, 139, 188, 229
Germany
  agricultural
    GDP, 33, 34t
    productivity, 35
  apple juice, 163
  beer exports, 260
  beverages
    exports, 259
    imports, 260
  and biotechnology mergers, 300t
  grain mill product imports, 259
  miscellaneous food exports, 260
  output per food worker, 262, 316
  peanut imports, 143t
  poultry imports, 222, 228
  TFP growth, 36f
  tobacco imports, 191t
  U.S. export destination, 309
  wheat, 62
  wine, 179f
Germplasm, 299
Gibson, J., 294
Gillespie, J.M., 238
Glade, E., 128
Glaze, 195
Global markets, 8, 243, 244
Globalization, 1, 134, 201, 214, 317, 322
Glucosinolates, 114
Gluten, 42f
GMO (genetically modified organisms), 151, 296, 318
GNP, 51, 56, 272
Golz, J., 70, 71
Goose meat, 219t
Gopinath, M., 14, 29, 30, 31f, 34f, 36f, 38n
Government regulation, 12
Graf, E., 214
Grain mill products
  exports, 253, 254t, 255
  imports, 256t, 257t
  output, 252
  SIC-204, 251
  U.S. supplier of, 259

Grain sorghum, 279, 280
Grapefruit
  exports, 159, 314
  import duties, 285
  production, 157, 158
  tariff reductions, 165
  trade, 155
Grapes, 161, 164, 165, 180
Gray, D., 166
Great Britain, 200
Greece, 62, 190, 191t
Grise, V.N., 189, 190
Grossman, G., 261
Grubel-Lloyd index, 261
Grupp, H., 19
GSM, 69, 82
Guatemala, 93, 132, 183t, 226t, 227t
Gulf ports, 60, 80
Gunter, D.L., 160
Guyana, 81, 82t, 84t, 86, 97, 183

HACCP, 213
Hafi, A., 187
Haiti, 75, 185
Haley, M., 236, 239
Haley, S.L., 67
Hand labor, 224
Handy, C., 261, 262, 317
Hansen, J., 85, 87
Hard red spring wheat, 62t, 65t, 72t, 74t, 75t. *See also* HRS
Hard red winter wheat, 62t, 65t, 72t, 74t, 75t. *See also* HRW
Harmonized Tariff Schedule, 183t
Harrison, R.W., 4, 17
Harwood, J.L., 63, 69
Hawaii, 6, 182
Hayes, D., 245
Hazard Analysis and Critical Control Points, 213
Health standards, 114
Heckerman, D., 125
Heckscher-Ohlin-Samuelson tradition, 261
Hecksher-Ohlin type, 27

Helmberger, P.G., 54
Henderson, D., 261, 262
Henneberry, S.B., 199
Henry, R., 219, 221, 222f, 223t, 224, 228, 229, 230
Herbicides, 14, 52, 96
Herndon, C., 122, 127
High fructose corn syrup, 186, 314
High fructose syrup, 288
High-oil corn, 53, 295t, 297f
High-value exports, 3
Hill, L., 104
Hillman, J., 171
Hirschberg, J., 261
Hoffman, L., 47, 52
Hog cholera, 275
Hog/corn ratio, 241
Holmes, P., 135
Holt, M.T., 241
Honduras, 93, 183t, 226t, 227t
Hong Kong
  imports of
    apples, 162
    cotton, 122t
    pork, 235, 247
    poultry, 217, 220, 225
    tobacco, 193
  U.S. exports to, 309
Hooker, N., 266
Horiuchi, G., 181
Hoskin, R., 103, 114
Houston, FOB, 80, 94
Houston, Jack E., 103
HRS (hard red spring wheat), 66, 70, 73, 76
HRW (hard white spring wheat), 66, 70, 73, 76
Hu, H., 262
Hubbell, B., 294
Hudson, D., 119, 122, 127
Hudson, W.J., 4, 6
Hughes, K.S., 14
Hung Seed, 300t
Hungary, 163, 226t, 245
Hwang, T.-C., 107

Hybridnova, 300t
Hydrogenation, 295

IBM, 301t
Idaho, 166
Illinois, 235
Imperfect competition, 18, 125
Importing
  licenses, 50, 69, 71
  quotas
    beef, 213, 321
    cotton, 286, 287
    dairy products, 277
    and NAFTA, 270
    peanuts, 146
    poultry, 227t, 276
    restrictive, 290
    rice, 83
    and Uruguay round, 290
Income elasticity, 56, 100
Incyte, 301t
India
  and biotechnology transfer, 304
  cotton, 123, 127, 131
  peanuts, 140, 141t, 142
  poultry, 226t, 227t
  rice
    in Arkansas rice model, 94, 96, 97
    and El Niño, 86
    production, 79
  sugar, 182, 192t
  wheat, 59
Indiana, 235
Indonesia
  apple imports, 162, 164
  cotton, 121, 122t
  and fruit quality, 164
  oranges market, 161
  peanuts, 143
  poultry, 226t, 227t
  rice
    in Arkansas rice model, 87, 88, 89, 95

Indonesia, rice *(continued)*
   and El Niño, 86
   import suppliers, 97
   state trading in, 85
   tariffs, 84
  soybeans, 113t
  tobacco imports, 191t
  wheat, 60
Industry structure, 17
Inflation, 6, 49, 93, 308
Infrastructure
  and agricultural TFP, 31f
  Cargill's, 301
  and competitiveness, 99, 107
  development, 175
  improvements of, 271, 272
  investments in, 27, 30, 37, 111, 228
  peanut, 151
  and soybean production, 116
  specialized, 286
  technology, 303
Integration of markets, 317
Intellectual property rights, 298, 299, 302, 303
Intermediate
  inputs, 30, 32, 37
  products, 181, 251, 286, 321
Internal supports, 70, 71
International Finance Corporation, 220
International Grains Council, 62t, 64t, 65t
International Labor Organization, 200
International markets
  and Arkansas Rice Model, 87
  citrus, 160
  and competitiveness, 4, 16, 21, 41, 307, 322
  corn, 41, 46, 56
  cotton, 120
  fruit, 166
  pork, 241, 248
  poultry, 219, 231
  and product characteristics, 262
  U.S. dependence on, 270, 290
Internet, 302

Iowa, 235, 245
Iowa State University, 245
Iran, 86, 88, 95
Iraq, 68, 86, 88, 95
Ireland, 179f
Irrigation, 7, 82, 96
Israel, 157, 189, 321
Italy
  beer imports, 260
  dairy, 259
  fruits, 157
  peanuts, 143
  pork, 235, 245
  rice, 86, 88, 97
  wheat, 62
  wine, 177, 178, 181

Jamaica, 84t, 183t
James, C., 293
Japan
  apple imports, 166
  and Arkansas rice model, 86, 94
  beef
    imports, 205, 206t, 208f, 209, 321
    market for, 210, 211
  bilateral trade agreement, 321
  citrus, 160
  corn
    high-oil, 53
    imports, 50, 279
    market access, 56
  cotton, 120, 122t, 126
  and economic growth, 49
  fruit tariff cuts, 165
  peanuts, 143t
  phytosanitary requirements, 165
  pork, 235, 236, 246
  poultry, 217, 220, 226t, 230
  processed foods
    exports to U.S., 256, 257t
    imports, 228, 259, 260
    U.S. exports to, 254t, 255, 262, 266, 272, 315, 316

Japan *(continued)*
  technology research, 53
  TFP growth, 35, 36f
  rice, 83, 93, 95
  soybeans, 105, 106, 111t, 113t
  sugar, 187
  tobacco, 190, 191t, 192, 193
  U.S. products destination, 304, 320
  wheat, 63, 64t, 66
  wine imports, 178
Johansen, A.E., 164
Johnson, P., 126
Joly, P., 299
Jones, E., 232, 263, 321
Jones-Costigan Sugar Act, 182
Jordan, K.H., 171
Josling, T., 50, 312
Juliano, B., 79
Jurenas, R., 146

Kalaitzandonakes, N.G., 52, 115, 262, 294, 299, 301, 302
Kapombe, C.R., 220
Kaylen, M.S., 238
Kennedy, P.L., 4, 16, 17, 153, 293
Kenya, 84
Kinsey, J.D., 199, 201, 241, 267
Kliebenstein, J.B., 238
Koo, W.K., 70, 71, 187
Korea
  beef, 209, 210, 211, 212, 213
  biotechnology mergers, 300t
  cotton, 121, 122t
  feed grain trade, 279
  financial crisis, 60
  fruit, 165, 166
  government involvement, 213
  pork, 235, 236, 246
  poultry, 226t, 227t
  rice, 83, 84, 86, 93, 95
  soybeans, 113t
  tobacco, 191t
  U.S. Agricultural exports to, 309
  wheat, 63, 64, 66

Krishna Valluru, 104
Krissoff, B., 166
Kropf, J., 186
Krueger, A.O., 6
Krugman, P.R., 4, 12, 25, 28, 38n

Labor
  costs
    and cotton, 131, 314
    and fruit, 160, 166
    high, 5, 313
    and peanuts, 145
    and rice, 81t, 96
  division of, 18
  efficiency, 313
  employment and NAFTA, 286
  family, 37
  growth in, 38
  hand, 224
  intensive, 29, 134, 175, 289, 290
  market, 200
  payments to, 18
  and poultry, 277
  productivity, 17, 262, 266, 316
  standards, 188
Lancaster, K., 261
Land prices, 43, 110
Landau, R., 15
Landell Mills, 182
Larson, D.W., 109t, 111t
Lasley, F.A., 229
Latin America, 93, 95, 97, 166, 180
Lau, L.J., 35
Law of comparative advantage, 12, 15
Lawrence, J.D., 238
Lee, J., 262
Leg quarters, 220, 224
Lemons/limes, 155, 157, 160
Lettuce, 170, 284
Level playing field, 13
Liberalization
  of agricultural trade, 95
  and CUSTA, 280
  economic, 132
  gains from, 116

Liberalization *(continued)*
  market, 51
  regional, 85, 99
  and sugar, 197, 188
  of trade, 13, 112, 117, 247, 270
  and wheat production, 76
  and wine exports, 181
  of world markets, 1
Libya, 259
Liu, K., 126
Live cattle, 273
Livestock
  and Clean Water Action Plan, 240
  consumption of, 115
  corn imports for, 50
  exports, 1, 271, 312, 320
  feed
    industry, 41
    foreign markets for, 42
    peanut meal, 114
    share of output, 204
    soybean meal, 100, 102, 104, 114
    wheat as, 63
  intensive operations, 239
  manure, 239
  producers, 106
  production, 56
  productivity, 29, 30
  protein using, 45
  trade, 8, 274
  U.S. sector, 289
LMC International, 182
Loan rate for
  corn, 46, 47, 48, 56
  cotton, 129
  rice, 91t
  soybeans, 116
Long-term agreements, 66, 69
Lord, R., 185
Louisiana, 182
Love, J., 177, 180
Low-fat recipes, 201
Lynham, M., 247
Lynx Therapeutics, 301t
Lysine, 52, 294, 295t

MacDonald, S., 127, 128
Macroeconomic policies
  and corn, 43, 46, 49, 51
  effects of, 6, 26
  and poultry, 231
Madagascar, 183t
Maize, 50
Malawi, 183t, 189, 190, 191t, 192, 194, 314
Malaysia, 84t, 86, 191t
Maltsbarger, R., 52, 115, 294
Mangoes, 285
Manitoba, 61
Manufacturing
  capability, 298
  cigarette, 189
  competition for, 14
  cotton, 134
  and developing countries, 51
  food industry, 251, 262, 263, 315
  peanut sector, 152
  phyto-, 115, 295t
  and terms of trade, 14
  textile, 314
Manure, 110, 239, 240
MAP (Market Access Program), 82, 181, 184
Marchant, M., 262
Marin, R., 166
Marion, B.W., 19
Marketing
  allotments, 188
  assessment, 189
  beef, 199, 200, 201, 214
  and biotechnology products, 318
  boards, 188
  Cargill's capabilities, 301
  costs
    rice, 82t, 97
    soybeans, 110, 111
    vegetables, 173
    wheat, 70, 71
  cotton, 120, 127
  efficiency, 87
  fruits, 166

Marketing *(continued)*
   grain, 313
   loan, 48, 130
   meat products, 214
   networks, 298
   pork, 241, 243, 248
   poultry, 200
   quotas, 188, 189, 193
   regulations, 263
   sugar, 194
   supply chain, 297
   systems, 72
   tobacco, 193, 194
   vegetables, 284
   wheat, 72
Marks, L., 299
Marra, M., 294
Martin, L.L., 16, 57, 144, 193
Mauritius, 183t
McCormick, A., 200
McCormick, I., 103, 114
McCorriston, S., 261
McDonald, S., 262
McElroy, R.G., 73
McGuirk, A.M., 200
Meat
   characteristics, 236
   chicken, 221, 230, 231
   of choice, 243
   competitive prices, 213
   competitiveness, 214
   consumer, 56
   consumption, 199, 202, 203, 243t, 248
   dark parts, 217, 224, 231
   demand, 51, 274, 277
   exports
      broiler, 312
      growth in, 203
      to Mexico, 271, 274, 320
      poultry, 217
      processed foods, 252, 253t, 254t, 316
      red, 309, 310
      turkey, 219

Meat, exports *(continued)*
      U.S., 219t
      value of, 266
      world, 219t
   imports, 105, 230, 255, 245t, 257t
   industrial elaboration of, 200
   leaner cuts, 201
   luxury, as a, 51
   marbled, 213
   marketing of, 214
   and nontariff barriers, 225
   output, 252
   packing, 18, 274, 275
   prepared products, 258
   prices, 224
   processing, 200, 201, 223
   production, 45, 105, 310, 314
   and productivity growth, 30t
   promotion, 200, 230
   quotas, 246
   red, 242
   requirements, 51
   SIC-201, 251
   tariffs, 225, 226t
   trade, 200, 272, 274, 289
   turkey, 217, 219, 230, 231
   variety, 235
   white cuts, 224, 242
Meilke, K.D., 106, 113t
Melons, 170, 285
MERCOSUR
   and cotton trade, 127, 132
   and poultry trade, 228
   and reduced tariffs, 116
   regional trade association, 112, 319
   and rice trade, 85
   and trading blocks, 6
Methyl bromide, 171, 172
Mexican peso, 114, 172, 270, 271, 273
Mexico
   beef
      consumption, 205, 206
      exports, 209
      imports, 212

Mexico, beef *(continued)*
    markets, 210
    production, 204
  biotechnology mergers, 300t
  broilers, 217
  corn, 50, 56
  cotton, 120, 121, 131
  as developing neighbor, 51
  fruits
    apples, 162, 164, 165, 167
    costs, 160
    grapefruit, 157
    limes, 160
    oranges, 157
    production, 157
    tariffs, 165
  and GATT, 271
  GNP increases, 51
  imports from, 271
  NAFTA impacts, 270, 272-289, 319, 320, 321
  peanuts, 141t, 142t, 143t, 148t, 149, 150
  pork, 235, 236, 245, 247
  poultry, 220, 226t, 227, 228
  processed foods, 253, 260, 265, 315, 317
  rice, 84t, 95, 113t
  soybeans, 113, 114
  sugar, 183t, 186, 187
  textile production, 120
  TFP growth, 35, 36f
  tobacco, 191t, 193
  turkey imports, 217
  vegetables
    costs, 173, 174f
    supplier of, 169, 170, 171, 172, 175
    tariffs on, 173
    trade, 319
  wage rates, 160
  wheat, 63, 64t, 69, 71, 72
  wine, 179
Meyer, L., 128
MFN, 51, 225, 228

Michel, K., 220
Michigan, 61, 161
Middle East, 67, 95, 96, 97, 190, 248
Midwest, 43, 235
Milberg, W.S., 22
Miles, R., 302
Milham, N., 16
Milk, 6, 259, 278, 279
Millennium Pharmaceuticals, 301t
Minnesota, 235
Mississippi, 79, 92, 110, 229
Missouri, 79, 92, 245
Mittelhammer, R.C., 230
MNE (multinational enterprises), 261, 262, 316
Mogen, 300t
Molecular farming, 295t
Monetary policy, 49, 56, 105, 116
Monke, E., 125
Monopsony, 54, 55f
Monsanto, 300t, 301, 303
Morocco, 64
Most favored nation, 51, 225
Mowery, D.C., 22
Mozambique, 183t
Multilateral trade, 13, 44, 95, 235
Munirathinam, R., 262
Myanmar, 88, 97
Mycogen, 300

Naegeley, S.K., 163
National Pork Production Council, 240
National Research Council, 214
Natural resource endowments, 107, 110
Nayga, R.M., 199, 214
Nebraska, 235
Neff, S., 260, 262, 317
Nehru, V., 35
Net returns, 89, 144, 145t, 146, 191t
Netherlands
  beef promotion, 200
  beer exports, 260

Netherlands *(continued)*
  biotechnology mergers, 300t
  dairy
    exports, 258
    imports, 259
  destination for U.S. exports, 309
  feed exports, 259
  free trade in E.U., 228
  high-cost feed, 230
  miscellaneous food exports, 260
  pork costs, 116
  poultry
    costs, 222, 223
    DRC for, 220
    exports, 231
    production, 219
  rice imports, 95, 96, 143
  tobacco imports, 191t
  tomato supplier, 170
  wine imports, 179f
New South Wales, 63
New York, 61, 161
New Zealand
  beef, 209, 210, 211, 212
  dairy, 258
  fruit, 161, 162
  sugar, 212
Newcastle disease, 225, 227
Newly industrialized countries, 255
Nicaragua, 93, 141t, 142, 150, 151, 183
Nigeria, 84t, 85
Ning, Y., 262
NIS (Newly Independent States), 203, 204, 205, 207
Nohria, N., 302
Nontariff barriers, 274, 290
  and agricultural trade, 221
  fruit trade, 164
  market share, 99
  pork trade, 246
  poultry trade, 225, 227t
  product standards, 264
  soybeans, 114
  tariffication, 312

Nontariff barriers *(continued)*
  technical barriers, 264
  U.S. product exports, 312
  vegetable trade, 172
  wine exports, 180, 181
North Africa, 66, 67
North American Free Trade Agreement
  agricultural trade, 13
  beef, 212
  citrus fruits, 160
  effects of, 320-321
  market access, 99
  regional trade association, 319
  sugar policy, 186
  trade determination, 6
  and U.S. agriculture, 269-291
  vegetable trade, 173
  wheat trade, 69
North Carolina
  cotton, 119
  peanuts, 139
  pork, 235, 237, 240
  poultry, 229
  tobacco, 188
North Carolina Pork Council, 237
North Dakota, 61
Northern Plains, 61
Norway, 165, 180
Novartis, 301t
Nutraceuticals, 295
Nuts, 272, 320

Oats, 279, 280, 281
OECD (Organization for Economic Co-operation and Development), 106, 112, 200
Offals, edible, 209, 212
Office of Technology Assessment, 44, 53, 58
Office of the U.S. Trade Representative, 186
Oilseeds
  and AGLINK model, 106
  biotechnology, 115

Oilseeds *(continued)*
  exports, 114, 271, 308, 309, 321
  peanuts, 139, 151
  and tariffication, 95
  trade, 116, 279, 282-283
  world
    markets, 112, 116
    supplies, 269
Oklahoma, 61, 139
Oleic acid, 294, 295t
Onions, 170, 283, 284
Ontario, 68, 179, 276
Oranges
  Brazil, 160
  costs, 160
  export share, 155
  exports, 155, 285, 314, 321
  import quota, 321
  juice, 158t, 159, 160, 165, 321
  price effects, 164
  processing economies, 160
  production, 157, 158t
  quality, 161
  tariffs, 165
  trade, 159
Orden, D., 264, 312
Oregon, 166
Ortmann, G., 63, 73
Overend, C., 262
Oxley, J.E., 22

Paarlberg, P., 43
Pacific
  Northwest, 61
  region, 85
  Rim, 11, 213, 321
Packaging, 201, 214, 260, 263
Pakistan
  in Arkansas Rice Model, 86, 94
  basmati rice, 96
  cotton, 121, 122, 127, 128, 129, 131
  rice tariff, 84t
Palm oil, 102, 103, 106, 116, 295, 313
Pampas, 110

Panama, 93, 183t, 187
Papua New Guinea, 183t
Paraguay, 100, 101, 110, 112, 121, 183t
Park, T., 125
Parkan, C., 17
Pasta, 54, 56
Patents, 298, 299, 303
PBI, 300t
Peaches, 155, 157, 161, 163, 164, 165
Peanuts
  butter, 288
  competitive position, 143-146
  and developing countries, 139
  exports, 314, 321
  farm program, 152
  legume, 139
  meal, 103
  oil, 102, 103
  oilseed, 151
  policy, 146-151
  price support, 152
  production, 140-141, 152
  regional crop, 139
  trade, 141-143, 288-289
Pears, 155, 157, 163, 165, 285
Pearson, J.L., 171
Pennsylvania, 161
Penov, I., 302
Perdue farms, 7
Peru, 68, 84, 183t
Pesticides, 30, 96, 166, 264
Pests
  and biotechnology, 46
  fruit fly, 166
  harmful, 263, 264, 312
  import of, 319
Peterson, H.C., 104
PGS, 300t
Pharmaceutical, 295, 301t
Philippines
  El Niño impacts, 86
  pork, 235, 245
  poultry tariff, 226t
  rice, 86, 97

Philippines *(continued)*
   state trading, 85
   sugar quota, 182, 183t, 187
Phytomanufacturing, 295t
Phytosanitary rules and
   apples, 165, 166
   biotechnology products, 318
   fruit to Mexico, 285
   peanuts, 151
   plants and animals, 263
   rice, 84, 85, 282
   soybeans, 114
   tobacco, 192
   trade, 6
   vegetables, 172
Pick, D., 125
Pioneer, 300t, 301t
Pluchnett, D., 53, 55
Plums, 155, 157, 161, 163, 164
Poland, 161, 226, 227t, 235, 244, 245, 246
Polopolus, L.C., 144
Population growth, 21, 242, 271
Pork
   in consumer diets, 102
   consumption
      changing dynamics, 241-243
      U.S., 244f
      world, 202t, 203
   and competitiveness, 243-248
   environmental costs, 249
   exports
      growth, 248, 271, 312
      to Mexico, 276
      processed foods, 262
      share, 2, 310t
      and TRQs, 275
      volume, 248, 310t
   imports, 236, 276
   industry, 235, 248
   meat share, 236, 276
   and NAFTA, 275-276, 320
   prices, 236, 242f
   production, 242f, 309, 310t
   promotion, 7

Pork *(continued)*
   rations, 45, 56, 102, 115
   and scale economies, 262
   trade, 275-276, 308
Porter, M., 12, 16, 17, 26
Potatoes, 170, 171, 284
Poultry
   and Brazil, 222, 223, 224, 231
   competitiveness, 221
   consumption, 115, 202, 203, 312
   costs, 221-224, 230, 331
   exports, 217, 218t, 219, 309, 310t, 312
   feed
      corn, 56
      demand, 45
      soymeal, 102, 114, 283
      supplies, 236
      wheat, 63
   market share, 45f, 310
   meat share, 45f, 202
   and NAFTA
      exports, 271, 272, 276-277, 320
      feed trade, 279, 283
      TRQs, 114
   prices, 224
   processed food exports, 308, 316
   production, 217, 224, 310t, 312
   promotion, 6
   trade
      agreements, 227-230
      barriers, 224-227
      research, 219-221
   and white meat, 242
Preston, W.P., 200
Price supports
   cotton, 129
   domestic, 319
   EU sugar, 186
   and FAIR act, 108
   peanuts, 146, 147, 151
   policy, 27, 28
   for small corn producers, 50
   sugar, 194
   tobacco, 189, 193, 194

Processed foods
  advantage in, 13, 14
  competitiveness
    environment for, 258-260
    and labor productivity, 316
    U.S. position, 260-263, 316
  efficiency gains, 14
  exports, 37, 252-255, 266, 315f, 316
  imports, 255-257, 315f
  and NAFTA, 289
  prices, 33, 37, 38
  product differentiation, 266
  rules and regulations, 267
  trade agreements, 263-266
Product
  differentiation, 15, 16, 17, 19, 20, 21
  quality, 19, 22, 23, 209, 213
Productivity
  advantage, 35, 262
  agricultural, 1, 5, 26, 29, 33
  and biotechnology, 317, 318
  capital, 17
  and competitiveness, 4, 28, 102, 107, 173
  determinants, 24
  factor, 26, 27, 107. *See also* TFP
  fruit, 156
  growth, 28, 30t, 33, 37, 38n
  labor, 17, 262, 266, 316
  and Land Grant System, 22
  in Mexico, 175
  and monetary policy, 49
  and public R&D, 37
  technology, 17, 43, 307
Public policy, 25, 27
Publicly funded research, 22
Purcell, W.D., 199
Putnam, J.J., 200, 241

Quality
  -enhanced, 294, 295, 296, 297, 301
  -enhancing, 17, 302
Quality of life, 239

Quarantines, 114, 276
Quiroz, J., 21
Quota rent
  peanuts, 144, 145, 146
  sugar, 187, 188
  tobacco, 190, 191t

R&D
  agriculture, 30, 31f, 32, 37, 38
  biotechnology, 295t, 303, 304
Rapeseed. *See also* Canola
  double low, 114
  duties on, 282
  exports
    Canadian, 105, 283
    growth, 103
  feed meal share, 103
  GRAS status, 103
  and health consciousness, 103
  Japanese imports, 105
  market share, 103
  production, 103, 139
  soyoil competitor, 102, 116, 313
  and trade liberalization, 106
Rask, N., 63, 109t, 111t
Red meat, 242, 312
Reed, M., 34, 264, 265, 270, 271, 308
Regmi, A., 127
Reich, R., 23, 35
Research and development, 9, 16, 18, 24, 70, 175
Rhone-Poulenc, 301t
Rice
  Arkansas Global Rice Model, 86-96
  biotechnology and, 295
  costs, 80-81, 96
  El Niño impacts, 85-86
  exports, 97, 280, 282, 310, 313
  import duties, 287
  long grain, 79
  market share, 2, 97, 309
  medium grain, 80
  milled, 80
  and NAFTA, 321
  phytosanitary ban, 282

Rice *(continued)*
  price, 80
  production, 79, 310, 313
  quality, 97, 313
  rough, 80
  trade agreements, 82-85
  and U.S. agricultural policy, 81-82
  world market for, 102
Richardson, J., 126, 127
Roberson, I., 127
Roberts, D., 195, 264, 265, 312
Roberts, I., 187
Robinson, S., 51
Rockingham, 230
Roe, T.L., 14, 26, 29, 30, 31f, 33, 36f, 38n
Romania, 226t
Rosson, P., 199
Rothwell, G., 219, 221, 222f, 223t, 224, 228, 229, 230
Round-Up Ready, 115
Russia
  cotton imports, 122t
  credit, 224
  imports
    dark meat, 221
    leg quarters, 224
    pork, 236, 237
    poultry, 217, 220, 231
    turkey, 217
    wheat flour, 259
  market for hogs, 235
  tariffs, 226t
  unstable growth, 56

Saguy, I.S., 214
Salin, V., 262
Salmonella, 225, 227
Salsgiver, J., 186
Sanitary and phytosanitary measures
  fruit trade, 166, 285
  and GATT principles, 265
  plants and animals, 263
  scientific evidence for, 312

Sanitary and phytosanitary measures *(continued)*
  standards, 85
  tobacco, 192
  vegetables, 172
Saskatchewan, 61
Saudi Arabia, 86, 88, 95, 96
Say, R.R., 229
Scandinavia, 67
Schmitz, A., 171
Schnittker, J.A., 52
Schuh, G.E., 46
Section 22, 140, 146, 147, 289
Senauer, B., 241
Shadow prices, 70, 74, 75
Shagam, S., 236
Shane, M.D., 14, 30, 32
Sharples, J., 16, 144
Sheldon, I., 261, 262
Simmons, G.L., 171
Sirhan, G., 126
Smith, E.B., 171
Solana-Rosillo, J., 262
Sorghum, 230, 279, 280, 295
South Africa (S. Africa), 141, 142, 162, 183t, 226t
Southard, L., 236
Southern states, 61, 229
Sows, 268, 276
Soybeans
  and competitiveness, 102, 107-111
  export taxes, 116
  exports, 99, 100, 310t
  and GMOs, 151
  high oleic acid, 294
  market share, 100, 101t, 116
  meal, 230, 231
  miracle crop, 115
  and oilseed trade, 282-283
  and peanuts, 139
  prices, 101
  processing, 102
  producers, 230
  production, 101, 102, 269, 310t
  quality-enhanced, 295t

Soybeans *(continued)*
  trade
    agreements, 112-115, 116
    research, 103-106
  transgenics, 294
  world market for, 115, 116
Soymeal
  and beef producers, 105
  exports, 106, 310t
  livestock feed, 102
  market share, 100, 101t, 102
  prices, 104, 105
  production, 310t
  protein market, 103
  substitutes for, 313
  tariffs on, 112, 113t
Soyoil
  competitors, 102, 104
  exports, 106, 310t
  imports by Japan, 105
  market share, 100, 101t, 102
  prices, 103
  production, 310t
  tariffs on, 112, 113t
Spain, 86, 157, 170, 177, 181, 259
Spinach, 170
Spreen, T.S., 172
Spring wheat, 54
Squash, 170, 173, 175f, 284
Sri Lanka, 77
St. Kitts and Nevis, 183t
Stanton, B.F., 73
Starch, 42, 51, 52, 295t, 296
Stewart, F., 230
Structural
  changes, 236, 237, 238, 239
  time series, 22
Stulp, V.J., 63
Suarez, N.R., 182, 186, 187
Subsidies
  agricultural product, 1, 313, 316
  and competitiveness, 76, 99, 311
  consumer, 50
  cotton, 129
  domestic, 73

Subsidies *(continued)*
  and EEP, 82
  export
    agricultural, 6, 263, 312, 313, 319
    corn, 109
    dairy, 278
    flour, 259
    fruit, 165
    oilseed, 106
    pork, 246, 275
    rice, 82, 85
    wheat, 59, 66-68, 71, 72, 75
    wine, 179, 180
  import, 6
  indirect, 127
  input, 128
  in Mexico, 51
  and prices, 8, 21
Substitutes
  high-oil corn (HOC), 297f
  product, 19, 293
  soybean, 313
  soymeal, 103
  sugar, 187
Sucrose, 294
Sugar
  competitiveness, 177, 182-185, 188
  costs, 185t
  dumped market for, 188
  exports, 253t, 255, 272, 314
  imports, 177, 255, 272, 314
  polices, 185, 186, 187
  prices, 184f, 188
  production, 177, 182
  program, 182
  quotas, 182
  substitutes, 187
  supplies, 194
  supports, 182, 194
  trade, 182, 194, 287, 288
  TRQs, 114
  and world economy, 187
Sunflower
  and AGLINK, 106
  and biotechnology, 295

Sunflower *(continued)*
  meal, 103
  production, 139
  saturated fat, 283
  and soybean exports, 283
  soyoil competitor, 102
  and vegetable oil, 295
Supply
  chain, 293, 297f, 302, 304
  control, 129, 182, 188, 276
  curve, 15, 17, 18, 54
  and demand
    agricultural products, 1, 269, 310
    constraints, 70
    farm prices, 290
    in rice model, 87
    shifts in, 43
    sorghum, 280
    in soybean model, 105
    and technology, 17
  management, 140, 144, 146, 151, 152, 277
Support prices
  agricultural, 3
  corn, 286
  EU, 103
  peanuts, 147, 152
  tobacco, 189
Swanson, E., 35
Swaziland, 183t
Sweden, 179f, 180
Swine, 53, 236, 247
Switzerland, 67, 165, 179f, 183t

Taiwan
  apple imports, 162
  Arkansas Global Rice Model, 86, 95
  corn market, 279
  cotton imports, 121, 122t
  financial crisis, 60
  foot-and-mouth disease, 236
  fruit market, 285
  land endowment, 239

Taiwan *(continued)*
  meat consuming, 56
  nontariff barriers, 180, 227t
  pork costs, 245
  rice, 83
  sugar TRQ, 183t
  tariffs, 226
  U.S. products destination, 309
  wheat imports, 63
  wine imports, 179f
Tangerines, 155, 157
Target prices, 46, 116
Tariffication, 95, 225, 312, 319
Tariffs
  barley, 280
  barriers, 6, 99, 172, 212, 213, 312-319
  conversion to, 69, 70, 312, 321
  cotton, 132
  dairy, 277, 278
  fruit, 164, 165, 166
  grain sorghum, 279, 280
  HFCS (HFS), 186, 288
  maize, 50
  NAFTA eliminates, 270
  oilseed, 106
  peanuts, 130, 147, 148t, 149t, 151
  phased elimination of, 320
  pork, 246, 247, 275
  poultry, 221, 224, 225, 226t, 227, 228
  quotas, 50, 83
  rate quotas, 50, 130, 182, 236, 275, 312
  reductions, 50, 271, 272, 291
  rice, 83, 84t, 85, 93
  textiles, 131
  tobacco, 182, 192, 193, 194
  vegetable oils, 50
  vegetables, 173, 175
  wheat, 59
  wine, 178, 179, 180, 181
Taxes
  Brazilian, 231
  export, 106, 116

Taxes *(continued)*
  ICMS, 106
  indirect, 117
  land, 81
  and macroeconomic policy, 6
  and soybean costs, 109t
  trade impacts on, 273
Taylor, T.G., 173
Technical Barriers to Trade, 263, 264, 270, 318
Technology
  advances, 4, 117, 214, 295
  breakthroughs, 238
  change, 29, 30, 38n, 87, 238
  development, 235
  efficiencies, 32
  environment, 11
  improvements, 175
  innovations, 22, 235, 312, 322
  platform, 293, 310
  progress, 35
  treadmill, 294
Teece, D., 298
Terms of trade, 26, 29, 35
Texas
  grapefruit, 158
  oranges, 157
  peanuts, 139, 147
  poultry, 229
  rice, 79, 80, 81t, 92, 96
  sugar, 182
Textiles
  bilateral trade agreement on, 135n
  competitiveness, 131
  and cotton fiber, 122
  export tax impacts, 127
  exports of, 127, 131, 286
  GATT provisions on, 133
  imports, 120, 314
  industries, 120, 122, 123
  markets, 122
  Mexican mills, 287
  policy, 123
  producing countries, 125, 134
  production, 120, 134

Textiles *(continued)*
  sector, 133
  tariffs on, 131
  U.S. manufacturing, 130
TFP (total factor productivity), 27, 107
  and agricultural growth, 28, 38
  components of, 34f
  EU growth in, 35, 37
  growth, 36f
  and R&D investments, 32, 37
  sources of, 30, 31f
Thailand
  cotton importer, 121, 122t
  financial crisis, 60
  labor costs, 313
  poultry
    costs, 222, 223, 224, 231
    DRC, 220
    production, 219
  rice
    Arkansas Global Rice Model, 86, 87, 94, 95, 96
    costs, 80, 81t, 82, 111
    exports, 282
    production, 79
    quality, 97
    tariffs, 83, 84t
    trade, 313
  sugar TRQ, 183t
  tariff reduction, 165
  tobacco, 191t, 192, 194, 314
Thigpen, E., 131, 134
Tilley, D.S., 160, 164
Tobacco
  assessment fee, 189
  competitiveness, 190
  competitors, 194
  costs, 191t, 314
  exports, 189, 309, 314
  imports, 189
  manufactured product exports, 314
  marketing quotas, 188
  production, 177, 188
  supply control, 188

Tobacco *(continued)*
   trade, 190-192
   Uruguay Round, 192-194
   U.S. program evaluation, 193
Tobacco Associates, 192, 195
Todd, M., 183
Tomatoes
   costs, 173, 174f
   exports, 170, 284
   imports, 170, 271, 284
   market integration, 171
   prices, 172
   production, 170
   returns to producers, 171
   trade, 172
Trade
   agreements
      agricultural, 12, 13
      beef, 212, 213
      citrus fruits, 160
      and competitiveness, 311, 318-321
      corn exports, 49-51, 56
      determining factor, 6
      multilateral, 41
      NAFTA, 269-291
      peanuts, 147-151, 152
      pork, 235, 246, 247
      poultry, 227, 228
      processed foods, 263-266
      rice, 82-85
      soybeans, 99, 100, 103, 112-115, 116
      sugar, 177, 185-187
      tobacco, 170, 180, 192, 193
      U.S. trading patterns, 269
      vegetables, 173-175
      wheat exports, 60, 66-70, 76
      wine, 177, 179-181
   associations, 7, 192
   barriers
      dismantle, 7
      EU, 260
      fruits, 314
      lower, 252

Trade, barriers *(continued)*
      peanuts, 151
      pork, 245, 246, 247
      poultry, 224-227, 277
      reduction of, 270
      rice, 83, 84
      soybeans, 103, 112, 114, 116
      sugar, 287
      and trade agreements, 13
      and U.S. agriculture, 289, 312
      wheat, 69, 70, 73, 76
      wine, 179, 180
   deficit, 11, 25, 252, 256
   flows
      and comparative advantage, 15
      cotton, 126, 127, 133
      vegetable, 177
      wheat, 59, 63, 71
   liberalization
      agricultural, 95
      and competitiveness, 13, 99, 270
      and economic growth, 117
      oats trade, 281
      oilseeds, 116
      pork exports, 247
      and regional groups, 112
      sugar, 188
      wine exports, 181
   patterns
      grains, 105
      manufactured food, 261
      NAFTA effects, 167
      poultry, 231
      and scale economies, 266
      shifts in, 1
      Sugar, 187
      WTO effects, 167
   policy
      anticompetitive, 99
      and competitiveness, 127, 312
      corn exports, 51, 56
      cotton, 120, 126, 127, 132
      framework, 322
      impacts, 6
      fruit, 164-166

Trade, policy *(continued)*
    meat trade, 203
    in Pakistan, 122
    peanuts, 146-151
    pork competition, 237
    wheat, 71, 73, 76
  secrets, 298
  surpluses, 11, 26, 277, 286
  theory, 26
Transgenics, 296
Trinidad and Tobago, 183t
Tropical products, 2, 6, 275
TRQ (tariff rate quota)
  barley, 280
  beef, 212, 213
  corn, 50
  cotton, 130, 286
  dairy products, 277
  peanuts, 143, 147, 149, 150, 288, 289
  pork, 275
  poultry, 225, 226t, 227, 228, 276, 277
  sugar, 182, 183t, 187
  tobacco, 189, 190, 192
  wheat, 281, 287
Trypsin, 296
Tunisia, 57
Turkey
  cotton, 121, 127, 128t
  fruit, 157, 161
  grain mill products, 259
  rice, 84t
  tobacco, 190, 191t, 194
  wheat, 59
Turkeys
  competitive position, 231
  cost, 230
  exports
    to Canada, 227, 277
    France, 219, 221
    to Mexico, 277
    share, 309, 310t
    U.S., 217, 218f, 219t, 310t
  import quotas, 276, 277

Turkeys *(continued)*
  production, 217, 219t, 310t
  tariffs, 225, 226t
Tweeten, L., 4, 12, 20, 88
Two stage least squares, 97

U.K. (United Kingdom)
  beverage exports, 259
  biotechnology mergers, 300t
  destination for U.S. exports, 309
  flour exports, 259
  market for hogs, 235
  peanut imports, 143t
  pet food exports, 259
  pork costs, 245
  poultry, 229
  prepared meat imports, 258
  processed foods, 316
  TFP growth, 33, 34f, 35
  tobacco imports, 191t
  wine imports, 178, 179f
Ukraine, 226t, 227t
United Brands, 7
United Nations, 258
Uruguay
  beef
    consumption, 206
    exports, 209
  and MERCOSUR, 112
  rice, 84t, 85, 86, 97
  sugar TRQ, 183t
Uruguay Round
  agriculture, 6, 318, 319
  beef, 209
  corn, 46, 49
  cotton, 127
  GMOs, 318
  market access, 312
  NAFTA, 270
  peanuts, 140, 144, 147, 148t
  pork, 246
  poultry, 225, 228
  processed foods, 263, 265
  rice, 83, 84, 95

Uruguay Round *(continued)*
    soybeans, 104, 106, 112
    sugar, 185, 194
    tobacco, 192
    wheat, 61, 70
    wine, 178, 180
U.S. Department of Agriculture, 67, 68, 188
U.S. Department of Commerce, 251, 252
U.S. International Trade Commission, 68
USDA
    cotton, 131
    data, 42, 45
    export estimates, 272
    FATUS, 308
    study, 246
    and world price adjustments, 134n
USSR, 113t
Uzbekestan, 121, 122

Valdés, A., 21
Van Duren, E., 16
VanSickle, J.J., 171, 172, 174t
Varangis, P., 131, 134
Varvello Vicente, M., 221
Vegetable oil
    consumption, 115, 295
    duties on, 282
    export market share, 102, 283
    prices, 106
    substitutes, 103
    tariffs, 53
Vegetables
    agribusiness and, 285
    competitiveness, 172-175, 314
    demand for, 201
    exports, 169, 309
    imports, 169, 320
    preserved, 251, 253t, 254t, 255, 256t
    trade
        Canada, 170, 272

Vegetables, trade *(continued)*
        increased, 175
        Mexico, 169, 170, 271, 272
        NAFTA impacts, 272, 283-284, 289, 290, 320
        research, 170-172
        Spain, 170
Venezuela
    pork, 247
    poultry, 226, 227t
    rice, 84t, 93
    wheat, 63, 64t
Victoria, 63
Vietnam
    poultry, 86, 88, 94, 96, 97
    rice, 79, 80, 81, 82, 86
Vineyards, 177, 181
Virginia, 139, 188
Vocke, G., 221
Vollrath, T., 26, 310
Voros, P., 261

Waggoner, O.B., 214
Wahl, T., 186, 230
Wailes, E.J., 87
Wall Street Journal, 237
Wampler Industries, 230
Warnken, P.F., 102, 105, 108, 115
Washington apples, 7, 161, 166
Water pollution, 239
Watermelon, 170
Watkins, P., 52
Wensley, M., 106, 113t
Western Europe, 71, 246
Western Hemisphere
    cotton, 120, 121
    FTAA, 321
    pork, 247
    rice, 85
    soybeans, 112
    trade agreements, 269
Westgren, R., 16, 144
WGTA (Western Grain Transportation Act), 67, 68

Index   357

Wheat
  classes, 60-61
  competitiveness, 70-75
  consumption, 63-65
  durum
    consumption, 63
    exports, 65, 66
    production, 60, 61, 62
    tariffs on, 69
    in wheat model, 70, 71, 72, 73, 74, 75, 76
  duties on, 281
  exports, 2, 65-66, 281, 308, 310t, 313
  flour, 259
  genetically modified, 295
  imports, 63-65, 281
  industry, 60
  markets, 59, 310
  output, 269
  policy, 116
  prices, 281
  production, 59, 61-63, 310t
  subsidies, 109
  trade
    agreements, 66-70, 76, 281
    exporting countries, 60
    importing countries, 59, 60-61
    volume, 75
  world market, 102
Wilkowske, G.H., 173
Wine
  competitive position, 194
  exports, 178, 179, 314
  grapes, 161, 177
  imports, 178, 256
  and MAP funds, 178t, 181
  promotion, 181, 194
  trade
    agreements, 179-181
    barriers, 181
    beverages, 260
    liberalization, 181
  world market, 177
Womach, J., 189

Women, 201, 202
World
  markets and
    agricultural policies, 116, 194
    beef, 214
    biotechnology, 114
    cigarette exports, 193
    comparative advantage, 35
    competitiveness, 32, 99
    cotton, 125
    environmental regulations, 104
    liberalization of, 1
    oilseeds, 114
    processed foods, 37
    productivity, 4
    soybeans, 100, 101, 102, 107, 108, 116
    soyoil, 102
    wine, 177
  prices
    agricultural component, 21
    in Arkansas rice model, 87, 89, 91t
    and cotton competitiveness program, 129, 130
    EEP bonus, 67
    EU support and, 103
    and export subsidies, 21, 66
    pork TRQ and, 275
    signals, 13
    and U.S. agricultural exports, 269
  trade
    agricultural products, 4, 310
    apples, 162
    Arkansas rice model, 89
    cotton, 119, 132
    and GATT/WTO, 112, 318
    and NAFTA, 112
    orange juice, 160
    peanuts, 142, 146
    rice, 79
    soybeans, 100
World Bank, 220, 272
WTO (World Trade Organization)
  agriculture, 6, 13, 164, 318, 319

WTO (World Trade Organization)
(continued)
  China accession, 83
  and conversion of trade barriers, 70
  created, 6
  competitiveness, 99
  dispute settlement, 265, 319
  fruit trade, 165, 167
  HFCS consultancies, 187
  market access, 312, 316
  and NAFTA restrictions, 270
  national treatment provisions, 266
  negotiating issues, 99
  other trade agreements, 321
  peanuts, 144, 151, 152
  pork, 246
  poultry, 225, 231
  rice, 83, 84, 85, 93, 95
  safeguard tariff, 236
  sanitary provisions, 318

WTO (World Trade Organization)
(continued)
  section XX, 264
  soybeans, 112, 113t, 114, 116
  and tariffication, 225
  Taiwan's membership in, 83
  trade barriers, 70, 76

Yang, S.-R., 70, 71
Yemen, 259
Young, K.B., 182

Zechendorf, B., 305
Zeneca, 300t, 301t
Zepp, G.A., 171
Zhang, P., 126
Zhong, F., 51
Zimbabwe, 183t, 189, 192, 194, 314

Order Your Own Copy of
This Important Book for Your Personal Library!

COMPETITION IN AGRICULTURE
The United States in the World Market

_____ in hardbound at $69.95 (ISBN: 1-56022-892-X)

_____ in softbound at $39.95 (ISBN: 1-56022-893-8)

COST OF BOOKS_____	☐ **BILL ME LATER:** ($5 service charge will be added) (Bill-me option is good on US/Canada/Mexico orders only; not good to jobbers, wholesalers, or subscription agencies.)
OUTSIDE USA/CANADA/ MEXICO: ADD 20%_____	
POSTAGE & HANDLING_____ (US: $3.00 for first book & $1.25 for each additional book) Outside US: $4.75 for first book & $1.75 for each additional book)	☐ Check here if billing address is different from shipping address and attach purchase order and billing address information.
	Signature _____
SUBTOTAL_____	☐ **PAYMENT ENCLOSED:** $_____
IN CANADA: ADD 7% GST_____	☐ **PLEASE CHARGE TO MY CREDIT CARD.**
STATE TAX_____ (NY, OH & MN residents, please add appropriate local sales tax)	☐ Visa ☐ MasterCard ☐ AmEx ☐ Discover ☐ Diner's Club
	Account # _____
FINAL TOTAL_____ (If paying in Canadian funds, convert using the current exchange rate. UNESCO coupons welcome.)	Exp. Date _____
	Signature _____

Prices in US dollars and subject to change without notice.

NAME _____

INSTITUTION _____

ADDRESS _____

CITY _____

STATE/ZIP _____

COUNTRY _____ COUNTY (NY residents only) _____

TEL _____ FAX _____

E-MAIL _____
May we use your e-mail address for confirmations and other types of information? ☐ Yes ☐ No

Order From Your Local Bookstore or Directly From
The Haworth Press, Inc.
10 Alice Street, Binghamton, New York 13904-1580 • USA
TELEPHONE: 1-800-HAWORTH (1-800-429-6784) / Outside US/Canada: (607) 722-5857
FAX: 1-800-895-0582 / Outside US/Canada: (607) 772-6362
E-mail: getinfo@haworthpressinc.com
PLEASE PHOTOCOPY THIS FORM FOR YOUR PERSONAL USE.

BOF96